Fluid Arguments

Fluid Arguments

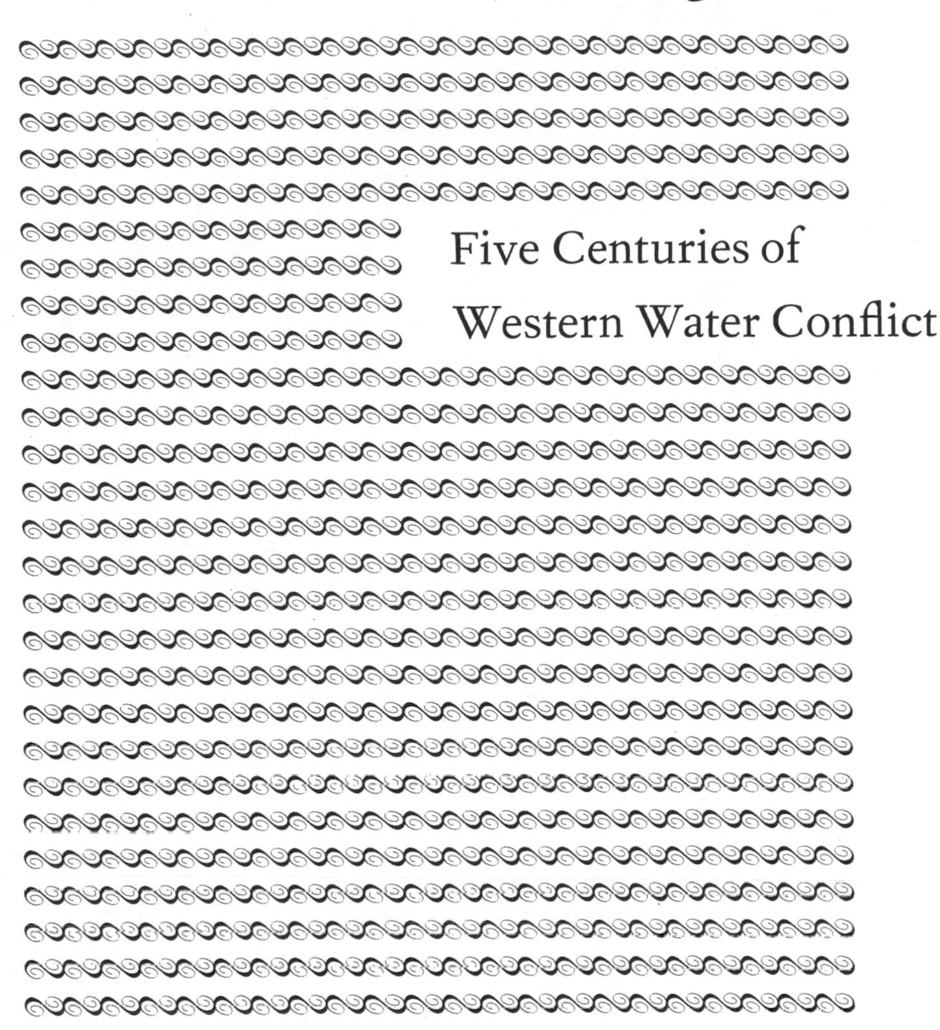

Five Centuries of Western Water Conflict

Edited by Char Miller

The University of Arizona Press, Tucson

The University of Arizona Press
www.uapress.arizona.edu

First paperback edition 2014

Printed in the United States of America
19 18 17 16 15 14 6 5 4 3 2

ISBN-13: 978-0-8165-2061-9 (cloth)
ISBN-13: 978-0-8165-0136-6 (paper)

Cover illustration: The Hayden-Rhodes Aqueduct—Reach 1.
Photograph compliments of the Central Arizona Project.
Original front cover design: Wilsted & Taylor Publishing Services

Library of Congress Cataloging-in-Publication Data
Fluid arguments: five centuries of western water conflict / edited by
Char Miller.
p. cm.
Includes index.
ISBN 0-8165-2061-5 (cloth : acid-free paper)
1. Water-supply—Southwestern States—History. 2. Water rights—
Southwestern States—History. 3. Water-Supply—Government
policy—Southwestern States. I. Miller, Char, 1951–
TD223.9.F58 2000
333.91'00976—dc21 00-010450

∞ This paper meets the requirements of ANSI/NISO Z39.48-1992
(Permanence of Paper).

For Eunice Herrington

Contents

Illustrations

Tables

Acknowledgments

Fluid Arguments has been a labor of love. The book grew out of an American Society for Environmental History–sponsored conference on water in the American West held at Trinity University in May 1998, a gathering that drew together an interdisciplinary array of scholars, water-management professionals, and citizens. The interaction between the audience and presenters was dynamic and led to considerable, often intense discussion of the intersection between past water policies and present dilemmas. The conversations continue and are embodied in the revisions that shape each of this book's chapters. I am grateful for all those who attended the conference and who, through their keen insights and careful commentary, helped refine the book's focus, argument, and structure.

My gratitude is as great for the many individuals, institutions, and organizations that helped underwrite the original conference and, indirectly, *Fluid Arguments*. Ronald G. Calgaard, then-president of Trinity University, was an early, enthusiastic, and generous supporter of my fledgling idea. The conference came into being with additional—and deeply appreciated—aid from the Texas Council for the Humanities, a state partner of the National Endowment for the Humanities; the Semmes Foundation, Inc.; the San Antonio Water System; the Center for Hazards and Environmental Geography of the Department of Geography and Planning at Southwest Texas State University; The Inter-American Legal Studies Program at The St. Mary's University School of Law; and the Departments of Geosciences, History, Physics, Political Science, and Sociology and Anthropology of Trinity University.

This book benefited as well from the kind support of many people at the University of Arizona Press: Acquiring Editor Patti Hartmann, who loved the idea of *Fluid Arguments;* Assistant Managing Editor Al Schroder, who helped me navigate the shoals of copyediting; and the anonymous reviewers, whose professional assessments were as rigorous as they were constructive. My fellow historians at Trinity University, as usual, have been wonderfully indulgent and helpful, as has our junior secretary, Mary Miller. But no mere words can convey what I (and my colleagues) owe Eunice Herrington, senior secretary of the History Department at Trinity University. For more than twelve years she has been an absolutely essential part of our scholarly lives; her literary instincts (always sharp), like her computer skills (ever accelerating), have helped us become better writers and more productive historians. Dedicating this book to Eunice is but a small way by which to thank her for her enormous contributions.

Divining the Past: An Introduction

Char Miller

Out of the nineteenth-century American West came this shrewd aphorism: "Whiskey's for drinking; water's for fighting." Attributed to Mark Twain, these pithy words suggest how crucial white gold was to the arid region's economic growth and development. Without water, earlier generations of European migrants to the West could not put plow to ground; graze cattle, goats, or sheep; mine gold or silver; harvest timber; or build cities. With all this at stake, why come to blows over a mere glassful of liquid pleasure?

The last one hundred years have only reinforced this argument: The drive to gain access to and then control the distribution of water has spawned any number of political brawls from Texas to California, Nevada to Colorado, Arizona to New Mexico. Over that time, however, there has been a marked shift in the source of these struggles. Since World War II, this once-rural landscape and the agrarian economy it sustained have given way to an increasingly urban and industrial society. Today, six of the nation's ten largest cities are located within the Southwest. This demographic transformation and the concentration of population it has produced in places as disparate as Houston, Phoenix, San Diego, Dallas, Los Angeles, and San Antonio has led to an intense demand for water to quench human thirst, keep suburban lawns green, and fill innumerable swimming pools.

This demand may escalate if projections of population growth for the Southwest during the coming century are accurate, meaning greater political conflict over the flow of water. It is a naive community and myopic governmental entity that does not understand what the City of Angels recognized at the beginning of the twentieth century—that whoever controls the spigot shapes the course of urban development. However, it is equally clear that contemporary communities cannot operate in the aggressive, imperial manner that once defined the Angelino pursuit of water. Initially constrained through a series of intergovernmental water compacts, cities are restrained as well by a conservationist political climate. In 1998, for example, East Bay MUD, a water utility district in Alameda County, California, was forced to ask voters if it could extend its services to a massive new housing development in a neighboring county. As both Tucson and Salt Lake City lap out into the surrounding desert, water restrictions are being imposed. Even the Texas Natural Resource Conservation Commission, in an aggressively pro-growth state, has been compelled to respond to the impact of drought on once-sanctioned sprawl. In the late 1990s, the commission proposed regulations that would govern the quantity of new homes that could be built over aquifer recharge zones; it has also advocated the monitoring of pollution runoff from commercial and residential development. If rigorously implemented, such regulatory action would be an important step in preserving the quality and maintaining an adequate supply of water for the Lone Star State's booming metropolitan areas.[1]

But every step forward will be challenged, a contentiousness that has been woven into the history of water in the American West. This rich, sustained, and combative historical context is the focus of *Fluid Arguments*. Through the interdisciplinary insights of ethnography, geography, history, political science, law, and urban studies, this book reveals the degree to which many of the late-twentieth-century arguments are imbedded within decades of accumulated debate. Penetrating the layers of these debates is crucial for understanding the often-tangled relationship between the history of and public policy concerning water use.

This complex of relationships is central to the scholarship of those writers—such as Helen Ingram, Donald Pisani, Donald Worster, Marc Reisner, Jim Sherow, Norris Hundley, William Kharl, and John Opie—who have forced us to reconceive the impact water and aridity have had on human cultures and ecosystems. Asking who controls the distribution of water and how those controls complicate its equitable flow across regions and social divisions are comparable to queries about the many environmental costs incurred with the construction of hydraulic systems that carry water from source to tap. As its authors and their endnotes suggest, *Fluid Arguments* builds off of these themes and extends their scope through its broad chronological coverage and geographical range, the variety of problems it addresses, and its case-study focus, checking the prevailing historiography against the reality of lives and landscapes.

Fluid Arguments opens with the northward thrust of New Spain into what is the now the borderlands region of Mexico and the United States.

Over this land, the Spanish exercised tentative political control, a partial consequence of the distance that separated these colonial outposts from the seat of imperial authority in Mexico City. Environmental factors were also responsible: As the sixteenth-century conquistadors quickly discovered, the absence of water made laughable their fantasies about the riches of this often-forbidding terrain; subsequent Spanish colonists learned soon enough that the size and significance of settlement and the extent of agricultural production depended on scarce water supplies. In the Tamaulipan Cession—a strip of territory running south from the Nueces River in present-day Texas to the Mexican state of Tamaulipas—the patterns of colonization were determined through a powerful formula; in this region, historian Jesús F. de la Teja observes, "land was plentiful, water was not." This shaped colonial and later Mexican observations of the region's political importance and economic potential. In the late 1820s, botanist Jean Louis Berlandier doubted that irrigation would ever become economical enough to transform this parched land and predicted that for "several centuries it will remain nothing but an immense prairie where herds can be bred." About irrigation Berlandier was wrong—within a century, as geographer John Tiefenbacher explains, a hydraulic empire was being constructed within the Rio Grande watershed—but from the seventeenth century on, aridity conditioned human endeavor. In this region, livestock grazing over large tracts was the preferred mode of production, a preference that over time determined property ownership under Spanish and Mexican law. This legal context, in de la Teja's words, developed a "cattle culture in which land, water, and animal were part of a whole." Much would change in the mid-nineteenth century with the arrival of "an alien set of rules and large amounts of capital from the United States," but not this. In how "land has been subdivided, used and transferred," water remained "the critical element in the equation."

Water's preeminence was just as pronounced in the area the Spanish dubbed Pimería Alta, south-central Arizona. Beginning with the Hohokam civilization, which before its disappearance in the late fifteenth century had used sticks to dig hundreds of miles of irrigation canals, indigenous peoples manipulated available water supplies to extend their agricultural productivity. Whatever the links between the Hohokam and later Pima Indians—Shelly Dudley doubts any direct connection—the Pimas apparently imitated their predecessors' hydraulic technology and later perhaps borrowed from the Spanish as well. Missionaries touring the region beginning in the mid-sixteenth century observed the cultivation of maize, cotton, and pumpkins; the use of irrigation canals; and the development, as one seventeenth-century visitor noted, of "much green land" in an otherwise brown world. But the extent of cultivation was limited by the amount of water that could be diverted from the Gila River and its tributaries. This was demonstrated during the years of contact with the Spanish, during which the Pima planted missionary-introduced crops such as wheat and legumes; these new foodstuffs could extend Pima agricultural output, but only if worked within the land's carrying capacity.

That balance provided plenty: Reports of late-eighteenth-century and early-nineteenth-century European and American travelers in the region indicate that the Pima regularly sold or traded surplus food supplies. Such agricultural success declined after the 1820s, not because the Pima intensified their demands on the land, but because recent arrivals—Euro-American residents—instituted what Dudley describes as a "complex set of environmental changes" that undercut the Pima's access to water and thus their ability to work and own their land.

In New Mexico, ecological change and claims of property ownership also were (and are) keyed to water rights. So argues Sandra Mathews-Lamb in her analysis of records from late-twentieth-century judicial proceedings concerned with the establishment of priority dates. Determining who *first* utilized water (and how continuously) on a particular site is central to contemporary legal battles over water availability and control of property. But the documentary research required to answer these concerns reveals a great deal about the confluence of environmental and social forces in the more distant past. Consider the Chama Basin in northern New Mexico, which ancestors of the Pueblo Indians called home more than a millennium ago. There, they (and their successors) used surface waters to grow a variety of vegetables and legumes. How productive they were became evident to early-sixteenth-century Spanish explorers, such as Coronado, and, at century's close, to those who sought to establish a settlement at San Gabriel. In letters and diaries, colonists marveled at the success of indigenous agriculture and were perplexed by their less-than-fruitful efforts to produce enough food to sustain their communities. Frustrated, these early colonists retreated into Mexico. Although later Spanish colonists and missionaries were more successful, always the difficulty of their life was measured in the amount of water at their disposal.

The same could be said for the Native American peoples who inhabited the western landscape. As the American military conquest of the trans-Mississippi West unfolded in the years following the Civil War and intensive migration rolled across the Plains, spilled into the intermountain basins, and pressed out to (or landed along) the Pacific Coast, these forces and settlers came into conflict with indigenous tribes and communities. As with the earlier Spanish intrusion, many of the battles revolved around control of resources, which in the process transformed long-standing relationships to place.

As Bonnie Lynn-Sherow reminds us, this power struggle, which invariably displaced Indian societies from the richest, most arable land and led to "the substitution of one environmental ethos for another," has generated a counterclaim that grants to the dispossessed "an intuitive relationship to 'nature' that is both sustainable for human life and beneficial to all life in general." This assertion—however understandable—hinders our understanding of the character and context of Indian resource use and thus minimizes our ability to secure their "unique place in American society," which Lynn-Sherow establishes in her study of the Kiowa's historic use of and relationship to water.

The Kiowa's perspectives have been complicated by the fact that the tribe migrated over time, first north along the eastern Rockies up into Wyoming, then southeasterly into the southern Plains, extensive movement across varied environments to which the tribe adapted each in turn. When by the eighteenth century the Kiowa had relocated to the grasslands of Oklahoma and Texas, their cultural depiction of water—most especially through the Saynday stories—reveal that it played a "nonanimate or malevolent role"; other tales link water negatively to the underworld. These supernatural perceptions continued throughout the reservation period of Kiowa history (post-1875), but moderated as well in response to the new demands of a more sedentary, agrarian life that intensified after allotment in 1901. Because groups were tied to particular sections of land, band identification began to flow along watercourses (the Elk Creek Kiowa, for example); in addition, these streams enabled the Kiowa to make the "transition from hunting and raiding" to a ranching economy. Ever since, water has remained largely what it has been for the Kiowa's white neighbors: an essential, practical resource.

But the Kiowa, like other Native American peoples, suffered a troubled and troubling transition into a capitalist economy. That broader shift is the subject of Donald Pisani's chapter on the federal government's Indian water policies. Those policies came to the fore with the passage of the General Allotment (or Dawes) Act of 1887, which spelled the end of the reservation period. By allotting 160 acres to each Indian and declaring the remaining lands to be "surplus," and thus open to sale to non-Indians, the act began the process of shrinking the amount of land under individual ownership. Many white reformers justified this rapid decline on the grounds that, with the introduction of irrigation to the arid portions of the West, indigenous peoples required less land for maintenance; these newly arable farmsteads, reformers theorized, would also help assimilate the Indians into the mainstream of American culture. As the irrigation plans went forward and hundreds of miles of ditches and canals were dug by Indian labor, the chief beneficiaries were white farmers who had purchased "surplus" allotments. By the late 1920s, after the Bureau of Indian Affairs had expended more than $27 million to irrigate approximately 700,000 acres, evidence indicated that two-thirds of that land was white owned; and because the Indians tended to use water for grazing, their "crops . . . returned only half the per-acre income of white farmers." Add to these depressing results the fact that Indian agriculture was badly undercapitalized, and it is no wonder the irrigation projects were failures. The Indian farmer was left to labor "with his bare hands," one U.S. senator asserted in 1914. "The white man could not make a success under similar circumstances."

That white farmers were considerably more successful than their Indian neighbors was not solely the result of their ability to take advantage of the General Allotment Act. They benefited as well from the West's curious water law, known as the doctrine of prior appropriation. Since 1866, notes Alan Newell, Congress had "recognized the paramount rights of states to adjudicate water rights," an adjudication that depended on a simple formulation: first in time, first in right. Those who could demonstrate prior manipulation of water resources could hinder others' abilities to divert stream flow or tap into underground supplies. The impact of this on Native American agricultural production, which in many cases did not begin in earnest until after the Dawes Act and subsequent allotment, was profound: White settlers could and did claim priority and were supported in the courts, further dooming western tribes to a precarious existence.

Complicating this scenario was a second, and later, legal finding, the Winters doctrine. Following a 1905 district court decision, which the Supreme Court affirmed in 1908, the federal government secured the responsibility "to assert its right to water for federal reservations," meaning that it could bring suit against nonfederal (that is, non-Indian) irrigators to insure adequate water on arid reservations. How much was adequate? The 1908 decision did not answer this crucial question. Ever since, as the analyses of Newell and political scientist Daniel McCool reveal, this issue of quantity has been at the center of a century-long series of courtroom battles. Finding a just answer has not been easy. What seems evident from the twenty-year-long case involving claims of the Mescalero Apache of New Mexico (the subject of Newell's research) and from McCool's overview of the complex, often bitter, late-twentieth-century negotiations involving numerous tribes, western states, and the federal government is that this new era of deliberations is as fraught with dangers as were the treaty settlements of the nineteenth century. As McCool suggests, there is also a glimmer of hope that through these negotiations "tribes may parley their reserved water rights into a demand for the return of lands lost through allotment, theft, and sale," thus offering a means to reverse the oppressive consequences of western water and land policies.

Politics and law could do only so much to set the conditions under which American settlement of the West occurred. Just ask any rancher, especially those who struggled to establish themselves in Texas in the late nineteenth century. They were confronted, as had been their Spanish and Mexican predecessors, with the pressing need to secure adequate water and food for their herds, a difficult task given the rough physiography of the south-central portions of the state. Their operations were confronted with another hazard, the vast distances between the lands on which their cattle grazed—a triangular shape of country from coastal Indianola to Laredo on the Rio Grande, with San Antonio at the northern apex—and the midwestern and eastern urban populations hungry for beef. The arduous movement of herds north to railheads in Kansas, more than seven hundred miles away, has become the stuff of popular legend. To that western lore, most especially about the famed Chisholm Trail, James Sherow has brought an ecological perspective. Following the cattle drives north, he is alert to changes in ecosystems, such as befouled water courses and the rapid grazing of the "nutritious little bluestem, the mainstay fuel for cattle and horses," which was replaced by the more sparse cover of buffalo grass. The loss of quantity and quality of water and food led drovers to widen the trail's path in search of rivers and forage; similar consequences resulted at the end of trail, when hundreds of thousands of thirsty and voracious cattle awaited transportation. The "ecodynamics of trail driving" thus shaped a series of biotic, physical, and cultural systems and meant that "certain consumers flourished at the expense of others."

In time, this ephemeral ecosystem itself would wither away and be replaced by an agricultural landscape. But establishing the new farmers' regime could be as difficult as anything the cattle industry had faced, as land speculators and irrigators in Grand Junction, Colorado, discovered. When the U.S. Army pushed the Utes out of the Grand Valley in the early 1880s, they opened for white settlement a remote region that, despite the presence of the swift and full Grand and Gunnison Rivers, was remarkably arid. "For Grand Valley residents the problem was not finding water but transferring it to the land," Brad Raley notes, a problem that led to a series of locally underwritten attempts to construct irrigation ditches to make this desert bloom. They all failed, due to a lack of funding and flawed planning; residents then sought outside capital and the technological expertise it could bring to bear. First Denver financiers and later Connecticut-based Travelers Insurance Company, lured by the potential profits from a well-watered land, invested in the scheme and undercut local control and participation, an old story in the history of western resource development. But to this much-told tale, Grand Valley offered a twist. Eastern investors found the project such a "financial sink hole" that although they completed the network of ditches, they never made a cent out of their investment; in time, the canals reverted to local ownership.

In another valley in a different part of the western frontier, the struggle to secure a steady flow of water for economic development also consumed the energies and capital of the first generation of Euro-American settlers. According to John Tiefenbacher, those arriving in the Lower Rio Grande

Valley in the late nineteenth century began to alter its historical environmental geography when they dug ditches first to irrigate sugar cane fields and, later and more effectively, land set aside for vegetables, rice, and citrus. Lacking transportation to major markets in the United States and Mexico inhibited the growth of an export economy until the railroad arrived in 1905. It kicked off a boom by laying down new towns, sparking a sharp speculation in real estate, and generating a host of new land and water companies. This boom lasted until the Depression, and until then its profits were plowed back into the soil. With the help of new fertilizers and insect controls, the agricultural use of the landscape intensified even as urban population, from Laredo to Brownsville, expanded rapidly. The competition for space has only accelerated, and one significant fallout is higher levels of toxicity in the water and air. A degraded environment compromises human health, Teifenbacher writes, especially among the poor, and there is no more impoverished population in the United States than in this valley, a loop that remains of critical concern in the post-NAFTA era.

That changes in "basic agricultural practices can produce dramatic alterations in the entire economy of a region," and therefore in human endeavor, is reinforced in Thomas Schafer's study of the evolution of cropping practices in southwestern Kansas. Identifying the regional shifts in specialization—from the nineteenth-century fixation on corn to intensive wheat production through the twentieth, and, at that century's close, to the development of a more diverse farming practice—leads him to conclude that this trend is more complex and variable than scholarship in agricultural geography would predict. The identification of these variations is possible due to the county-level analysis Schafer adopts, which reveals the manner in which farm communities adapted to and transformed local environmental conditions. The move away from corn, a humid crop, to wheat, which was better suited to a climate that produced a mere fifteen to twenty-two inches of rain annually, reflected farmers' growing awareness of the world in which they operated. Alter the source of water, however, and the contours of the world change. Technological innovations in the 1940s enabled farmers to pump water out of the Ogallala aquifer, inaugurating a "wetter," more diverse form of agriculture cast over a larger expanse. With the subsequent development of the center-pivot irrigation system, a new environment was constructed that would have befuddled earlier generations of farmers; best glimpsed from above, this new verdant vista was built of "giant lawn sprinklers, circular fields, and underground rain."

Yet is an aerial snapshot the best way to comprehend this terrain? John Opie suggests not, proposing instead a perspective from beneath the surface, one that emphasizes the critical role that the water resources contained within the Ogallala aquifer have played in the transformation of the "old Dust Bowl region of western Kansas and the Texas-Oklahoma panhandle." In so arguing, Opie also advances a broader claim: It is time to remap, or resize, the High Plains along lines of natural resources and ecosystems. This is an argument John Wesley Powell had advanced in the late nineteenth century. Opie refines the geographer's claim by observing that when we jettison the traditional demarcation of western space along political subdivision—particularly counties—and rectangular sections, we will secure more accurate information about the extent and quality of resources set within competing human pressures and environmental needs. Maps that groundwater management districts in the region use may locate each and every center-pivot and aquifer boundary, but when they ignore "natural systems like watersheds and soil types," such maps are of considerably less utility. New methods of mapping may be timely for another reason. During the mid-1990s, the numbers of industrial hog confinement operations exploded, placing enormous pressures on the land use, befouling the air, and threatening water supplies, one result of which was the division of rural communities into "habitable and uninhabitable zones." Precisely charting these consequences will provide citizens and their governments with a clearer sense of what has happened and with clues for how best to respond. Slipping loose of the now-dominant map conventions will allow us to see ourselves and the physical world we inhabit in a new, more accurate light.

It is also time to reevaluate the significance of dams in the history of water development. Donald Jackson asserts that one aspect of this reevaluation must include the recognition that the private sector played a crucial role in the creation of these public-works projects. This claim, which modifies the long-standing historiographical emphasis on the federal government's heavy investment in the massive dams constructed during the New Deal, turns on the erection in the 1920s of a modest flood control dam to the northwest of Phoenix, Arizona. Although built in response to local flooding problems, the Cave Creek Dam, by its engineering, design, financing, and political support, reflects a broader, regional concern with the contours of urban growth. Believing that a spur line connecting with the Santa Fe Railroad well to the north would facilitate the nascent community's development, Phoenix boosters were eager to construct the Cave Creek Dam to protect the track from washouts. Because it was well-versed in exploiting the connection between the extension of its lines and profitable real estate investments and commercial expansion, the railroad company supported the flood control project. Its powerful presence enabled the company to swing support behind the hiring of dam engineer John Eastwood, whose multiple arch design not only was as safe, but used less materials and was less expensive than Boulder, Grand Coulee, and the other soaring constructions of the next decade. A viable, if ultimately forgotten alternative to a "federally dominated system of water development," Cave Creek Dam stands as a reminder of the road not taken.

The path followed turned the massive New Deal dams into classic expressions of American technocracy. As Mark Harvey observes, these monumental objects were richly symbolic of the cultural anxieties that rattled the nation during the Depression and the Cold War and have had profound environmental and social consequences. The stunning complex of dams along the Colorado, Platte, Snake, Columbia, and Missouri Rivers degraded riparian ecosystems, inundated natural landmarks, uprooted communities, and turned fast-flowing watercourses into placid reservoirs. They also produced considerable work in a region of high unemployment, generated cheap hydroelectricity to power new industries on the West Coast, and sparked the emergence of a potent political coalition that channeled federal spending into these ambitious projects. But these dam boosters had their critics too. By the mid-1960s, activists using new federal laws such as the National Environmental Policy Act and the Wild and Scenic Rivers Act had begun to challenge these technological marvels and the presumption that underlay their construction, namely that impounding western rivers would resolve forever the problem of western aridity.

That shimmering vision is not exclusive to the United States, of course. China and many other Third World nations have gambled similarly that the potential economic payoff derived from the damming of rivers is too great to ignore. So the government of Carlos Salinas de Gortari believed when in 1988 it pressed for the construction of the El Cuchillo Dam in the

Mexican state of Nuevo León. In his close analysis of the history of the dam, legal scholar Raúl Sánchez uncovers the political forces that insured its rapid development, reveals the trade-offs involved in diverting water away from agricultural production to urban consumption, and traces the environmental deterioration and social dislocation that resulted when the floodgates were closed. As the reservoir began to rise behind the forty-three-meter dam, "a slow-motion disaster began to unfold downstream as the life-sustaining waters of the San Juan River stopped flowing." These disastrous consequences were not contained within Mexico, however, because the San Juan is an important tributary of the Rio Grande. Although the impoundment of its waters do not violate a 1944 treaty between Mexico and the United States, the sharply diminished flow affected water consumers on both side of the border. However, neither Mexican nor American citizens had any institutional recourse to protest this substantial loss, an impotence that Sánchez believes can only be remedied through the assertion of an enforceable, international claim of human rights.

The prospects for that legal strategy are as uncertain as the future development and distribution of water is unclear. But this much is certain—the economic development of the American West in the twenty-first century will not be able to depend solely on vast quantities of cheap water stored behind big dams. Certainly, new dams will not be constructed, due to their environmental and financial costs. This more pragmatic reality does not mean that the mirage of finding a new, inexpensive source of ready water does not dazzle still. Proponents of desalinization, such as former Senator Paul Simon, have convinced themselves that the construction of a massive complex of water filtration plants will provide for all our future needs. As with an earlier generation of dam-builders who exaggerated the virtues of an interlocking series of dams, reservoirs, and canals, Simon quotes approvingly this early 1960s projection from President John F. Kennedy: "Before this decade is out, we will see more and more evidence of man's ability at an economic rate to secure freshwater from saltwater, and when that day comes, then we will literally see the deserts bloom." Desalinization is the last best hope.[2]

That this latest lure—like its predecessor's now-bankrupt claims—deftly ignores a multitude of environmental problems in its single-minded pursuit of a wet Holy Grail, ought to give us pause. So too should the claims that water conservation is, in Simon's words, only a "big short-term payoff." Recent data not only suggest the payoff is long term, but that concerted efforts to cut consumption have beneficial environmental outcomes. Several innovations in conservation are especially noteworthy—the low-flush toilet, industrial recycling, and drip irrigation have had an impressive impact on water consumption in urban and rural areas in the United States. According to a U.S. Geological Survey report released in 1997, Americans' use of water between 1980 and 1995 dropped about 9 percent, despite a 16 percent increase in the nation's population.

Nowhere has this change been more evident than in Los Angeles, whose history of water piracy and profligacy is legendary. Since 1970, as its population increased by one million, the city's consumption of water has remained roughly at the same level. The Southern California Metropolitan Water District, covering Ventura in the north to San Diego in the south, estimates that it has reduced the region's annual need for imported water by 710,000 acre-feet. The "huge new source of water for the city of Los Angeles," argues S. David Freeman, director of the Department of Water and Power, "was the water we were wasting."[3]

That the same can be said about Las Vegas, Tucson, and San Antonio; about irrigation (the average quantity of water used for irrigated land dropped by 16 percent between 1980 and 1995); and about industrial consumption (overall use is down 35 percent) suggests the depth of the commitment to conservative measures. These steady declines also have had environmental consequences, the most striking of which is reflected in the rising levels of water in Mono Lake in eastern California; once nearly drained so that Los Angeles faucets and irrigation systems would be flush, the lake has risen more than ten feet since the mid-1990s, a restoration of considerable significance, given the conflicted history of southern California water politics. There also has been a confounding economic ramification. The success of water conservation may signal a dramatic break from the conventional wisdom that holds that "water use inevitably rises along with economic and population growth," a link often used to justify the spending of large amounts of money to establish new sources of water. Contemporary conservation efficiencies highlight the inefficiencies that were built into past water-delivery systems.[4]

Conservation alone will not resolve all future water requirements, but subsequent spikes in demand may be met and new supplies of water made available through another significant alteration in the West's water regime. Hal Rothman argues, for instance, that we are on the cusp of a new age in water marketing and consumption patterns. Tracking the remarkable changes in western demography and income and noting the sharp shift in the West's economies—from rural and agrarian to urban and service—he points out that one result will be (indeed, already has been) a move to reallocate the waters captured within the region's reservoirs and aquifers. Farmers and ranchers, whose livelihoods were built on ready access to low-cost water, are now in direct competition for water with powerful and much more profitable urban economies. The latest struggle over this vital western resource will produce a new group of winners and losers, Rothman predicts, who will usher in a newly reconfigured West, a region in which once-dominant agricultural interests will play an increasingly diminished role. As for the victors, their triumph will not depend on their ability to throw a punch, a pugilistic fantasy that so captured Mark Twain's nineteenth-century imagination, but on the depth of their pockets. To buy their way into dominance, they will need to throw money around as if it were water.

Notes

1. Elmer Kelton, "Bone Dry," *Texas Monthly,* July 1996, 75, 101–2; Nicole Foy, "Water Capture before High Court," *San Antonio Express-News,* 19 November 1998, 18A; Lisa Vorderbrueggen, "Growth Curb Faces Big Board Vote," *Contra Costa Times,* 12 September 1999; Tony Davois, "Desert Sprawl," Special Issue, *High Country News,* 18 January 1999.

2. Paul Simon, *Tapped Out: The Coming World Crisis in Water and What We Can Do About It* (New York: Welcome Rain Publishers, 1998), 87; for an extended critique of Simon's arguments, see Char Miller, "Water Works," *Texas Observer,* 22 January 1999, 32–33; Wayne B. Solley, Robert R. Pierce, and Howard A. Perlman, "Estimated Use of Water in the United States in 1995," U.S. Geological Survey Circular no. 1200, (Washington, D.C.: U.S. Geological Survey, 1998); Wayne B. Solley, "Estimates of Water Use in the Western States," report to the Western Water Policy Review Advisory Commission, August 1997, 1; *Water in the West: The Challenge of the Next Century,* report of the Western Water Policy Review Advisory Commission, June 1998, ch. 5.

3. Simon, *Tapped Out,* 125–46; William K. Stevens, "U.S. Use of Water Fell in 1980–

95 Even as Nation's Population Rose," *Houston Post*, 10 November 1998, 10A; Douglas P. Shuit, "Water Conservation Efforts Begin to Pay Off in Southland," *Los Angeles Times*, 15 June 1999, A1, 22–23; Solley et al., "Estimated Use of Water in the United States in 1995," 1; Solley, "Estimates of Water Use in the Western United States," 3–17.

4. Ibid.; Char Miller, ed., *Water in the West: A High Country News Reader*, (Corvallis: Oregon State University Press, 2000) tracks some of these issues.

Land and Water on

New Spain's Frontiers

I

"Only Fit for Raising Stock"

Spanish and Mexican Land and Water Rights in the Tamaulipan Cession

Jesús F. de la Teja

In Mexico, land and water have been in constant flux since pre-Columbian times. The aridity of much of the American Southwest and Mexico has always been a limiting factor on land use. The farther north New Spain extended, the more important an issue water became in determining the uses and value of land. The history of this land-granting activity is the accumulation of customs and laws, ancient practices and modern necessities, both in Spain and the New World. Texas is a direct heir of this history, as is evident in the rich and complex history of its trans-Nueces region.[1]

Spanish Colonial and Land-Granting Practices

When Hernán Cortés completed his subjugation of the Aztec empire in 1521, the Spaniards in America already had a generation's worth of practice distributing the spoils of victory. Because the greatest fortune and fame rested in conquest and plunder, the early Spanish colonists showed little interest in acquiring and working their rural estates. The conquistadors had entitled themselves to the product of the Indians' labor on village lands, and this was, by and large, enough. The Spanish crown, which held dominion over the territory and people of the Americas under a papal bull of 1493, acted through various agents to grant land and water to colonists, always maintaining that such *mercedes* had to be perfected through confirmation from the crown.

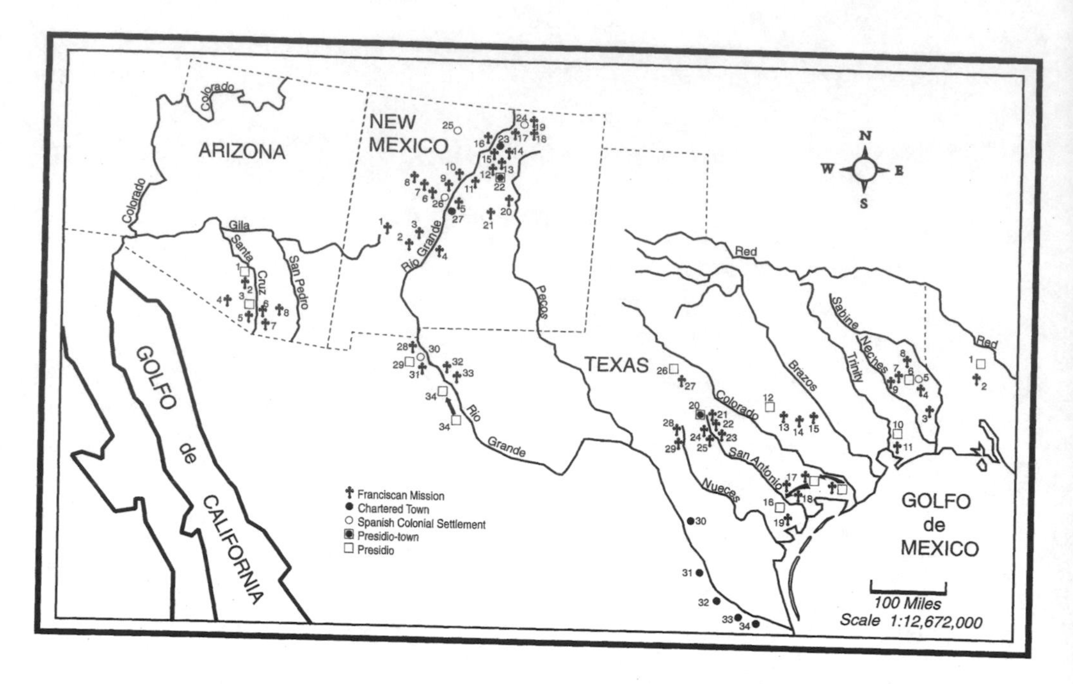

FIGURE 1.1 Eighteenth-century Spanish foundations of the American Southwest. *Arizona:* (1) San Agustín del Tucson (1776–); (2) San Xavier del Bac (1756–); (3) San Ignacio de Tubac (1752–1776), San Rafael de Tubac (1787–); (4) Arivaca (Visita, 1730s–); (5) San José de Tumacácori (1751–); (6) Calabazas (Visita, 1730s–1787); (7) Los Santos Angeles de Guevavi (1732–1775); (8) Sonoíta (Visita, 1730s–1774). *New Mexico* (dates are for post–Pueblo Revolt of 1680 Spanish resettlement): (1) Nuestra Señora de Guadalupe, Zuni (1699–); (2) San Esteban, Acoma (1699–); (3) San José, Laguna (1699–); (4) San Agustín, Isleta (1690s–); (5) Nuestra Señora de los Dolores y San Antonio, Sandía (1760s–); (6) Santa Ana, Tamaya (1695–); (7) Nuestra Señora de la Asunción, Zia (1694–); (8) San Diego, Jemez (1694–); (9) San Felipe, Koots-cha (1694–); (10) Santo Domingo, Khe-wa (1730–); (11) San Buenaventura, Cochiti (1694–); (12) San Diego, Tesuque (1695–); (13) Nuestra Señora de Guadalupe, Pojoaque (1700s–); (14) San Francisco, Nambe (1694–); (15) San Ildefonso, Po-woh-geh (1690s–); (16) Santa Clara, Kha-po (1694–); (17) San Juan, O'kang (1694–); (18) San Lorenzo, Picuris (1730–); (19) San Gerónimo, Taos (1695–); (20) Nuestra Señora de los Angeles, Pecos (1694–1838); (21) Nuestra Señora de los Remedios, Galisteo (1706–1794); (22) Santa Fe (1692–); (23) Santa Cruz de la Cañada (1695–); (24) Ranchos de Taos (1760s–); (25) Abiquiú (1754–); (26) Bernalillo (1698–); (27) Albuquerque (1706–); (28) Nuestra Señora de Guadalupe (1659–); (29) Paso del Norte (1684–;

Presidio, 1684–1773); (30) San Lorenzo (1682–); (31) San Antonio, Senecú (1682–); (32) Corpus Christi, Ysleta (1682–); (33) Nuestra Señora del Socorro (1682–); (34) Presidio San Elizario (first location, 1774–1789; second location, 1789–). *Texas:* (1) Nuestra Señora del Pilar de los Adaes (1721–1773); (2) San Miguel de Linares de los Adaes (1716–1773); (3) Nuestra Señora de los Dolores de los Ais (1716–1773); (4) Nuestra Señora de Guadalupe de los Nacogdoches (1716–1773); (5) Nacogdoches (1779–); (6) Nuestra Señora de los Dolores de los Tejas (1716–1729); (7) Nuestra Señora de la Purísima Concepción de Acuña (1716–1730); (8) San José de los Nazonis (1716–1730); (9) San Francisco de los Neches (1716–1730); (10) San Agustín de Ahumada (El Orcoquisac, 1756–1771); (11) Nuestra Señora de la Luz (1756–1771); (12) San Francisco Xavier (1751–1755); (13) Nuestra Señora de la Candelaria (1749–1755); (14) San Xavier (1746–1755); (15) San Ildefonso (1749–1755); (16) Nuestra Señora de Loreto (La Bahía; first location, 1722–1726; second location, 1726–1749; third location, 1749–); (17) Nuestra Señora del Espíritu Santo de Zúñiga (first location, 1722–1726; second location, 1726–1749; third location, 1749–); (18) Nuestra Señora del Rosario de los Cujanes (1754–); (19) Nuestra Señora del Refugio (1790–); (20) Presidio San Antonio de Béxar (1718–), Villa San Fernando de Béxar (1731–); (21) San Antonio de Valero (Alamo, 1718–1793); (22) Nuestra Señora de la Purísima Concepción (1731–1794); (23) San Juan Capistrano (1731–1794); (24) San José y San Miguel de Aguayo (1720–1794); (25) San Francisco de la Espada (1731–1794); (26) San Luis de las Amarillas (1757–1768); (27) Santa Cruz de San Sabá (1757–1758); (28) San Lorenzo de la Santa Cruz (1762–1769); (29) Nuestra Señora de la Candelaria del Cañon (1762–1766); (30) San Agustín de Laredo (1755–); (31) San Ignacio de Loyola de Revilla (now Guerrero, 1750–); (32) Lugar de Mier (1753–); (33) Nuestra Señora de Santa Ana de Camargo (1749–);(34) Nuestra Señora de Guadalupe de Reynosa (1749–).

Cortés first set the pattern for land distribution in New Spain. He organized land grants into *peonía* units (approximately 21 acres) for foot soldiers and *caballería* units (approximately 105 acres) for mounted conquistadors. Cortés also distributed Indians in *encomienda* (trusteeship in which Indians owed labor in return for the trustee's "protection") in disregard of royal orders, bowing instead to the wishes of his most important followers. These three units of measure underwent transformations as the Mexican frontier moved northward. Encomiendas became extremely difficult to come by and their exploitative capacity was circumscribed by royal decrees, thus limiting their overall importance to the economy. Spaniards of even the lowest status refused to accept the second-class peonía grants, making the caballería the standard grant unit. Over time, both terms came to refer to units of land area, rather than measures of entitlement, in an increasingly complex land distribution system.[2]

In the early 1530s, the second *Audiencia* authorized grants to all Spaniards who founded new towns and settled as farmers. The crown authorized governors and the viceroy to make or authorize grants under certain conditions, although final confirmation remained with the crown. Each original settler received a grant of one or two caballerías on condition that he establish a homestead and improve the land for a period of ten years (later reduced to six), after which time the grantee could dispose of the land. Officials also made caballería grants from the large tracts that became part of the royal domain as their Indian owners disappeared due to epidemics and *congregación* (relocation).[3]

During this early period, grazing stock shared the land with crops. Following European medieval tradition, livestock grazed freely on all vacant land and in *agostaderos* (fields after the harvest). As a matter of fact, with the exception of lands specifically appropriated, all pastures, woods, and waters were subject to common use. As herds proliferated and conflicts arose between stockmen and farmers, town councils began to grant stockmen reserves where they could maintain their herds and flocks. By 1567, the practice was so extensive that the crown issued ordinances establishing the grant of *estancias* (or *sitios*) *de ganado mayor* (approximately 4,300 acres) and *menor* (approximately 1,900 acres). Eventually, these stockmen also extended their dominion by buying out Indians, occupying vacancies between established estates, and appropriating lands without respecting legal forms.[4]

As the frontier moved northward into drier regions, the size of estates grew. The absence of sedentary Indian populations for labor, along with limited water resources, favored animal husbandry. In well-watered valleys crop-oriented agricultural enterprises thrived, and many large estates became mixed-use operations. These *haciendas* were both economic and social units. Land owners or their agents exercised considerable economic and judicial power over their workers and attempted to make their estates as self-sufficient as possible. Some estates, for instance that of the Marqueses de Aguayo, became so large that they were composed of a number of haciendas. *Hacendados* collected the labor forces on these properties into villages of permanent workers or made use of work gangs from neighboring Indian communities.[5]

Below the hacienda in size and prosperity was the *rancho,* a midsized to small property typically of a mixed-use nature, worked directly by the owner and a few hands. Ranchos tended to be located on more marginal lands or at the leading edges of the frontier. In Nuevo Santander and Texas, "rancho" referred to even the largest estates because of the limited extent of improvements. In general, these properties tended to suffer from the same defects of title as the larger estates, that is, their owners did not have confirmation of title from the crown.[6]

The farther north the frontier expanded, the more important a role water played in land acquisition. Land was plentiful, water was not. As crown property, flowing water was available for common uses such as washing, cooking, fishing, and watering livestock. Particular uses of water, for example, operating a mill race or an irrigation system, required a specific grant of water rights. The only time water rights accompanied a surface grant was when the property in question contained a spring. Also, from earliest times the extralegal use of water paralleled the extralegal and illegal use of land, which in turn allowed the crown to apply the same remedies toward the former that it did toward the latter. *Composición,* auction, and outright grant created water rights separate and distinct from land rights. As a result, by the mid-sixteenth century land titles for agricultural estates specified whether or not they carried irrigation rights. At Saltillo, for example, late-sixteenth-century mercedes combined one or two *ojos de agua* (springs) with two to five caballerías and an amount of pasture land.[7]

Consequently, from the earliest days of Spanish settlement, land granting along the northward expanding frontier was on a large scale, a process of negotiation between the crown and grantees, intimately tied to livestock grazing, and, to a considerable extent, conditioned by the availability of water. It placed Spanish livestock raisers in competition, first with sedentary Indian peoples and then with nonsedentary hunter-gatherers. Over time, these dynamics produced a flexible system of real property rights on which the land acquisition and tenure practices of Nuevo Santander rested.

The Title Process

Throughout the colonial period the acquisition of property out of the royal domain consisted of the same basic steps. First, a claimant made a *denuncia* (claim) before a local authority having jurisdiction and stated the purpose and need for the land or water. The local official before whom the claim was filed forwarded the matter to the appropriate court for an opinion on the suitability of the request. An authorized official, acting on the competent attorney's opinion, then ordered a *vista de ojos* (visual inspection), survey, and appraisal. Following these actions, the tract was placed at auction and the authorized official approved the winning bid. The local authority in charge of the proceedings then placed the grantee in possession. The grantee received a *testimonio* (witnessed copy) of the proceedings. Finally, the grantee sought confirmation of his title from the appropriate royal authority. Thus, grants issued by *cabildos* (town councils), *audiencias* (high courts), governors, and viceroys had to be confirmed by the crown or whichever authority the crown entrusted with confirmation.[8]

The costs and bureaucratic hurdles associated with following such a complicated procedure caused widespread noncompliance. At various times during the colonial era, and for various reasons, the crown attempted to reform and streamline the confirmation process. In need of a new fleet following the 1588 Armada disaster, Philip II decreed in 1591 that defective titles should be cured through composición, that is, payment of a fee to the royal treasury based on an appraisal of the property in question. Also, the crown ordered that grants and composiciones should be made only through auction to the highest bidder. Until 1618, the viceroy had confirmation authority; thereafter until the 1690s, the king, either directly or through the viceroy, appointed *jueces de ventas y composiciones* (judges for sales and compositions) to oversee the title process and make confirmation.[9]

Although the language remained that of merced (grant), all transfers from the crown to private ownership of estates were sales.[10] In the north, where land was of limited value without water, the more powerful ranchers used composición to acquire title to sources of water. New towns and their settlers and missions continued to receive outright grants of crop land and irrigation water and communal grazing lands. Over time there developed a patchwork of landholding that included large and small estates to which the owners sometimes held title and sometimes not; estates held in trust by missionaries for their Indian communities; town tracts that were subdivided into private home lots *(solares)*, individually owned agricultural fields *(suertes)*, municipal rental lands *(propios)*, and communal grazing lands *(ejidos)*; and a vast expanse of untitled royal domain that might be used by all subjects until acquired by an individual through composición.[11]

The Spanish monarchy's decline in the late seventeenth century created a crisis in government, eventually led to the War of the Spanish Succession and transfer of the crown to the House of Bourbon, and affected the distribution of the royal domain in the New World. Already in the 1690s the power to designate a competent land judge fell to a member of the Council of the Indies, who appointed a member of each Audiencia to serve as *juez privativo de ventas y composiciones.* This land judge, in consultation with the Audiencia's *fiscal* (attorney), was to confirm titles and oversee composición and sale proceedings. Once Philip V established a secure hold on the Spanish crown, he began a reform process that extended through the reign of his son, Charles III. In 1735, Philip reasserted overall control over land matters by restoring the confirmation authority to the crown. This policy proved a failure, however, because it once again led to long and costly proceedings that resulted in noncompliance. Consequently, in 1754 Ferdinand VI returned confirmation authority to the Audiencias and delegated to the viceroys and Audiencias responsibility for naming *jueces privativos de composiciones y tierras.*[12]

The first colonists of Nuevo Santander received grants under rules that date from the reign of Ferdinand's successor, Charles III. He ordered a new administrative system, the *intendencia,* which was instituted in New Spain during the 1780s. The new organization divided New Spain into twelve jurisdictions, the head of each being responsible for fiscal and administrative matters within his jurisdiction, including sales and composiciones. The system included a Junta Superior de Real Hacienda, a treasury council headed by the viceroy, that held final confirmation authority.[13]

The last important colonial-era reform effort with regard to transfers of royal domain to private ownership occurred at the beginning of the nineteenth century, when the crown severely restricted the size of mercedes. This ruling came as a result of an investigation conducted by the *intendente* of San Luis Potosí, whose jurisdiction included Texas and Nuevo Santander, into the size and value of rural grants within his territory. Reports of vast tracts selling for ten pesos per sitio or less led to an 1802 order of the Junta Superior de Real Hacienda limiting grants to four sitios per wealthy claimant and two sitios per commoner. The junta also fixed land prices based on the property's relationship to water: sixty pesos per sitio for *tierra de regadío* (irrigable), thirty pesos per sitio for *tierra de temporal* (dry-farming), and ten pesos per sitio for *tierra de pasto* (pasture). In 1805 the crown backed up this unrealistic plan by issuing it as a *cédula.*[14]

Following adoption of the Federal Constitution of 1824, the Mexican government transferred jurisdiction over public land issues to the states. The national government retained only the authority to prevent settlement by immigrants within close proximity to the United States and to grant lands in federal territories such as California and New Mexico. In California, the secularization of missions and colonization efforts led officials to grant large tracts of former mission lands to civilian settlers. In New Mexico, efforts to stimulate economic activity led to the distribution of large grazing land tracts as far north as present-day Colorado.[15]

Coahuila y Texas, which had a pressing need to regulate Anglo-American settlement in the province, passed its first colonization law in 1825. It allowed immigrants up to one sitio for farming and livestock raising, pricing the land according to the traditional categories: irrigable, dry-farming, and pasture. Additionally, Mexican nationals could purchase up to eleven sitios. To regulate and encourage orderly settlement by the immigrants, Coahuila y Texas employed an *empresario* system, whereby it granted exclusive colonization rights over a defined area to individuals who agreed to bring a stated number of families to Texas.[16]

These legal procedures involved in gaining government land were not set in stone, but varied according to changes in the administrative philosophy, economic concerns of government, and local conditions. Officials made grants on an individual basis and, along the northern frontier, usually long after the land had been occupied and put in production. On the one hand, individuals claiming a tract of land exposed themselves to losing it in the auction process, but, on the other hand, those using land without gaining title exposed themselves to having the land denounced by a competitor. In the last decades of the colonial era, authorities took into consideration the limited water resources of northern New Spain in designing regional land-granting practices. Mexican independence accelerated the pace of land distribution throughout the frontier, including Tamaulipas, which attempted to find a way of occupying its desert northern reaches.

Seno Mexicano: *The Desert by the Sea*

Perceived from the earliest colonial days as an inhospitable and dry region, the area along the Gulf of Mexico coast began to receive serious attention from colonists only near the end of the seventeenth century. Indian hostilities and the harshness of the country between the Sierra Madre Oriental and the coastal Tamaulipas Range prevented effective Spanish occupation of the entire region until the mid-eighteenth century.

During the late sixteenth century and throughout the seventeenth century, occasional incursions into the area did take place: slaving expeditions, shipwrecks, and probings northward from Tampico and eastward from Monterrey in search of new pastures. Also during the seventeenth century, settlers from Nuevo León began making irregular trips to the salt deposits north of the Río Grande. After La Salle landed at Matagorda Bay in 1685, the Spanish sent land and sea expeditions in search of the French colony. However, none of these left a good description of the region between the Río Grande and the Nueces River.[17]

French incursions into the northern gulf region required action, however, and the result was Spanish occupation of strategic points in what came to be known as the province of Texas. Beginning in 1716, the Spanish built presidio-mission complexes at three strategic locations, Los Adaes (present Robeline, Louisiana), San Antonio, and Bahía del Espíritu Santo (Matagorda Bay), which became the basis for the only successful Texas colonial settlements. From these essentially religious-military communities, the Spanish tentatively occupied the countryside with open-range livestock operations (called ranchos) dependent on the abundant water courses from the San Antonio River Valley northward. Distance from markets; the presence of Apache, Comanche, and other native peoples who successfully resisted subjugation; and the availability of ample natural and human resources closer to the interior of the viceroyalty meant underdevelopment for Texas throughout the colonial period.[18]

Spanish occupation of Texas, precipitated as it was by foreign encroachments along the present Louisiana-Texas border region and the missionary zeal of the Franciscans, left a huge expanse of land south of the Nueces River to be explored and occupied. Serious exploration of the region did not occur until the mid-eighteenth century, when increasing urgency in dealing with Indian raiders who fled to the safety of the mountains and fears of foreign encroachment forced the crown's hand. José de Escandón, an accomplished soldier and Indian fighter, became responsible for the project.[19] As a first step to colonizing the region, he organized a series of exploratory expeditions to canvas the entire territory. Capt. Joaquín de Orobio Basterra of the Texas presidio of La Bahía (then situated in Mission Valley, near Victoria) conducted one of these expeditions in early 1747. He discovered, among other things, that the Nueces River did not flow into the Río Grande. Orobio also determined that there was "no place suitable for settlement from the area of the Nueces River to the Río Grande, essentially because of the great absence of fresh water."[20]

A decade after Orobio made these observations, another inspector, military engineer Agustín López de la Cámara Alta, reached a similar conclusion: "In the direction of Texas, all the northern part of the countryside is deserted and sterile, even wanting water, which is only available in the Rio Grande and Nueces River and a few small water holes."[21] Cámara Alta's advice, therefore, was that the crown should save its money for more worthy projects. According to him, only along the Río Grande could settlement be undertaken, "this territory being worthless except for raising livestock."[22]

In the course of the next half-century all inspections of the region reached the same conclusion. Trusted with determining the veracity of Indian reports of Englishmen on the coast, in September 1766 Diego Ortiz Parrilla commanded a small troop that explored along what later became known as Padre Island and also reported the same absence of water and hostility of the environment.[23] Twenty years later, José de Evía, a Spanish naval officer, conducted the most detailed reconnaissance of the Spanish gulf coast up to that time. Within two days of departing from the mouth of the Río Grande in August 1786, Evía's journal begins recording the dry and hostile character of the coastline: "The land is extremely low, and from the few dunes there are many small mouths *(boquillas)* . . . . Our drinking water drawing short, people were distributed on the land to search for some . . . which proved useless, because not even by digging did they find any."[24] Traveling as far north as the southern terminus of Matagorda Bay, the expedition turned back for lack of supplies. Due to the absence of water between the Nueces and Río Grande, Evía found little evidence of Hispanic settlement. Not surprisingly, the only livestock activity he encountered was along the Nueces River, which he described as "populated with cattle haciendas."[25]

Observations during the Mexican period, at the very moment when settlers from the riverine communities were establishing ranches throughout the region, confirm colonial-era judgments. Jean Louis Berlandier, botanist for the Mexican frontier inspection tour of 1828–1829, noted the limited potential of the territory between San Antonio and Laredo: "All that space between the banks of the Nueces and the Rio Bravo is comprised of large, little-wooded plains, seldom broken by streams. Some pools spaced at intervals determine the day's march. The further one goes to the south, the less fertile the earth becomes, due to its dryness."[26] Berlandier's judgment was unequivocal: "Agricultural industry will never be able to flourish between the Nueces and the Rio Bravo del Norte. The countryside does not lend itself to the irrigation which is absolutely necessary in that region. For several centuries it will remain nothing but an immense prairie where herds can be bred."[27]

The consistency in the descriptions of the coastal region's physical characteristics suggest reasons why economic development was so limited. Shallow and exposed anchorages, problematic winds, lack of adequate fresh water, unnavigable rivers, and remoteness from population centers all served as disincentives to development during the colonial and Mexican periods.

Nuevo Santander–Tamaulipas–South Texas

Beginning in 1749, José de Escandón led the most concerted colonization project in New Spain since the sixteenth century. The region corresponds to today's Mexican state of Tamaulipas and South Texas from Corpus Christi to Laredo to the mouth of the Río Grande. On submitting his report on the state of affairs in Nuevo Santander in 1755, Escandón could boast of founding more than thirty settlements—towns and missions—and establishing agricultural and livestock operations as far as the north side of the Río Grande.[28]

Escandón did not initially make individual grants to the colonists. He gave three general reasons: shortage of preferred sites, which would lead to disputes; continued arrival of good settlers, some of whom might be more worthy of good grants than the original colonists; and a shortage of time in which to carry out all the activities entrusted to him. Escandón also contended that there was no one he would trust with the job.[29]

His carefully planned colonization strategy, which included a string of strategically located settlements and missions, effectively occupied much of the province within a decade. Not surprisingly, the only part of his plan that failed was the occupation of the trans-Nueces region. Although Escandón managed to obtain relocation of the Texas outpost of La Bahía to the San Antonio River, he was not able to raise the necessary settlers to establish a corresponding town on the Nueces.[30]

Inability to fully develop the trans-Nueces region during the colonial period is not surprising, especially in light of the Texas experience. The absence of precious metals and convenient trade routes, combined with the success of some the region's nonsedentary peoples in retaining their hunting and raiding lifeways, stymied expansion from the outposts of San Antonio and La Bahía. Additionally, the Franciscan missions associated with these two settlements had preempted the best pasture lands along the best streams in the region, leaving slim pickings for the civilian population until secularization began in the 1790s (a process not unlike that in California until the 1820s). The only organized colonization effort in Texas worth mentioning is the 1729–1731 effort to bring Canary Islanders to settle in the province. Prohibitive costs, the unsuitability of these Old World peasants to the hostile frontier conditions, and the absence of economic incentives doomed the project to failure. Although their coming resulted in establishment of the only chartered town in Texas and the systematic distribution of solares and suertes, outside the town proper an informal system of land tenure ruled. San Antonio's 1801 town council explained the situation eloquently:

> Since this province was founded we do not know of any sale other than the eighteen sitios for one hundred pesos made to the late Captain Don Luis Antonio Menchaca. . . . The other ranch possessions have been made gratuitously to some of the more meritorious *vecinos* of the place; some of these lands to this day remain settled and others entirely abandoned without the legitimate owners making any petitions regarding them, with neither they nor anyone else attempting new possessions. From all of which can be determined the little or no value that may be assigned to these lands.[31]

Ironically, given the availability of plentiful water supplies in the region, the need for compact settlement created by chronic Indian raiding produced the only water disputes of colonial Texas. With five missions, a presidio, and a civilian town clustered within a ten-mile stretch of the upper San Antonio River, the available water supplies from the river and its tributary, San Pedro Creek, had to be carefully husbanded. The settlement site had originally been selected because of its facility for irrigation. Between 1718 and 1778 an irrigation system was developed, composed of weirs, canals, aqueducts, and mill runs, some of which continue to operate to this day. The result was the development of a productive agricultural complex that met both local needs and allowed for limited exports to other parts of region. Downstream from San Antonio, at La Bahía, and in eastern Texas, where the terrain precluded the gravity-flow irrigation systems preferred by Spanish colonists, the land required dry farming, strictly subsistence agriculture, and an even greater reliance on animal husbandry for income.[32]

Unlike Texas, Nuevo Santander was the product of a more systematic and secular colonization effort. Although the trans-Nueces region shared some of the same obstacles to development with Texas, in particular distance from markets and an absence of precious metals, it did have the advantage of having fewer Indian problems and no competition for resources from an aggressive mission system. It seems that Escandón must have learned the lessons from other areas, Texas in particular, where the missionaries had been given the lead in colonization, with the result that conflict between the religious and civilian communities retarded development.[33] Given that Escandón had not made individual grants, when José Tienda de Cuervo carried out his inspection of the Río Grande settlements in 1757, he found the colonists dissatisfied with the land tenure situation. Some areas were becoming overcrowded, at others there were boundary disputes. Therefore, Tienda de Cuervo recommended that lands should promptly be distributed on an individual basis. It took ten years for royal officials to respond to Tienda de Cuervo's recommendation and Escandón's pleas with the appointment of a special commission.[34]

Juan Armando de Palacio, a military man, and José de Ossorio de Llamas, counsel to the Audiencia, conducted the *Visita General* (general inspection tour), during which they made the first land grants in the region. Their mission was to organize land distribution by appointing surveyors and experts and make sure everything was properly recorded. They arrived in the Río Grande Valley at Laredo on 15 May 1767 and proceeded downstream over the next four months. Their legacy was the distinctive *porción* survey pattern that continues to dominate landholding along the border between Laredo and Reynosa.[35]

The commission faced two important challenges in carrying out its mission. First, the limited availability of water meant that they had to devise some system that allowed maximum access to what water did exist, hence, the long, narrow tracts of approximately 9/13 by 11 to 14 leagues (although, depending on the number of porciones needed, the width could be shortened and the length stretched). Second, at some settlements disputes had arisen between the original settlers and later arrivals over the amount each individual was to receive. The solution here was to give roughly the same amount of pasture land, two sitios, but vary the amount of farmland, measured in caballerías, according to date of arrival. Also requiring measurement were the town tracts, each of which was to consist of four sitios, including plazas, solares, propios, and ejidos. In return for the grants, settlers agreed to improve their properties, organize for self-defense, and take possession within two months of the grant.[36]

By the time the Visita General began its work, a number of families already conducted considerable ranching activity on the eastern side of the Río Grande. Cámara Alta described the 1757 settlers of Reynosa, Camargo, and Revilla as "rich" on the livestock and salt trade. About thirty miles downstream of Laredo, José Vázquez Borrego established his Nuestra Señora de los Dolores ranch in 1750. When Escandón visited the ranch in 1753, it showed promise and he made a verbal grant to Borrego. Eventually Dolores consisted of fifty sitios de ganado menor and twenty-five sitios de ganado mayor. In 1755, another important rancher, Tomás Sánchez, established himself to the northwest of Dolores, where he occupied fifteen sitios that eventually became the town of Laredo.[37]

Escandón had also allowed settlement on the eastern side of the Río Grande downstream at Camargo. The town's founder, Blas María de la Garza Falcón, had a ranch across from the town. When the porciones were surveyed, his son succeeded in getting the tract numbered as porción 80, a double porción to which Garza Falcón was entitled as founder. A group of families that in 1752 had established themselves away from the river at the site of some wells called El Sauz managed to get their lands included in the 1767 distribution of Camargo porciones.[38]

Clearly, the ranching character of the region was well established by the time the Visita General completed its work. Grants made from that time forward specifically include language characterizing them as pasture lands. A copy of the title to Vicente Ynojosa, "Las Mesteñas, Pititas y La Abra," for thirty-four sitios de ganado mayor, one sitio de ganado menor, eight caballerías, and 273,625 square *varas,* made in 1798, affords a glimpse at the thinking behind these grants. Ynojosa's petition of 11 June 1794 stated that he was a resident of Reynosa, the son of Juan José Ynojosa, a founding resident of the town. He petitioned for the tract in view of his father's and his own merits and "by virtue of having some field chattels and not having my own land to graze them on, and there being vacant next to my father's land, lands . . . which are not inhabited by anyone except some gentile Indian nations, and by me."[39]

Even when the possibility existed for a broader type of grant, the residents of the region did not think in terms other than ranching. Following the flood of 1802, the long-suffering residents of Reynosa asked for and received permission to move the town. Francisco Ballí, who gave up the twenty-five sitios of higher ground east of the original site where the town was relocated, received, among other compensation, a right to pick out an equal amount of land on the banks of the Río Grande. He found two parcels, each undersize, still available. One was "very overgrown and not suited for stock raising," and the other was forty leagues away. The land he did want, although fifty leagues (150 miles) distant, did have two advantages over the others: it was twenty-five sitios in one piece; and, as his agent put it in the petition, "it is the one he considers best suited for the purpose of raising large stock, which is the function for which he needs it."[40]

The two decades immediately preceding the Mexican War of Independence was a time of great activity in the north. Peace with the Comanches and a decline in depredations by other Indian peoples led to renewed interest in the countryside. The population continued to grow and the government seemed interested in making land available not just to the powerful, but to anyone interested in pursuing stock raising who had the seed stock to establish a claim.[41] According to available evidence from the San Patricio first class land-grant records at the Texas General Land Office, Spanish officials completed about sixteen grants and at least fourteen individuals established occupation or made an original denuncia between 1800 and 1815. Certainly, other claims fell by the wayside and remain lost to history as a result of the Mexican War of Independence.

Once the political situation began to settle down following independence in Tamaulipas, the new name of the former Spanish province of Nuevo Santander, land distribution resumed. The continuity between postindependence proceedings and colonial ones is striking. The steps to obtaining a grant were essentially the same: limitations on grant size remained in place, the price for the land remained effectively unchanged, and the purposes claimed for the sought-after land remained identical.

The state's first colonization law, issued on 15 December 1826, had two important objectives: first, to rekindle interest in occupying the available public domain by issuing terms for settlement as generous as those offered by neighboring Coahuila y Texas (in some instances the terms are identical); and, second, to clear up ownership questions by making abandoned lands subject to colonization and forcing owners of excessively large tracts to alienate the excess. Lawmakers were also optimistic, at least as far as the region where most of the public domain was available—the trans-Nueces—in calling for land to be categorized as grazing, dry-farming, or irrigable agricultural land.[42]

One of the earliest surviving documents from the period is the title to "Santa Ysabel, San Martín, and Buena Vista." The complicated story of these three tracts, all adjacent to the 1781 grant to José Salvador de la Garza, "Espíritu Santo," can be reduced to the efforts of Manuel Prieto, heir to the eastern part of Espíritu Santo, to gain control of lands between the Laguna Madre and his. When José Treviño, a rancher with cattle and no land, applied for the tract known as "San Martín," Prieto stepped forward to challenge not only Treviño, but to claim "Santa Ysabel" and "Buena Vista" as well. The authorities sided with Treviño, but because no one stepped forward to claim the other two tracts, Prieto received them. When Manuel de la Garza and Rafael García, who had been using those lands since 1823, heard about the government's action, they filed their own petition to prevent the grant from being made. The response of the Matamoros city council when it was asked to comment on the case is equally instructive on the attitudes toward the use of land in the region:

> The stocks of cattle owned by said Garza and García are considerable, and for that reason sufficient to occupy more lands than those solicited; to which may be added that being just, in proportion they may likewise stock them, sell and cultivate them and benefit them [in] other ways. If the foregoing should appear in favor of the Citizens Garza and García so that to them may be allotted the lands which they ask for, it will not appear in favor of Citizen Manuel Prieto, who occupied an inverse attitude from them, because if the first mentioned do not own one league, the latter owns as many as ten leagues, and still another ten leagues which he is enjoying without having sufficient stock with which to occupy one or the other; in fact not enough to use even the half of the lands, and

for which reason he has on them a number of renters with a great many cattle, who Prieto allows to occupy the same upon the payment to him of corresponding rents, simply because he cannot alone occupy them with his own cattle.[43]

Two principles emerge from the above assessment. One is that all parties agree on the paramount nature of livestock as the determining factor for who should get land and in what quantity. The other is that the ideal in land tenure is a correlation between the number of stock and the amount of land. These beliefs made their way into the land law of 17 November 1833, which made the relationship between cattle and land in Tamaulipas even clearer. Article 18 states:

> Denouncements made to the government for stock raising land will be admitted: (1) if the claimant is not the owner of large estates; (2) if, in the judgment of the government, the necessity of the tract is proven and the claimant obligates himself to occupy each sitio with at least one hundred head of cattle or horse stock, within the precise term of four months counting from the date on which the title is issued; but even in such cases, the government shall not be able to grant more than six sitios to a single individual.[44]

What the law did was formalize a long-standing practice that already relied on the claimant establishing a need for the land on the basis of livestock interests.

That same year, Tamaulipas state government further protected the settlers of the Río Grande frontier by issuing a special order that applied only to them. In view of the abandonments occasioned by the Indian wars that occurred during the war-of-independence period, the governor provided for grants of up to five sitios at a cost of ten pesos per sitio. To be eligible, a settler had to prove residence and ownership of sufficient stock to justify the grant and commit to retaining the property for twenty years following the concession. Article 3 of the decree required "the respective Ayuntamiento [to] inform itself by previous justification, whether the interested party who may present himself, has lands of his own and stock with which to occupy them."[45] In this action also, state authorities considered vital both the equity issue and livestock ownership.

Tamaulipas governments made more than one hundred grants in the trans-Nueces region before sovereignty transferred to Texas. Although most of these grants have no surviving documentation of the original proceedings, there is sufficient evidence to show that the patterns established in the colonial period continued to function in the early national period. These patterns included: the consistent description of tracts as agostadero; the persistent use of the phrase "sitio de ganado mayor" (when the sitio had been standardized and the distinction between ganado mayor and ganado menor no longer mattered with relation to the size of the grant); and continued efforts to make ranching accessible to a broad socioeconomic spectrum of the population. These are important indicators of the development of a cattle culture in which land, water, and animal were part of a whole.

Conclusion

The South Texas case demonstrates a number of features of a dynamic, flexible, and open ranching society. It was a combination of frontier responses to local conditions and centuries-old Hispanic legal principles. The locally evolved land system transcended Spanish and Mexican sovereignties and lasted until the latter half of the nineteenth century, when its demise was brought about by the arrival of an alien set of rules and large amounts of capital from the United States. Among the continuities from the colonial to the Mexican period were limitations on land-grant size, demonstration of need as basis for a grant, and expectations for land use limited to livestock purposes. Also surviving as a matter of course were granting procedures including division of the granting and confirmation authority between local and state-level authorities and separate visual inspections and surveys.

The continuities from the Mexican to the state period have been social, economic, and demographic in character. From 1848 until the early twentieth century, South Texas remained a sparsely settled, ranching dominated, mostly Hispanic region, a condition that remains true for much of the area today. Richard King and the cattle barons who followed recognized the suitability of many of the adaptations made by *rancheros* and *vaqueros* and continued to rely on these practices into the twentieth century. Also surviving and, in fact, remaining the central theme in regional development, is the issue of limited water availability.[46] In how the land has been subdivided, used, and transferred, water has been the critical variable in the equation.

Another important element has been Spanish-Mexican land and water law, important aspects of which remain at the root of present Texas jurisprudence, even though Anglo-American jurists have not always done a good job of interpreting that law. Hans Baade's thorough monograph on the evolution of Texas water law points out what some of these errors were and how the *Valmont Plantations* Texas court decision of 1961 finally recognized the complexity of the Spanish-Mexican law on the matter. The court's recognition that all grants out of the public domain had to emanate from the crown, including irrigation rights, overthrew the simplistic notion that Spanish water rights were riparian. Neither were these rights based on prior appropriation, as New Mexico and Arizona courts had held. In studying the history of the relationship between land and water in the trans-Nueces region, the rights of present owners, some of whom hold title from their original-grantee ancestors, are better served.[47]

As the Hispanic character of South Texas refuses to dissolve in the American "melting pot," there is a renewed urgency to study the history of Spain and Mexico in those parts of Texas (approximately one-seventh the area of the state), where Euro-American settlement extends back a century and a half before 1848. By studying how Spanish and Mexican settlers and government officials dealt with the distinctive circumstances of the South Texas region, we might gain useful insights for successfully managing existing land and water resources throughout the Southwest.

Notes

1. Hans W. Baade, "The Historical Background of Texas Water Law: A Tribute to Jack Pope," *St. Mary's Law Journal* 18, 1 (1986): 1–98; Betty E. Dobkins, *The Spanish Element in Texas Water Law* (Austin: University of Texas Press, 1959); Enrique Florescano, *Origen y desarrollo de los problemas agrarios de México, 1500–1821* (1971; reprint, Mexico City: Ediciones Era, 1976); Ricki S. Janicek, "The Development of Early Mexican Land Policy: Coahuila and Texas, 1810–1825" (Ph.D. diss., Tulane University, 1985); Guadalupe Rivera Marín de Iturbe, *La propiedad territorial en México, 1301–1810* (Mexico City: Siglo Veintiuno Editores, 1983); Michael C. Meyer, *Water in the Hispanic Southwest: A Social and Legal History, 1550–1850* (Tucson: University of Arizona Press, 1984); Martha Chávez Padrón, *El derecho agrario en México*, 9th ed. (Mexico City: Editorial Porrúa, 1988).

2. Florescano, *Origen y desarrollo* 29, 38–41; Marín de Iturbe, *La propiedad territorial*, 179–81; Lesley Byrd Simpson, *The Encomienda in New Spain: The Beginning of Spanish*

Mexico, rev. ed. (Berkeley: University of California Press, 1966), 56–64; for information on the numbers, origins, and social and economic distribution of the encomenderos, see Robert Himmerich y Valencia, *The Encomenderos of New Spain, 1521–1555* (Austin: University of Texas Press, 1991).

3. Enrique Florescano, "The Hacienda in New Spain" in *Colonial Spanish America,* ed. Leslie Bethell (Cambridge, U. K.: Cambridge University Press, 1987), 255–56; Marín de Iturbe, *La propiedad territorial,* 179.

4. Florescano, "The Hacienda," 258–59; François Chevalier, *Land and Society in Colonial Mexico: The Great Hacienda* (Berkeley: University of California Press, 1963), 84–114; Philip W. Powell, *La guerra chichimeca (1550–1600)* (Mexico City: Fondo de Cultura Económica, 1977), 20–21; Elinor G. K. Melville, *A Plague of Sheep: Environmental Consequences of the Conquest of Mexico* (Cambridge, U. K.: Cambridge University Press, 1994), 116–50; Marín de Iturbe, *La propiedad territorial,* 183.

5. Chevalier, *Land and Society,* 148–84; Charles H. Harris III, *A Mexican Family Empire: The Latifundio of the Sánchez Navarros, 1765–1867* (Austin: University of Texas Press, 1975), 5–10; Florescano, "The Hacienda," 261–68.

6. Chevalier, *Land and Society,* 287–88; Eric Van Young, *Hacienda and Market in Eighteenth-Century Mexico: The Rural Economy of the Guadalajara Region, 1675–1820* (Berkeley: University of California Press, 1981), 107–9; Jesús F. de la Teja, *San Antonio de Béxar: A Community on New Spain's Northern Frontier* (Albuquerque: University of New Mexico Press, 1995), 97–99.

7. Baade, "Historical Background of Texas Water Law," 37–38; Janicek, "Development of Early Mexican Land Policy," 13–14; Michael C. Meyer, *Water in the Hispanic Southwest,* 117–20, 133–44.

8. This outline of the title process is based on examination of hundreds of land grant files in the Texas General Land Office (hereafter TGLO) dating from the Spanish and Mexican periods and on Janicek, "Development of Early Mexican Land Policy," 22. A simplified description of the process is found in John Sayles and Henry Sayles, *A Treatise on the Laws of Texas Relating to Real Estate, and Actions to Try Title and For Possession of Lands and Tenements* (St. Louis, Mo.: The Gilbert Book Co., 1890), 1: 60–62, 124–25, 157–58, 169–70.

9. The principal laws governing the distribution, granting, and selling of royal lands are found in Book 4, Title 12, *Recopilación de leyes de los reinos de las Indias.* See also Chevalier, *Land and Society,* 266–70; Florescano, *Origen y desarrollo,* 32–33; Janicek, "Development of Early Mexican Land Policy," 16.

10. Baade, "Historical Background of Texas Water Law," 35, notes that the misleading terminology has led even U.S. Supreme Court justices astray.

11. General laws regarding public land uses are found in Book 4, Title 17, *Recopilación de leyes de los reinos de las Indias;* see also Baade, "Historical Background of Texas Water Law," 57–60; Harris, *A Mexican Family Empire,* 5–6; Janicek, "Development of Early Mexican Land Policy," 16–18; Meyer, *Water in the Hispanic Southwest,* 29–30.

12. Florescano, *Origen y desarrollo,* 34; Janicek, "Development of Early Mexican Land

Policy," 16; Marín de Iturbe, *La propiedad territorial*, 185–87; Victor Westphall, *Mercedes Reales: Hispanic Land Grants of the Upper Rio Grande Region* (Albuquerque: University of New Mexico Press, 1983), 16. The text, with a gloss, of the 1754 instructions is found in Wistano Luis Orozco, *Legislación y jurisprudencia sobre terrenos baldios* (1754; facsimile reprint, Mexico City: Ediciones El Caballito, 1974), 59–95.

13. Chávez Padrón, *El derecho agrario*, 169; Janicek, "Development of Early Mexican Land Policy," 33–35; Orozco, *Legislación y jurisprudencia*, 96–99.

14. Baade, "Historical Background of Texas Water Law," 39–40; Janicek, "Development of Early Mexican Land Policy," 38–39.

15. David J. Weber, *The Mexican Frontier, 1821–1846: The American Southwest under Mexico* (Albuquerque: University of New Mexico Press, 1982), 180–81, 190–206.

16. Sayles and Sayles, *A Treatise on the Laws of Texas*, 1: 86–99; Virginia H. Taylor, *The Spanish Archives of the General Land Office of Texas*, (1955; reprint, St. Louis, Mo.: Ingmire Publications, 1983), 31–68.

17. Armando C. Alonzo, *Tejano Legacy: Rancheros and Settlers in South Texas, 1734–1900* (Albuquerque: University of New Mexico Press, 1998), 16–17; Peter Gerhard, *The North Frontier of New Spain*, rev. ed. (Norman: University of Oklahoma Press, 1993), 359–63. The history of Gulf of Mexico exploration through the end of the Spanish colonial period has been ably synthesized in the trilogy by Robert S. Weddle, *Spanish Sea: The Gulf of Mexico in North American Discovery, 1500–1685* (College Station: Texas A&M University Press, 1985); *The French Thorn: Rival Explorers in the Spanish Sea, 1682–1762* (College Station: Texas A&M University Press, 1991); *Changing Tides: Twilight and Dawn in the Spanish Sea, 1763–1803* (College Station: Texas A&M University Press, 1995).

18. The best available monograph on the history of Spanish colonial Texas is Donald Chipman, *Spanish Texas, 1519–1821* (Austin: University of Texas Press, 1992). A more brief synthesis focusing on demographic, social, and economic aspects is Jesús F. de la Teja, "Spanish Colonial Texas" in *New Views of Borderlands History*, ed. Robert H. Jackson (Albuquerque: University of New Mexico Press, 1998), 107–30.

19. On Escandón, see Hubert J. Miller, *José de Escandón: Colonizer of Nuevo Santander* (Edinburg, Tex.: Nuevo Santander Press, 1980); Francis Lawrence Hill, *José de Escandón and the Founding of Nuevo Santander: A Study in Spanish Colonization* (Columbus: Ohio State University Press, 1926).

20. José de Escandón, *Reconocimiento de la Costa del Seno Mexicano* (Mexico City: Archivo de la Historia de Tamaulipas, 1946), 72.

21. Agustín López de la Cámara Alta, *Descripción general de la nueva colonia de Santander* (Mexico City: Archivo de la Historia de Tamaulipas, 1946), 159. All translations by author.

22. Ibid., 16.

23. Weddle, *Changing Tides*, 29–31.

24. Jack D. L. Holmes, ed., *José de Evía y sus reconocimientos del Golfo de México, 1787–1796* (Madrid: Ediciones José Porrúa Turanzas, 1968), 162.

25. Ibid., 182–83.

26. Jean Louis Berlandier, *Journey to Mexico during the Years 1826 to 1834* (Austin: Texas State Historical Association, 1980), 2: 420.

27. Ibid., 2: 423.

28. The best study examining the founding and early development of far South Texas is Florence Johnson Scott, *Historical Heritage of the Lower Rio Grande* (San Antonio, Tex.: The Naylor Co., 1937).

29. Scott, *Historical Heritage*, 45; Oakah L. Jones Jr., *Los Paisanos: Spanish Settlers on the Northern Frontier of New Spain* (Norman: University of Oklahoma Press, 1979), 67; Alejandro Prieto, *Historia, geografía y estadística del estado de Tamaulipas* (facsimile reprint, Mexico City: Manuel Porrúa, Librería, 1975), table facing p. 190.

30. Carlos E. Castañeda, *Our Catholic Heritage in Texas, 1519–1936* (1936–1958; reprint, New York: Arno Press, 1976), 3: 150–55, 177–88; Jones, *Los Paisanos*, 65–67.

31. de la Teja, *San Antonio de Béxar*, 100–101. In chapter 5 of this book, de la Teja discusses the development of San Antonio's countryside as a ranching frontier in the eighteenth century. Similar work remains to be done on La Bahía.

32. de la Teja, "Spanish Colonial Texas," 120–21.

33. Ibid., 121.

34. Scott, *Historical Heritage*, 59–60; Alonzo, *Tejano Legacy*, 35–36; Gilberto M. Hinojosa, *A Borderlands Town in Transition: Laredo, 1755–1870* (College Station: Texas A&M University Press, 1983), 6–8.

35. Scott, *Historical Heritage*, 60–98; Alonzo, *Tejano Legacy*, 36–39.

36. The activities of the commissioners in the five Rio Grande Valley jurisdictions, Camargo, Laredo, Mier, Revilla, and Reynosa, are recorded in the proceedings *Auto de la general visita*, which were filed in the archives of each town. Certified copies made in the late nineteenth century are on record in the Archives and Records Division of the TGLO, under the title of "Act of the Visit of the Royal Commissioners of 1767 to the town of . . ." These records form the basis of this and succeeding paragraphs covering the work of the *visita*.

37. Alonzo, *Tejano Legacy*, 30–32; Cámara Alta, *Descripción general*, 25–26; Gilberto R. Cruz, *Let There Be Towns: Spanish Municipal Origins in the American Southwest, 1610–1810* (College Station: Texas A&M University Press, 1988), 91–93.

38. Cruz, *Let There Be Towns*, 90–91; Texas General Land Office, *Guide to Spanish and Mexican Land Grants in South Texas* (Austin: Texas General Land Office, 1988), entry 64.

39. Vicente Ynojosa, Spanish Collection 136: 10, TGLO, Austin, Texas.

40. José Francisco Balli, "La Barreta," San Patricio 1–831, TGLO, Austin, Texas.

41. This and succeeding paragraphs discussing the distribution of land are based on the author's examination of more than one hundred title files in the Archives and Records Division, TGLO. See also Alonzo, *Tejano Legacy*, 55–61.

42. "Colonization Law of the State of Tamaulipas, 15 December 1826" in *The Laws of Texas, 1822–1897*, comp. and ed. H. P. N. Gammel (Austin, Tex.: The Gammel Book Co., 1898), 1: 454–59.

43. Copy of abstract of title, English translation to Santa Ysabel, San Martin, Buena Vista, Cameron County, Texas, p. 3, James Wells Papers, Center for American History, University of Texas at Austin.

44. *Estado de Tamaulipas: Ley de colonización. Espedida por el honorable congreso de este estado, y sancionada por el ecsmo. sr. gobernador D. Francisco Vital Fernández* (Ciudad Victoria, Mexico, 17 November 1833).

45. Scott, *Historical Heritage,* 128.

46. The obstacles to applying technological solutions to the water problems of South Texas through the beginning of the twentieth century were so insurmountable that it is impossible to find historical sites associated with the issue. In their study of sites involving water works, T. Lindsay Baker, Steven R. Rae, et al., *Water for the Southwest: Historical Survey and Guide to Historic Sites,* ASCE Historical Publication no. 3 (New York: American Society of Civil Engineers, 1973), the authors did not designate a single site south of San Antonio for inclusion in their project.

47. Baade, "Historical Background of Texas Water Law," esp. 91–98.

2

Water, the Gila River Pimas, and the Arrival of the Spanish

Shelly C. Dudley

Following the Gadsden Purchase in 1853 and the subjugation of the hostile Western Apaches, Euro-Americans traveled to the Arizona Territory to transform the land into cultivated fields and to develop new communities. Many of these pioneers came to the Middle and Upper Gila River Valleys to establish homesteads, create small towns, and use the flow of the Gila River. The actions of these settlers diminished the water supply of the Gila River Pimas and forever changed their way of life. But these American settlers were not the first Europeans to come into contact with the Pimas living along the Gila River. Although evidence is inconclusive about whether the expeditions of Father Marcos de Niza (1539) or Francisco Coronado (1540) came into direct contact with the Pimas, by the late seventeenth century the Spanish had extended their frontier from Mexico into Pimería Alta, into what is now southern Arizona. Missions in southern Arizona were established as far north as San Xavier del Bac near Tucson, but the Spanish influence went beyond these communities, reaching to the Pimas and Maricopas living along the Gila River.

Just how significant was the Spanish impact on the Pimas and their environment? Historian Robert MacCameron, using colonial New Mexico as his model, asserts that environmental change intensified from Native American to Spanish to Euro-American settlement. One of the ways to determine the consequences of the Spaniards' presence is to examine their impact on the farming practices and water use of the Indians on the Gila River; although the Spanish introduced new seeds, tools, and animals, the

Pimas were able to maintain a rough balance with their natural surroundings until the arrival of the Americans.[1]

Although central Arizona is a desert, its environment has supported people for more than a millennium. The Hohokam civilization adapted to the dry climate by constructing irrigation canals using only wooden sticks. Archaeologists have found more than 100 miles of prehistoric acequias in the Florence–Casa Grande area and between 125 and 300 miles of Hohokam canals in the Salt River Valley. Estimates of irrigated acreage from these early canals range between 65,000 and 250,000 acres, with the inhabitants growing corn, beans, cotton, and squash. Scholars still dispute the reasons behind the disappearance of the Hohokam, but all agree that the communities and irrigation structures they had constructed were no longer in use by the end of the fifteenth century. There is also considerable debate whether there was a direct link between the Hohokam and the later Pima peoples. Some archaeologists and a few of the younger, late-twentieth-century members of the tribe advance this argument; most local, contemporary Gila Pimas do not. Nor does Pima tradition and oral history support such claims. According to storyteller Juan Smith, when the Pimas first reached the Gila River Valley they did not irrigate the land, but farmed only using the rainfall. "But they saw all about them a great canal system left by the Vipishad, a Puebloan people." After viewing the ditches, the Gila River Pimas decided to construct irrigation canals to water their fields.[2]

Written evidence documenting the extent and significance of the Pima irrigation practices before the time of the first Spanish exists in the correspondence and diaries of the early Spanish missionaries and travelers. One of the most thorough explorers was Father Eusebio Francisco Kino, who arrived in Sonora in early 1687 as part of the Jesuit mission program to bring the aboriginal inhabitants of the New World into full control of the Spanish empire. Although the Spaniards still feared attacks by the Indians in northern Sonora following the Pueblo Revolt of 1680, Father Kino soon established more than two dozen missions and *visitas,* including Nuestra Señora de los Dolores, San Xavier del Bac, and Tumacácori. The southern Pimas gathered to the padre, listened to his teachings, and received his gifts, including several new types of crops; at the missions, the Indians planted fruit trees and wheat, first introduced into the Southwest by Kino.[3]

Making the first of his four visits to the Gila River Pimas in November

1694, Father Kino arrived at the first *ranchería* of El Tusonimo, which he renamed La Encarnación. Kino did not record the existence of any Pima-built irrigation canals, but he mentioned the "large aqueduct with a very great embankment" near the Hohokam ruins at Casa Grande on his trip there in 1697. Lt. Juan Mateo Manje, acting as Kino's military escort, reported on the agricultural fields of the Gila River Pimas, but did not mention the use of irrigating structures. The Spanish soldier wrote that the Indians were friendly and provided the group with corn and beans, indicating the Pimas had a surplus of food to give to the travelers. Manje, however, noted that if the Cocomaricopas near Gila Bend constructed irrigation canals, they would have been able to cultivate more land.[4]

Because neither Kino nor Manje recorded seeing irrigating ditches among the Gila Pimas, archaeologist William Doelle suggests that canals were not used by these Indians until after the initial Spanish contact. Doelle asserts that the "early Gila Pima agriculture was dependent primarily on flood waters, but may in some cases have employed minimal techniques of artificial irrigation." Although Doelle also states the Pima may have used small ditches to carry the river water to fields, a "sophisticated irrigation system" was in all likelihood not possible. More complete evidence of the existence of irrigation canals does not appear until 1744, when Father Jacobo Sedelmayr of Bavaria reported the Indians at Sudacson growing wheat using irrigation.[5]

When did the Gila River Pimas start practicing irrigation to water their fields? Although Doelle believes it started within a fifty-year period after the Spanish first visited the Pimas, it is also possible that the Pimas learned of irrigation from their contacts with the Indians in Pimería Baja (southern Sonora) before Kino's arrival. This scenario can be sustained by examining the environmental conditions of the region that the Gila River Pima inhabited. According to anthropologist Paul Ezell, the area is "characterized by the most highly developed irrigation system in the Southwest," a point ethnologist Frank Russell extended: "At the first appearance of the Pimas it may be presumed that they used the canals already constructed by their predecessors, hence they would be dull indeed if they could not maintain irrigation systems sufficient for their needs." Edward Castetter and Willis Bell also indicated that *aboriginal* Gila Pimas had used canal irrigation. Other documentary evidence of irrigation in southern Arizona before the arrival of the Spanish, not including the structures left by the Hohokam, exists in the writings of Manje and in the 1694 diary of Cristóbal Martín Bernal. Bernal, who traveled down the San Pedro River to the Gila with Father Kino, wrote about San Augustín, which had "very good lands all under irrigation and in the great abundance of grains which the natives harvest."[6]

Although the Jesuits established irrigation at their missions in southern Arizona, none of the Spanish fathers reported teaching irrigation techniques to the Gila River Pimas, nor did they spend enough time at the villages to help construct the acequias. Although Paul Ezell asserts that the early Spaniards were not irrigation engineers, he also argues it is difficult to accept the idea that acculturation of irrigation without direction could have arisen within a generation, especially when the Opas and Cocomaricopas were in the same situation and they did not start irrigating by the time of Sedelmayr's visit in the mid-eighteenth century. Seeming to accept both theories, John Ressler contends that although the Spanish immigrants who came to the borderlands were already familiar with the scarcity of water and brought with them methods of irrigation and water storage, they encountered Indians who had learned to conserve water and practice irrigation. Whether introduced or assisted by the Spanish entrada, by the middle of the eighteenth century the Gila River Pimas had an irrigation system containing dams of brush construction in the riverbed as well as laterals and ditches delivering water to their fields.[7]

Scholar Douglas Hurt asserts, "After contact with the Spanish explorers and missionaries, Indian agriculture in the Southwest changed dramatically," but in fact this transformation of the landscape occurred over hundreds of years. Before the arrival of the Spaniards, the Gila Pimas were already cultivating maize, beans, pumpkins, and cotton. Father Kino introduced wheat, chickpeas, peppers, and melons, and the Gileños (as the Spaniards called the Gila River Pima) selected the new crops that would adapt best to their needs and environment. Although aboriginally maize was the leading crop of the Pimas and could be used as a food in many different forms, wheat soon became the major crop of the Gila River Pima. Their adaption of it was also shrewd: Wheat complemented the growing cycle because it was a winter crop, planted in December and harvested in June, whereas maize was first sown in April and gathered in June; a second crop could be planted in July and reaped in October, before the wheat was ready for planting. Thus, wheat provided a crop when cultivated foodstuffs were not as plentiful and helped distribute the farm labor over the growing season, while expanding needed food supplies. The addition of wheat, which produced a heavier yield of grain and more than doubled the capacity of an irrigated acre to support the Pimas, expanded the amount of irrigated land the Gila River Pimas controlled. This supplement to their production also enabled the Pima to trade for different goods with neighboring tribes or assist those whose supplies might not be as substantial.[8]

Spanish visitors recorded the agricultural productivity of the Gileños in their journals and letters over the next century. Among them was the Jesuit Father Ignacio Xavier Keller from Moravia, who arrived in Sonora in 1731 and later received jurisdiction over the region of northern Pimería Alta. He traveled along the Gila River between 1736 and 1737, visiting the Pima villages, baptizing the young babies, "where he beheld the fertile lands of that valley." One of his occasional companions, Father Jacobo Sedelmayr, traveled to the Gila River looking to establish a mission and recorded his observations on the land between the settlements of Tuquisan and Sudacson. Sedelmayr noted that "on both banks of the river and on its islands have much green land. The Indians sow corn, beans, pumpkins, watermelons, cotton, from which they make garments, and the Indians of Sudacson plant wheat with irrigation canals."[9]

Other visitors commented on the fertility of the Gila River Valley and the agricultural production of the Pimas. In the "Rudo Ensayo," the anonymous Jesuit writer provided a geographical picture of Sonora between 1761 and 1762. The author described the ranches on the bottom lands along the Gila as "fruitful and suitable for wheat [and] Indian corn" and the cotton left in the field after harvesting as more than that raised in Sonora. Father Kino had earlier remarked that the natives of Pimería Alta produced very good cotton fabrics. Miguel Venegas, a Mexican Jesuit writing on the natural history of California, noted the Pima ranchería of Judac was "a pleasant fertile country, well watered by means of trenches . . . from the Gila."[10]

Following the expulsion by royal decree of the Jesuit priests from Mexico in 1767, the gray-robed Franciscans came to Sonora and Pimería Alta to aid the military in the pacification of the Indians and to extend the mission program. A number of these friars left extensive diaries of their journeys in southern Arizona. Father Francisco Garcés reported that all of the pueblas near the village of La Encarnación del Sutaquison "raise large crops of wheat, some of corn, cotton, calabashes, etc., to which end they have constructed good irrigating canals." On another trip, Garcés described leaving the road near San Simón y Judas de Uparsoytac (Gila Bend) and descending to the Gila River, where water was being taken out for irrigating the fields. Father Pedro Font, who accompanied Garcés on a number of journeys, went to the fields with the Pima governor of San Juan Capistrano de Uturituc, approximately twenty miles west from Casa Grande, and recounted that, "These milpas are enclosed by stakes, cultivated in sections, with five canals or draws and are excessively clean." On that same trek, Capt. Juan Bautista de Anza remarked on the vast cultivated lands of the Pima near the village of Sutaquison (Vah Ki). On an expedition to open a route to Monterey, California, in 1774, Capt. de Anza traveled across southern Arizona and recorded that the Gila "fields of wheat which they now possess are so large that, standing in the middle of them, one cannot see the ends, because of their great length."[11]

At the time of the Spanish entrada, the Gila River Pimas already employed several agricultural implements. They used the digging stick, considered to be the most important farm tool, for planting, clearing the ground of bushes, and digging irrigating canals. (It is surmised that the Hohokam dug their massive canals with the digging stick as their only tool.) The hoe, made of ironwood with a sharpened blade, cleared away the weeds and loosened the dirt around the plants. The wooden shovel, carved from cottonwood or mesquite, was used to throw out the dirt loosened by the digging stick in the construction of the irrigation canals. The origin of the shovel is unclear, however; whereas Russell contends that it was aboriginal, others, such as Castetter and Bell, report that Pima informants state it came from the Spanish and that their tribesmen initially removed the loosened dirt in baskets. Other post-Spanish farm tools that the Pimas employed were a stick with an iron blade, called a dibble, and a three-pronged rake fashioned from a mesquite tree. The Spanish also introduced the wooden plow to the Indians, but according to Russell, its use began only after the influx of the Euro-Americans created the need for surplus crops in the mid-1850s; this first plow used a wooden yoke similar to the type found in southern Europe. Prior to its introduction, however, the traditional farming tools of the Pimas had been adequate for their agricultural production.[12]

Although the Gila River Pima planted crops to supplement their food supply, the natural resources of the region constituted an essential part of their diet. Among the most important plants were the mesquite and screwbean. The Indians gathered the fruit from the mesquite tree, ground the pods into flour, and baked it into loaves of bread. Although Father Font complained that the Pimas had an "evil odor" because they ate *péchita* ground from the mesquite pod, which when mixed with other seeds had a strong smell, he admitted they were "gentle and of good heart." Although wild gourds were not eaten, the Pima crafted them into utensils and rattles. Father Kino recorded that the Indians at the ranchería of San Andrés, close to Casa Grande, were fishermen and used nets and tackle to capture fish throughout the year. The fish helped sustain the Pimas along with the maize, beans, and calabashes. While the wild plants and cultivated crops provided a major portion of their diet, the Indians also hunted jackrabbits, deer, and mountain sheep. Lt. Manje noted a large pile of bighorn sheep horns at Tusonimoho and the early American travelers also reported that the Pimas ate fish caught from the Gila River and hunted bighorn sheep.[13]

Gila River Pima probably did not keep any domestic animals, except possibly dogs, until after contact with the Spaniards. The majority of Gileños saw their first horse when Father Kino came riding into their villages. Initially fearful of the animals, the Cocomaricopa youngsters thought the horses might eat them and were relieved to learn they ate grass. When Kino lost a few ponies along the Gila River in 1702, the local tribesmen gathered up the animals and returned them to him. Although Kino provided cattle and horses to the missions and settlements in southern Sonora, there is no record of his bringing livestock to the villages along the Gila River. By 1740, however, the Gila Pimas were raising horses and small livestock, probably having acquired them through trade with the Spanish settlements to the south. The horse was considered a prize possession and was used for traveling and for trading, but not for eating. In 1775, Father Francisco Garcés noted the Indians at Ranchería de San Juan Capistrano had poultry and horses, some of which were bartered with the soldiers. Almost seventy-five years later, Lt. William Emory reported that the Pimas still held the ownership of horses and mules in high regard.[14]

Cattle were not acquired by the Gila River Pimas until the Mexican period, sometime between 1820 and 1830. One reason for the lack of cattle might have been the poor range conditions along the Gila River, a fact that many of the Spanish travelers noted. Father Font, in reporting that he thought the region occupied by the Gileños would be good for missions, observed there were two drawbacks—the ongoing harassment by the Apaches and the shortage of pasturage. According to Font, the country could barely sustain horses let alone the breeding of cattle. Capt. Juan Bautista de Anza also recorded that although there was "abundance of water and good lands for crops . . . there is little or no pasturage." It was not until the California gold rush of the late 1840s that the Pimas purchased from Mexican settlers and the Tohono O'odham the cattle to increase their agricultural production and for trade with overland migrants. While additional animals were obtained by the tribe, a single ox might have been used for plowing the fields of a whole village or family cluster. A few of the Americans remarked on eating beef at the Pima villages during their overland journey, but cattle usually were not kept for food and were only eaten after the owner's death or the demise of the animal.[15]

At almost the same time as Mexico achieved its independence from Spain in 1821, the Euro-Americans started to make their appearance in Arizona. The early trappers and explorers moved along the Gila River hunting beaver or traveling to California, and a few of these individuals left accounts of the kind treatment they received after visiting the Pima villages. During the Mexican War, Gen. Stephen W. Kearny and his army traveled from Fort Leavenworth through Arizona on their way to the Pacific Coast to aid in the conquest of California. As a member of the advance guard, Lt. Emory of the Topographical Engineers recorded his meeting with the Pimas in November 1846: "The town was nine miles distant" from the army encampment, "yet in three hours our camp was filled with Pimos loaded with corn, beans, honey, and zandias (watermelons). A brisk trade was at once opened." Emory went on to describe the luxuriant crops, irrigated fields, and the generosity of the Pimas in providing corn, wheat, and beans to the soldiers. "We were at once impressed with the beauty, order, and disposition of the arrangements for irrigating and draining the land." Emory noted that the fields were subdivided with ridges of earth in plots of land two hundred feet by one hundred feet for ease in irrigating; he commented that the cattle were few in number and used for plowing the fields.[16]

When the Mormon Battalion under Col. Philip St. George Cooke arrived at the Gila River in December 1846, he too benefited from the Pimas' willingness to trade with the soldiers, with whom they exchanged corn, watermelons, and other produce for clothing or beads. As he traveled along the Gila River toward California, Cooke wrote: "The march was fifteen miles. The whole distance was through cultivated grounds, and a luxuriantly rich soil; there is a very large zequia [acequia] well out from the river."[17]

Other members of the Mormon Battalion kept journals and recorded the exchange of shirts and clothing for pumpkins and other food. Christopher Layton wrote, "We traded buttons cut from our clothes for cakes of bread, and also some old clothing for corn, beans, molasses, squash." Later, at the Maricopa villages, Layton noted the rich cultivated fields and remarked that Col. Cooke suggested that the Saints return to the region because it "would be a good place." Another member of the group, Henry Standage, noted that the Indians at the Pima villages were "well provided for with grain &c. Purchased some beans and meal of them so that we have full rations once more." Some of these same Mormons would return to Arizona and settle in the Salt and Gila River Valleys.[18]

After the treaty of Guadalupe Hidalgo in 1848 and the absorption of California and the northern Arizona Territory into the United States, the Pima villages became a haven for the traveling Americans. Numerous Forty-niners left diaries of their journeys across the southwestern desert in which they recorded the help they received from the Gila Pimas. Gen. Frederick Townsend reported in 1849 that the Pimas "supply the Government troops with all their surplus grain" and the Indians appeared to be pleased to have the Americans among them. George Evans, traveling through southern Arizona during the summer heat, recorded that the Gila River Pimas "molest no one, and look to trade to the emigrants such articles as corn, wheat, beans, peas, and an occasional trade of horses or mules." The Indians wanted clothing in exchange for cornmeal, the first of that foodstuff Evans had seen since leaving Mexico. Before leaving the region, Evans also noted that the Pimas had constructed a dam that diverted water through canals to the irrigated fields.[19]

While surveying the boundary between the United States and Mexico in the summer of 1852, John Bartlett wrote that the Pimas brought food and water to the Americans and acted as guides. He noted the extensive irrigation system and cultivated fields, reporting that the lands were "better irrigated, their crops are larger, and the flour which they make from their wheat and maize is quite as good as the Mexicans make." Bartlett concluded that no tribe in the United States was more deserving of the attention of philanthropists than the Pimas. Unfortunately, with the settlement of the Arizona Territory by the Euro-Americans, even the philanthropists could not have helped the Pimas when the waters of the Gila River ran dry in the late nineteenth and early twentieth centuries.[20]

During the Hispanic period, the Pimas had had a permanent source of water in the Gila River. Some scholars believe there may have been periodic droughts that cut the river's flow. Others dispute that belief by noting that the river sank underground at La Encarnación, only to resurface at Sudac-sson, and later join with the Salt River; the Spanish and other contemporaries might have perceived this "disappearance" as the river having dried up. Later, in the early nineteenth century, Euro-American travelers also noted this phenomenon, but thought it due to the Indians diverting all the water from the Gila. The Pima were not, but they had dug wells near where the river went underground to boost their water supply. Later, they would dig additional wells when the increased salinity of the Lower Gila River compelled them to locate other sources of potable water.[21]

With the arrival of the Euro-Americans trappers in the 1820s along the Gila River in Arizona and later the Mormon settlers in the Upper Gila River Valley, the demands on and the quality of that water supply changed drastically. Fur trappers, when they destroyed the indigenous beaver population, also destroyed a network of small dams along the Gila and its tributaries from which the Pima had benefited. The introduction of livestock throughout the watershed also marred water quality. Large herds thoroughly grazed native grasses, leading to increased soil erosion and thus silted river waters. This, in turn, escalated the number of floods and attendant damage, so much so that the floods altered the natural course of the river. This complex set of environmental changes, along with the diversion of the Gila that the new settlers instituted to meet their needs, diminished the Pimas' access to a stable water supply. This had a dramatic impact on their agricultural productivity: By the last quarter of the nineteenth century, the Indians on the Gila River no longer were able to furnish a surplus food supply to the American migrants, let alone provide for their own consumption.[22]

The arrival of the Spaniards had altered the life of the Gila River Pimas. Although it is debated whether the Pimas practiced irrigated agriculture before the Spanish entrada, the new plants, such as wheat, and tools, such as a rudimentary plow, the missionaries and the military introduced to the region increased the amount of land the Pima put into production and thus aided in the creation of a surplus of crops. But these alterations were readily absorbed into the Gileños way of life. The same would not be true following the Americans' arrival in the 1840s. The American presence would rapidly increase a few years later when, in the aftermath of the Mexican-American War, the United States purchased the land south of the Gila River from Mexico. By 1860, the Pimas were still able to produce more than 440,000 pounds of surplus wheat for sale to the Butterfield Overland Mail Stage Line, but this fuller use of the new seeds and tools introduced by the Spaniards a hundred years earlier started to destroy the Pima environment and the culture it once had sustained. Moreover, and more to the point, the fertility of the land and cessation of hostilities with the Apaches allowed American migrants to settle the Middle and Upper Gila River Valleys and to divert its stream flows onto their new fields. The Pimas, who had expanded their subsistence agriculture into a commercial enterprise with the arrival of these new settlers, were no longer able to maintain power over their water supply and land. Losing hold of these, the Pimas lost control of their future.[23]

Notes

1. Robert MacCameron, "Environmental Change in Colonial New Mexico," in *Out of the Woods: Essays in Environmental History*, ed. Char Miller and Hal Rothman (Pittsburgh: University of Pittsburgh Press, 1997), 79–97.

2. Thomas E. Sheridan, *Arizona: A History* (Tucson: University of Arizona Press,

1995), 12, 19; Amadeo M. Rea, *At the Desert's Green Edge: An Ethnobotany of the Gila River Pima* (Tucson: University of Arizona Press, 1997), 6, 147.

3. Donald Cutter and Iris Engstrand, *Quest for Empire: Spanish Settlement in the Southwest* (Golden, Colo.: Fulcrum Publishing, 1996), 120–24; Bernard L. Fontana, *Entrada: The Legacy of Spain and Mexico in the United States* (Tucson: Southwest Parks and Monuments Association, 1994), 93–94.

4. Eusebio Francisco Kino, *Historical Memoir of Pimeria Alta*, ed. Herbert Bolton (Berkeley: University of California Press, 1948), 127, 171; Juan Mateo Manje, *Unknown Arizona and Sonora, 1693–1721* (Tucson: Arizona Silhouettes, 1954), 120–24; William Doelle, "The Gila Pima in the Late Seventeenth Century" in *The Protohistoric Period in the North American Southwest, A.D. 1450–1700*, ed. David R. Wilcox, Arizona State University Anthropological Research Paper no. 24 (1981): 62.

5. Doelle, "The Gila Pima in the Late Seventeenth Century," 63; Jacobo Sedelmayr, *Jacobo Sedelmayr: Missionary, Frontiersman, Explorer in Arizona and Sonora, 1744–1751*, ed. Peter Dunne (Tucson: Arizona Pioneer's Historical Society, 1955), 23; Paul H. Ezell, "The Hispanic Acculturation of the Gila River Pimas" *American Anthropological Association* 63, no. 5, pt. 2 (October 1961): 33, 36.

6. Paul H. Ezell, "The Conditions of Hispanic Piman Contacts on the Gila River," *America Indigena* 17, no. 2 (April 1957): 164; Ezell, "Hispanic Acculturation," 36–37; Frank Russell, *The Pima Indians*, ed. Bernard L. Fontana (Tucson: University of Arizona Press, 1975), 87; Edward F. Castetter and Willis H. Bell, *Pima and Papago Indian Agriculture* (Albuquerque: University of New Mexico Press, 1942), 40; Cristóbal Martín Bernal, "Diary, November, 1694," trans. from Spanish to English, copy from National Archives, Laguna Niguel, Record Group 75, Bureau of Indian Affairs, Phoenix Area Office, Gila River Indian Reservation.

7. Ezell, "Hispanic Acculturation," 37. Ezell also contends that the Spaniards who came in the seventeenth century were miners, missionaries, and military explorers, but not farmers or irrigation engineers. It is difficult to support this hypothesis considering the work done at the Spanish missions in Sonora. John Q. Ressler, "Indian and Spanish Water-Control on New Spain's Northwest Frontier," in Oakah Jones, *The Spanish Borderlands: A First Reader* (Los Angeles: L. L. Morrison, 1974), 233, 237.

8. R. Douglas Hurt, *Indian Agriculture in America: Prehistory to the Present* (Lawrence: University Press of Kansas, 1987), 43; Tonia Horton, personal interview, Tempe, Arizona, 14 May 1998; Russell, *Pima Indians*, 50, 75, 90; Robert A. Hackenberg, "Pima and Papago Ecological Adaptations," in *Handbook of North American Indians: Southwest*, vol. 10 (Washington, D.C.: Smithsonian Institution Press, 1994), 168.

9. Cutter and Engstrand, *Quest for Empire*, 149–51; Carl Hayden, *A History of the Pima Indians and the San Carlos Irrigation Project*, 89th Cong., 1st sess., S. Doc. 11, reprinted 1965, 8–10.

10. Russell, *Pima Indians*, 103; Hayden, *History of the Pima Indians*, 11–13; Miguel Venegas, *A Natural and Civil History of California* (London, 1759; reprint, New York, N.Y.: Readex Microprint, 1966), 2: 184.

11. Herbert Eugene Bolton, *Anza's California Expeditions: Opening a Land Route to California*, vol. 2 (New York: Russell & Russell, 1966), 127, 387–88; Hayden, *History of the Pima Indians*, 12–14; Herbert Eugene Bolton, *Anza's California Expeditions: The San Francisco Colony*, vol. 3 (New York: Russell & Russell, 1966), 19–20, 217–18. Bolton places Uturituc between the current communities of Sacaton and Sweetwater.

12. Russell, *Pima Indians*, 97–99; Castetter and Bell, *Pima and Papago Indian Agriculture*, 138; Hurt, *Indian Agriculture in America*, 44.

13. Doelle, "The Gila Pima," 64; Russell, *Pima Indians*, 87, 91; Kino, *Historical Memoir*, 194; Herbert Eugene Bolton, *Anza's California Expeditions: An Outpost of Empire*, vol. 1 (New York, N.Y.: Russell and Russell, 1966), 236; Rea, *At the Desert's Green Edge*, 31; John Russell Bartlett, *Personal Narrative of Explorations and Incidents in Texas, New Mexico, California, Sonora, and Chihuahua*, vol. 2 (New York: D. Appleton & Co., 1854), 240.

14. Ezell, "Hispanic Acculturation," 44–45.

15. Hayden, *History of the Pima Indians*, 12; Ezell, "Hispanic Acculturation," 44–47; Bolton, *An Outpost of Empire*, 264; Bolton, *Opening a Land Route to California*, 122; Russell, *Pima Indians*, 85.

16. Lt. Colonel W. H. Emory, *Notes of a Military Reconnaissance, from Fort Leavenworth, in Missouri, to San Diego, in California* (Washington, D.C.: Wendell & Van Benthuysen, 1848), 82–84.

17. Philip St. George Cooke, *The Conquest of New Mexico and California* (Albuquerque: Horn and Wallace, 1964).

18. Christopher Layton, *Christopher Layton: Colonizer, Statesman, Leader* (Kaysville, Utah: Christopher Layton Family Organization, 1966), 62; Frank Alfred Golder, *The March of the Mormon Battalion from Council Bluffs to California, Taken from the Journal of Henry Standage* (New York: The Century Co., 1928), 197–99; Hayden, *History of the Pima Indians*, 21–23.

19. George W. B. Evans, *Mexican Gold Trail: The Journal of a Forty-Niner* (San Marino, Calif.: Huntington Library, 1945), 152; Hayden, *History of the Pima Indians*, 23–24.

20. Bartlett, *Personal Narratives*, 219, 232–33, 263.

21. Ezell, "Hispanic Acculturation," 10–11.

22. Henry F. Dobyns, "Who Killed the Gila?" in *Water in a Thirsty Land: Sonoran Desert and Mountain River History* (Prescott, Ariz.: Pinon Press, 1978), 17–19.

23. Shelly Dudley, "Pima Indians, Water Rights, and the Federal Government, *U.S. v. Gila Valley Irrigation District*," (master's thesis, Arizona State University, 1996), 8–9; MacCameron, "Environmental Change." MacCameron also discusses the dramatic impact of the Anglo-Americans in relation to the Spanish influence on agriculture in the Rio Grande Valley.

3

"Between This River and That"

Establishing Water Rights in the Chama Basin of New Mexico

Sandra K. Mathews-Lamb

New Mexico Pueblo Indian land and water rights, first officially recognized by the Spaniards as early as the sixteenth century, plague today's U.S. court system. Difficulties concerning water rights continually resurface, although "virtually all of the state's surface waters are fully appropriated."[1] In other words, almost "every acre-foot of surface water already belongs to someone."[2] Whereas ownership of water rights seems clearly defined, priority dates are still hotly debated. This chapter will explain the historical background for the recent fight over priority dates in northern New Mexico, the home of the first Spaniards who ventured deep into the heart of Pueblo lands to establish a permanent Spanish frontier settlement.[3]

Early New Mexico Settlements

This region was occupied more than one thousand years ago by the Anasazi, or, more properly, the Pueblo ancestors, who had inhabited the mesa tops and canyons in Arizona, New Mexico, Utah, and Colorado for centuries. These ancient people learned how to use existing surface water to grow maize, introduced by the early inhabitants of the Valley of Mexico. Until the 1270s, these early New Mexicans built homes and an elaborate culture based on an animist religion that depended on the seasonal cycles of the earth and the changes they produced. By 1300, the effects of a severe twenty-five-year drought forced the Pueblo ancestors to leave their mesa-top and cliff homes and migrate to the river bottoms and valleys to the south. They continued cultivating beans, squash, maize, and pumpkins and altered their lifestyle to incorporate the changes that living in valleys demanded.[4]

The arrival of Spaniards during the 1520s in southern Arizona and perhaps New Mexico disrupted the Anasazi lifestyle. Some of these intrusive Spaniards, including those led by Nuño de Guzmán, invaded southern regions of today's American Southwest to take Native Americans as slaves for work in Spanish mines, haciendas, and homes. Word spread quickly about the Europeans who destroyed communities and wrenched people from their homes. Thus, the 1539 Zuni pueblo of Hawikuh's fatal reception for Estebanico, a black slave from Azimore, Morocco, was not surprising.[5] Spanish Viceroy Antonio de Mendoza had sent Fray Marcos de Niza and Estebanico on a quest to discover whether the tales of rich trading cities to the north were true.[6] Upon his return to Mexico City, Marcos de Niza dutifully reported that he had seen one of the Seven Cities of Cíbola.[7] Shortly thereafter, Viceroy Mendoza authorized a full-scale exploration of the northern province to be led by Francisco Vásquez de Coronado.

Leading the first Spaniards to explore the northern Río Grande Valley, Coronado arrived in New Mexico in 1540 and remained until 1542. Coronado intended to find out if Marcos de Niza told the truth regarding the Seven Cities of Cíbola, so with the guidance of Niza, Coronado retraced the friar's footsteps from the south and west into New Mexico. He quickly realized that Niza had stretched the truth about seeing one of the Seven Cities of Cíbola. Not deterred and hoping to find a city to match Mexico City in wealth and glory, Coronado decided to press on. He traveled east into the Río Grande Valley by the early winter of 1540. Determining it to be an excellent location for wintering, Coronado set up camp and began organizing exploratory expeditions. After less than two years, Coronado realized that New Mexico's lands held no great value and he ordered the expedition back to New Spain. For more than a generation, New Spain ignored the northern frontier as a result of Coronado's failure to find wealth. Further exploration was also hindered by two important events to the south. Spaniards had recently discovered La Bufa, the rich silver mine near Zacatecas, into which Spanish investors poured great energy and resources to develop riches. The Spanish government also became engaged in a struggle with the Chichimecs of the Sierra Madre Range, who found their new Spanish neighbors to be a nuisance. However, despite these distractions, the northern frontier had not been entirely forgotten.[8]

After several illegal expeditions into New Mexico in the late 1580s and early 1590s, the viceroy reconsidered settlement of the northern frontier.[9] Antonio de Espejo, who led an unapproved expedition into New Mexico in 1582, begged King Philip II to entrust him with exploring and settling the northern frontier. He promised to serve the king with greater advantage than any others who vied for the contract.[10] Yet, Spain had just issued a decree giving New Spain's viceroy the power to select the recipient of the contract, and the viceroy ignored Espejo's application.

Two powerful rivals for the New Mexico contract soon emerged. Juan Lomas y Colmenares, an Andalusian, had the wealth necessary for financing such an undertaking, including entire "villages, mines, vast wheat fields, vineyards, and livestock innumerable."[11] In 1589 he submitted his proposal to the viceroy, the Marquis of Villamanrique. Lomas y Colmenares demanded concessions beyond what the law allowed. Because the Colonization Laws of 1573 intended to curb the power of colonial nobles, the viceroy rebuffed his offer.[12]

In 1594, Viceroy Don Luís Velasco summoned Francisco de Urdiñola to submit a proposal for colonizing the northern frontier. Urdiñola arrived in the New World hoping to find greater wealth. By all accounts, he succeeded, for he owned millions of acres in eastern Vizcaya. Before he could complete his New Mexico contract negotiations, however, the *audiencia* (high court) of Nueva Galicia indicted Urdiñola for poisoning his wife and killing some servants. Lomas y Colmenares, his accuser, knew that a trial would force Urdiñola to spend his time and wealth defending himself in court and therefore Lomas y Colmenares could regain the court's favor and the New Mexico contract. By the time the audiencia absolved Urdiñola of the charges, however, Don Juan de Oñate, son of Cristóbal de Oñate, one of the rich founders of the Zacatecas mine, had already won the proprietary contract to settle the northern frontier and had begun his journey into New Mexico.[13] Oñate's contract, however, did not get quick or unchallenged approval.

Viceroy Velasco, who had encouraged Oñate, was replaced on 15 September 1595 by Don Gaspar de Zúñiga y Acevedo, the Count of Monterrey, three days before Oñate met with Velasco to have his contract approved. The count did not consider Oñate a close friend, as had Velasco. Zúñiga y Acevedo modified Oñate's contract, reporting that others had confided to him that Oñate was "unfit to head a colonizing effort, and that his wealth was insufficient and his estate overburdened with debt."[14] The following year, Don Pedro Ponce de León requested that he be allowed to replace Oñate as colonizer of New Mexico because he had far more resources and qualifications. Ponce de León argued that Oñate had already expended a great deal of money and creditors now dogged him. He further stated that no one would sign up with Oñate because his expedition was doomed to fail. Oñate's mission was suspended in September 1596 so that King Philip II could sort through the requests and complaints. The path looked clear for Ponce de León to explore New Mexico, until the king finally ruled in Oñate's favor. On 2 April 1597 the king issued the order rescinding Oñate's suspension, which Oñate received later that summer. Oñate then began rebuilding the numbers of his recruits, as many of the soldier-colonists had defected during the previous, uncertain months. Although his contract obligated him to bring and provision 200 soldier-colonists and their families to New Mexico at his own expense, Oñate could only convince 129 soldier-colonists, their families, and nearly a dozen Franciscan friars to travel into the northern frontier.

Oñate's contract included several provisions, rights, and responsibilities. The crown gave Oñate the titles governor and captain general, thereby giving him administrative and military control over the province of New Mexico. He acquired economic control of the province as *adelantado*. The crown also granted Oñate an *encomienda* as well as the right to grant colonists lands, encomiendas, and the title of *hidalgo*.[15] In return, Oñate had to recruit and provision the soldier-colonists and their families to travel into New Mexico. Providing them with provisions would be expensive and forced Oñate to find investors for the proprietary expedition. To please the investors, Oñate would have to make New Mexico pay. He knew he must search the New Mexico mountains for a mother lode that would make the expedition profitable. Similar to the Virginia Company, which formed nearly ten years later on the east coast of British North America, Oñate felt great pressure to succeed. Also like the Virginia Company, he would ultimately fail, thus paving the way for New Mexico to become a royal colony.[16]

The three-year delay had caused many who had previously enlisted to desert the expedition. Exasperated, Gov. Oñate wrote that "regardless of my efforts I shall not be able to hold the people together."[17] Finally allowed to proceed, on 8 January 1598 a relieved, yet anxious Oñate mustered the colonists and began the journey north into Pueblo territory, thereby establishing the Camino Real de Tierra Adentro.[18] Oñate describes his ordeal and goals for his New Mexico venture:

> The obstacles placed in the way of the New Mexico expedition . . . have been so great that it may be judged to my advantage that I was able to tolerate them. These I have reported to your majesty, enumerating the damages and losses that have resulted from the delays, how much it is still costing me, and what I have endured because I would not abandon my purpose to serve your majesty. . . . I took eighty carts and wagons, with the necessary equipment and supplies, confident that the mercy of God favored my endeavor and that I would attain the object of the expedition, which is the increase of the holy Catholic faith and the enrichment of your royal crown.[19]

Oñate's perseverance and determination trained him well for the hardships that he foresaw.

Those civilians and soldiers who joined him on this journey did so with hopes of finding great material wealth; for the Franciscan missionaries, the gains would be spiritual—they hoped to Christianize the many native people whom Coronado reported had lived in the numerous agricultural communities. By the end of April 1598, they reached the Río Grande just below El Paso. There Oñate took possession of the entire colony of New Mexico for Spain. Forging northward, Oñate was forced to divide the caravan and travel in groups more than a mile apart to allow the springs along the Camino Real to refill. By 11 July 1598, the caravan began to dribble into the Indian pueblo called Ohke, located at the confluence of the Río Grande and Río Chama in northern New Mexico.[20] The military itinerary of the expedition describes how the carts continued to arrive for many more days:

"We waited for the carts until August 18 of the said year, 1598, when they arrived. This was the eve of the feast of the blessed San Luis Obispo, on whose day, a year before, they had arrived at San Bartolomé after a long wait at Casco, harassed by the children of this world in the prosecution of this blessed expedition."[21] Their journey had not been easy and their lives in the remote province would continue to challenge even the most dedicated of women and men. At the village they named San Juan Bautista (also known as San Juan de los Caballeros), the Spanish colony prepared itself for the upcoming winter and the spring planting season. Spanish crops, including wheat and barley, were not resistant to the harsh, semi-arid climate of the New Mexico frontier, which received less than seventeen inches of precipitation annually. For their crops and their community to survive, on 10 August Oñate issued a call for volunteers to build an acequia, or irrigation ditch. The military itinerary describes the day: "On the 11th we began work on the irrigation ditch for the city of our father, Saint Francis. . . . Some fifteen hundred barbarian Indians gathered on this day and helped us with our work."[22] The acequia would water fields to support the proposed community (and headquarters), which the Spaniards intended to call San Francisco de los Españoles, named to honor the Franciscan fathers who accompanied the expedition.

Scholars have disputed the exact location of this ditch and the Spanish "city."[23] Marc Simmons, in his exhaustive biography of Juan de Oñate, asked just where was the site of this municipality and the irrigated fields that were intended to provide it food? That is a question that continues to plague scholars because historical documents neglect to give the answer. The logical assumption is that the town site was very close to San Juan Pueblo and on the same side of the river. If that was the case, then the newly irrigated farms would have been separated from it by the Río Grande because the San Juan Indians insist, even today, that most of their agriculture has always been on the west side of the river. That raises the possibility, of course, that San Francisco too may have been set across the Río Grande, either above or below the existing San Gabriel Pueblo (as the Spaniards called the village of Yunge).[24] A more exact description of the community site, as reported by contemporary Fray Gerónimo de Zárate Salmerón, was "between this river [the Río Grande] and that of Zama [Chama]."[25] The fact that Oñate succeeded in establishing a colony at the confluence of the

Río Grande and Río Chama is significant and highly disputed. Benjamin M. Read, a late-nineteenth-century New Mexico scholar and antiquarian who called into questioned the exact location of San Francisco de los Españoles, believed that the first Spanish settlement was located at Yugwinge on the "east side of the Río Grande and that the first church was built at the pueblo of San Juan."[26] To establish an accurate priority date, modern scholars have attempted to determine from these conflicting sources and opinions the precise locale of the acequias and how long the Spaniards continuously occupied the lands.

With the acequia under construction, Oñate began preparations to explore the Río Grande Valley and the mountains beyond for precious metals to help finance the colony. Because of his experience as a miner, he led these numerous excursions to collect and assay the minerals they unearthed. Oñate's regular absences, however, provided an opportunity for the colonists left behind to commiserate, complain, and plot. They suffered tremendously and complained because the colony ran short of supplies including food, clothing, and even shelter. Increased Spanish demands for food and supplies from Pueblo Indians through the abused encomienda system angered the previously friendly natives. When nomadic Apaches and Navajos began raiding the Spanish community in an attempt to acquire supplies, as they had from the Pueblo Indians for generations, tension and frustration within the small Spanish settlement reached new heights. Without adequate food, shelter, and protection, many plotted to desert New Mexico. Oñate responded:

> Thus, by the end of August, I began to prepare the people of the army for the rigorous winter of which the Indians and the nature of the land warned us. And the devil, who has always tried to prevent the great loss that he would suffer through our coming, resorted to one of his usual tricks with the mutiny of more than forty-five soldiers and officers, who in anger at not finding bars of silver on the ground right away and resentful because I did not allow them to abuse the natives either in their persons or property, became dissatisfied with the land, or rather with me. They tried to band together and escape to New Spain, or so they said, but their intention, as became clear later, was rather to take slaves and clothing, and to commit other outrages.[27]

Oñate's supporters admonished those willing to give up, but the die was cast. Arid conditions, the desperate need of Oñate to find wealth, and the growing need for security from nomadic raids combined to make life in San Francisco unbearable.

Perhaps the soldier-colonists and their families had a valid complaint. Because Oñate authorized and led many expeditions between 1598 and 1601, very rarely did the small outpost have a full compliment of soldiers. In the early days of 1599, Oñate sent seventy soldiers to battle Acoma pueblo and punish them for their uprising the previous year, leaving fewer than fifty men to protect San Francisco.[28] He knew he could not continue the colony without assistance from the king. In Oñate's letter to the viceroy on 2 March 1599, he wrote, "I humbly beg and entreat you, since it is of such importance to the service of God and his majesty, to send me all the help possible, both to colonize and pacify."[29] Hoping to provide proof of the province's great wealth before reinforcements arrived before Christmas in 1600, Oñate went on at least three shorter expeditions looking for mineral resources. He left only a small complement of soldiers to protect the women, children, and the infirm.[30] Gaspar Pérez de Villagrá, chronicler for the Oñate expedition, described how women of the colony resorted to dressing in men's clothing and patrolling the rooftops to guard against nomadic Indian attacks.[31] The colony desperately tried to hold on, while it became apparent that the old community at Ohke simply could not provide for the Spanish community.

Sometime between March 1599 and late December 1600, Oñate moved his capital across the river. Although he made no reference in letters to the king regarding this move, he intended to build a standard Spanish city complete with a city square with government buildings, town lots for homes, garden plots on the outskirts of the town, and water rights to provide daily sustenance for the settlers. The colonists, however, refused to participate in the building of a "proper" city. Years later, a soldier testified that he believed "the reason for this [their refusal] was their dissatisfaction at remaining and their desire to abandon the land because of the great privations they were suffering."[32]

After the Spaniards moved across the river to the Pueblo of Yunque, they renamed it San Gabriel. They established a new irrigation system. Spaniards used the water from the Río Chama to "irrigate wheat, barley, corn, and other things that they plant in gardens."[33] With great hope for future successes, Oñate reported on Pueblo Indian crops in the area as well, saying "their corn and vegetables . . . are the best and largest . . . in the world."[34] Other Spaniards also commented on the surplus of corn, cotton, beans, squash, and melons. While some Pueblo Indians irrigated, others depended upon seasonal rainfall.[35] Regardless of the apparent abundance in the valley, the colony suffered tremendously. Oñate himself described New Mexico as having eight months of winter and four months of hell.[36]

When reinforcements arrived from the south in 1600, the colonists were only temporarily pacified. The following year, when Oñate led an expedition of more than seventy hand-picked men northeast onto the Kansas plains looking for Quivira, many colonists and friars deserted. When the deserters arrived in Mexico City, they told their stories of Oñate's abuses of the Acoma Indians, the lack of financial rewards in New Mexico, the long and cold winters, the miserably endless and dry summers, and the constant lack of supplies and food for the colonists.[37] The increased demands on Pueblo Indians for food might indicate a lack of willingness by or inability of Spaniards to utilize the acequias, instead depending upon the Pueblo Indians to provide them sustenance. Fray Juan de Escalona's report to the viceroy on 1 October 1601 argues that the lack of food production was at the heart of the settlement's problems:

> The governor did not want to sow a community plot to feed his people, although we friars urged him to do so, and the Indians agreed to it so that they would not be deprived of their food. This effort was of no avail, and now the Indians have to provide everything. As a result, all of the corn they had saved for years past has been consumed, and not a kernel is left over for them. The whole land has thus been reduced to such need that the Indians drop dead from starvation wherever they live; and they eat dirt and charcoal ground up with some seeds and a little corn in order to sustain life.[38]

This seems to be an indication that the irrigation ditch, and the fields it would have watered, was not used properly to feed Oñate's colony and the neighboring Pueblo people, a point Escalona returned to when he wrote to his prelate the same day: "If God had not directed some individuals to sow a small amount of wheat that could be irrigated we would all have starved and died."[39] Who in the colonial outpost made use of the ditch and its adjacent fields? Most likely those loyal colonists who intended to stay the course; those who plotted desertion probably would invest neither the time nor the energy to plant fields they intended to abandon.

Many colonists who remained loyal to Oñate later testified that the situation was not as dire as reported by those who fled the colony. The investigation was written in questionnaire form to which the soldier-colonists were to reply:

> If he knows that there have never been such abundant supplies as at present, for of the two thousand fanegas of corn that may be required at this camp each year, there is now a crop of fifteen hundred fanegas of wheat from our own harvest . . . that there are more than three thousand head of cattle and sheep, and that the gardens are filled with fruits and vegetables, with which we could support ourselves . . . if he knows that in the three years we have been here we have planted wheat and vegetables and that both the planting and harvests have increased.[40]

Each of the colonists responded that each question was true. One of them elaborated that, "when [I] was in the land last year [I] saw that there were fewer fields planted than now, so it is obvious that production is increasing."[41] That colonists placed more fields under production is a clear indication that the colony also increased the irrigation. That some colonists were terribly dissatisfied and frustrated is irrefutable, however, and it worried Oñate.

Until 1604, Oñate likely stayed closer to home to guard against more desertions. In that year, however, he headed an expedition for the Sea of Cortez and remained absent until 1605. It is likely that during the same year, Oñate began to move the colony from San Gabriel to Santa Fe. Custodian of New Mexico Fray Alonso de Posadas, "who doubtless had access to the provincial archives prior to their destruction in the revolt of 1680, states explicitly in his *Informe*, . . . written about 1684, that Oñate descubrióla [la Villa de Santa Fé] el año de 1605 [that he discovered Santa Fé in 1605]."[42] Other than this tidbit of information, little else is known about Oñate until his resignation as governor and captain general of New Mexico on 24 August 1607.[43] Oñate failed to make New Mexico pay and therefore relieved himself of the burdensome and expensive New Mexico venture. Yet, even in his darkest hour, he begged the crown not to give up on New Mexico: "If we should all leave the land, it will be necessary to take along more than six hundred Christian Indians. The result of this will be . . . [that] the natives will not even dare to welcome the Spaniards in future years. . . . It would be less harmful to cease making more Christians than to allow those who have already been converted to be lost."[44]

Establishing Community and Water Rights

Hoping not to lose their monopoly in New Mexico, Franciscans soon reported to the crown that they had baptized more than seven thousand Indian souls.[45] Perhaps they exaggerated the numbers to prove the value of continued settlement of New Mexico and to convince the crown of their importance in the venture. They argued that the crown's duty as head of the Catholic Church in the New World obligated Spain to continue providing Christian instruction to the neophytes; the Pueblo Indians, excellent farmers with their use of irrigated fields, undoubtedly could provide for the sustenance of the friars. Apparently these appeals were successful because New Mexico became a royal colony, and the first royal governor, Pedro de Peralta, began his trek north. During his governorship, Peralta officially founded the capital of Santa Fe. France V. Scholes, a well-known authority on seventeenth-century New Mexico, wrote:

> To this new site and the settlement which slowly grew up there was given the name of Santa Fé. Laying out of the villa and the sites for the principal buildings was probably done in the spring of 1610, and the Spanish population of San Gabriel was transferred to it during the succeeding months as rapidly as buildings could be erected. From that time to the present Santa Fé has remained the capital of New Mexico. The Villa of San Gabriel was abandoned, and during the remainder of the period prior to the Pueblo Revolt of 1680 Santa Fé was probably the only organized community in the province with a full local government.[46]

Although Scholes argued that the entire Spanish community left San Gabriel, it was likely that those who had received land grants near San Gabriel remained behind, or at least refused to give up their claims, visiting the land periodically to plant, irrigate, cultivate, and harvest. Whether they continued to inhabit the area until 1680 is central to the question of establishing a priority date for the Río Chama. Stanley M. Hordes, a contract historian, has argued that the Spaniards by no means abandoned the old capital: "A 1608 report by Captain Pedro Martínez de Montoya [sic], outlining the services that he had performed for the royal authorities over the previous years, contained references to his having spent time in Santa Fe, while building a house at 'la Villa de San Gabriel,' serving there as *alcalde ordinario*."[47] Although Hordes located a document indicating that a Villa de San Gabriel existed, San Gabriel was not formally established as a villa. That Martínez de Montoya served as an alcalde ordinario at San Gabriel did not indicate that he permanently resided in there.[48] What Martínez de Montoya's report indicates is that the government believed that San Gabriel needed an administrator—an indication that a population base existed in San Gabriel and the surrounding area. By definition, however, an alcalde ordinario administered one of six or eight rural districts:

> Their most important duties were probably police and judicial, viz., to administer petty justice, to adjust differences concerning lands and water rights, to assist the friars in the maintenance of mission discipline, to oversee the employment of Indians as house servants, farm laborers, and herdsmen by the Spanish and caste ranchers, and to supervise the routine of pueblo life, working with and through the petty Indian magistrates and officials.[49]

Martínez de Montoya's appointment as alcalde ordinario was not proof of a permanent Spanish settlement at San Gabriel, only that Spain needed an administrator for at least one of these many functions. Because at least one pueblo resided within the jurisdiction of San Gabriel (San Juan), the alcalde perhaps oversaw only Pueblo Indians. That he built a house at "la Villa de San Gabriel," is significant, however. This house served as a visible sign that Martínez de Montoya had no intention of abandoning his claim in San Gabriel. On the frontier of New Spain, continuous occupation did not necessarily mean daily residence. Local customary law permitted temporary abandonment. Only when flagrant or intentional abandonment occurred did the crown revoke the right of settlement.[50] Whether to serve as a permanent or temporary home, this house is an indication of at least periodic settlement, but not necessarily irrefutable proof of continued beneficial use of water.

Other historical sources provide evidence that the Spanish did not permanently occupy San Gabriel. Because establishment of priority dates demands proof of continued beneficial use of water, all documentary evidence must be consulted to determine whether a settlement remained viable and used its water for agricultural or other purposes. Historians must thus consider Franciscan Superior Fray Alonso de Benavides's extensive and detailed description of New Mexico, *The Memorial of 1630*. In this report, written first for the Spanish king in 1630 with another edition for the pope in 1634, Benavides hoped to convince either man to create a New Mexico bishopric and appoint him as its bishop. He described the great advances that the Franciscans had made in Christianizing the Pueblo Indians, as well as the rapid population growth in the northern province (as required for a bishopric). Benavides hoped to illustrate the great development New Mexico had achieved and thus promoted all corners of the frontier that had Spanish and Pueblo Indian settlements. Although Benavides did not succeed in acquiring a bishopric, his report provides vital clues regarding locations of Pueblo and Spanish communities. Interestingly, Benavides did not even mention San Juan Pueblo or a Spanish presence at Río Chama.[51] Some scholars have pointed to Benavides's omission of a settlement as proof that the region had been abandoned by Spaniards.

Yet various Spaniards intended to inhabit the Río Chama region, as other evidence suggests. In the 1660s, the Inquisition confiscated, embargoed, and inventoried the property of a New Mexican Spaniard named Francisco Gómez Robledo. The documents included a petition for an *estancia* (small ranch) by his father, Francisco Gómez, at Yunque and a land title to an estancia at San Juan de los Teguas Pueblo.[52] This may indicate that the land was continuously occupied by father and son. Due to the necessity of irrigation for the survival of Spanish crops, the father and son likely used the water rights in the valley to water fields, pastures, and livestock.[53]

Suggestions of continued settlement and probable water use can be found in a 1665 Spanish document that referred to Francisco García, then fifty years old, as a "native of the villa of San Gabriel de Iunque."[54] According to Hordes, this statement indicated that a Spanish woman had given birth to a son at San Gabriel in 1615. If in fact García had been born in San Gabriel, it is likely that he lived there and was not simply passing though. Whether he was actually born there, the significance of his identification as "native" of that community is paramount; while his presence at

San Gabriel does not prove beneficial use of the acequia, it was very likely the family used water from the Río Chama to grow crops for sustenance.[55] What this suggests is that not everyone had completely abandoned San Gabriel by the early 1610s.

These documents, although not land titles, provide important evidence not otherwise accessed easily, for almost all Spanish documents dated before 1680 were destroyed in the Pueblo Revolt. After New Mexico became a royal and missionary colony, Pueblo Indians rose up and forced the Spaniards from their territory in 1680, killing more than four hundred Spanish citizens and almost all of the Franciscan friars. Furious at their forced conversion and Spanish demands for tribute and labor, the coalition of Pueblo Indians determined to eliminate everything Spanish. Pueblo Indians burned or otherwise destroyed Spanish buildings—and in these structures lay documents like grants, wills, and letters that might have indicated beneficial use as well as land ownership.[56] As a result, scholars and lawyers have become dependent upon Inquisitional court proceedings, church records, personal correspondence (of which very little existed), and the reports of Villagrá; Benavides; Juan de Torquemada; and the postrevolt governor, Don Diego de Vargas.[57]

From 1680 until 1692, the Spaniards lived far to the south in El Paso del Norte, while the Pueblo Indians remained in control of New Mexico proper. The Indians tried to return to their traditional lifestyle, which they had lost gradually over the previous eighty-two years, but their hopes in this regard were for nought.[58] Increased deprivations by nomadic raiders forced some Pueblo Indians to travel to El Paso and ask the Spaniards to return to protect their villages. In 1692, Gov. Diego de Vargas began the reconquest of New Mexico, and the following year he and a contingent of soldier-settlers forcefully retook the capital of Santa Fe by cutting off the water supply to the well-defended city. While on his military expeditions throughout New Mexico, Vargas took note of the location of communities, abandoned haciendas, and irrigation ditches. Stopping at El Paso prior to his reconquest of the Middle Río Grande Valley, he commented that the "few fields were late and lightly planted, because of the time spent repairing and cleaning the acequias."[59] Vargas also observed that the acequia did "not have a fixed location, since the course of the river moves. The land is such that it cannot be sown continually, because by the third year, it is infested with sandburs and other briers."[60] At Senecú, a native settlement near El Paso, Vargas took possession, indicating that he gave "the father president possession . . . in the land I indicated and gave to him in his majesty's name, any persons and the water from their acequia."[61] In a dispute between the two native communities near El Paso, Socorro and Ysleta del Sur, Vargas ordered that "Thus, the lands indicated, as well as water from the acequia, are subject to their will, and likewise the Indians."[62] At Socorro, a Spanish village just north of a ninety-mile stretch of the Camino Real called the *jornada del muerto* (journey of the dead man) that was devoid of sufficient water, Vargas described the extensive and fertile lands, "There are acequias and a church, whose walls are sound."[63] While he commented on some of the agricultural developments of the native people, Col. Luis Granillo, Vargas's lieutenant governor and captain general, made a reconnaissance of the area north of Santa Fe to the old settlement at the confluence of the Río Chama and Río Grande. Vargas sent Granillo to reconnoiter the locations of "the farms and places separately, making a map showing the names of the places and the names of their former owner, the quality of the lands, and the distances, and he will examine personally and will confer with the above mentioned as to the number of persons who can be settled on the same."[64] Granillo reported that the physical remains of homes and even a tower still existed. Interestingly he described a meadow in the middle of which stood "the farm on which Agustin Romero was settled during the planting season because he had his cultivable land on said tract."[65] That Romero brought his family to live on the farm during the growing season is no surprise, due to the nomadic Indian raids and other uncertainties that farm life produced. Romero stayed to watch over his crops and livestock, as did many other Spaniards in the frontier—perhaps even at San Gabriel.

As New Mexico became more populated after Vargas quelled the last Pueblo uprising by 1696, more Spaniards began to request land grants at or near the newly established Villa de Santa Cruz de la Cañada and the Río Chama area. Priority dates for the Río Chama, therefore, reflect this more recent move into the valley. That does not sit well with late-twentieth-century water rights advocates, who demand instead priority dates that reflect their family's longevity in the valley, in some cases dating back to the prerevolt settlement of Gov. Oñate. Because few records from that period exist, continual settlement and beneficial use must be proven through other available sources, such as wills, extant land grants and transfers, civil and religious court proceedings, census reports, and various clerical records.[66]

Conclusion

The evidence presented in introductory form in this chapter provides a basis for understanding the current dispute over priority dates in the Río Chama Valley. In August 1996, Special Master Vickie L. Gabin held an evidentiary hearing in which the descendants of the original settlement at the confluence of the Río Chama and the Río Grande disputed the priority date of 1714 that the New Mexico State Engineer Office imposed. They argue that their ancestors had not abandoned the region, but had continued to occupy the land adjacent to and had used the Chamita, Hernandez, and Salazar acequias since they first occupied the community of San Gabriel in 1600. If the descendants did not receive the earlier priority date requested, their lawyer Fred J. Walz argued, the communities of Bernalillo and Albuquerque (which had been established after the revolt, yet prior to 1714) could get an earlier priority date; if this happened, the older communities on the Río Chama would be deprived of their priority. Walz argued that the state had the responsibility to prove abandonment, while the state argued that the plaintiffs had to prove continued beneficial use of water.[67]

In the proceedings, historians John O. Baxter and Stanley M. Hordes offered reports, depositions, and testimony indicating their beliefs as to the location of the acequia and community as it was represented in historical evidence. Baxter and Hordes depended on the previously mentioned Spanish documents that they separately translated, then disputed each other's translations on key phrases. The discrepancies in translations did not provide the only stumbling block in understanding the modern water rights at Río Chama.[68] Evidence supported both sides of the argument, for example, the priority of water rights, ancestry, continual habitation of region, beneficial use of water, abandonment of the area, and the exact date of the construction of the acequias. The lack of documents that resulted from the Pueblo Revolt's destructiveness, conflicting translations of extant primary sources, and opposing interpretations of the significance of evidence have greatly confused the issue of water rights in New Mexico.

The greatest difficulty in assessing Río Chama water rights emanates from the lack of extant land grants and transfers. It is also very difficult to trace the ancestry of modern Hispanos who claim rights to land and water rights because ecclesiastical records were destroyed during the Pueblo Revolt. Contemporary researchers continue to delve into the historical record in hopes of finding the link that claimants so desperately seek—perhaps it will be found in a will in Mexico City, in a letter in Sevilla, or even in an old trunk in Guadalajara.[69] In the meantime, scholars continue to debate extant sources and to search for previously unknown documents, hoping one day to find the key to understanding the history of New Mexico's water rights so that we can better comprehend its complicated present.

Notes

1. Linda G. Harris, *New Mexico Water Rights,* Miscellaneous Report no. 15 (Las Cruces: New Mexico Water Rights Resources Research Institution, n.d.), 33.

2. Ibid.

3. Continual settlement and beneficial use cannot be proven without establishing ancestral ties to the original Spanish inhabitants. The many ecclesiastical collections at the New Mexico State Records Center and Archives assist researchers in this regard. This ancestral connection is not enough to prove ownership of water rights. Regarding the difficulties of researching these land grants and water rights, many of the original records were lost during the great Pueblo Revolt of 1680–1696, when the Pueblo Indians rose up and threw the Spanish out of New Mexico for twelve years. Because their goal included the destruction of everything Spanish, church records and land grants were burned or otherwise lost. These lost documents might have provided the missing links to many land grant and water rights cases now hotly debated. For a partial guide to New Mexico water issues, see Stanley Crawford, *Mayordomo: Chronicle of an Acequia in Northern New Mexico* (Albuquerque: University of New Mexico Press, 1988); Michael C. Meyer, *Water in the Hispanic Southwest: A Social and Legal History, 1550–1850* (Tucson: University of Arizona Press, 1984); Daniel Tyler, *The Mythical Pueblo Rights Doctrine: Water Administration in Hispanic New Mexico,* Southwestern Studies Series, no. 91 (El Paso: Texas Western Press, 1990); José A. Rivera, *Acequia Culture: Water, Land, and Community in the Southwest* (Albuquerque: University of New Mexico Press, 1998).

4. Carroll L. Riley, *Rio del Norte: People of the Upper Rio Grandè from Earliest Times to the Pueblo Revolt* (Salt Lake City: University of Utah Press, 1995).

5. According to Pueblo Indian oral tradition, Estebanico arrived with a gourd deco-

rated with a red and white feather—indicative of an enemy gourd. Some Pueblo Indians believed he was a spy or an advance guard for the slave raiders who surely followed close behind. He was killed, reportedly while attempting to escape. Marc Simmons, *New Mexico: An Interpretive History* (Albuquerque: University of New Mexico, 1988); see also Pedro de Castañeda de Najera, *The Journey of Coronado, 1540–1542,* trans. and ed. George Parker Winship, with an introduction by Donald C. Cutter (Golden, Colo.: Fulcrum Publishers, 1990).

6. In 1536, Alvar Núñez Cabeza de Vaca arrived in Mexico City with three other survivors of a failed expedition to Florida, which had begun nine years earlier. See Cyclone Covey, ed., *Alvar Núñez Cabeza de Vaca's Adventures into the Unknown Interior of America,* (New York: Collier Books, 1961).

7. Purportedly, during the Moorish conquest of Iberia, the Portuguese bishops decided that instead of giving up the wealth of the Catholic Church, they would build seven cities of gold. After several voyages without seeing the cities, the Spaniards believed the seven cities were in the New World. Throughout the 1700s, Spanish expeditions had someone who spoke Portuguese, just in case they encountered the cities.

8. Herbert Eugene Bolton, *Coronado: Knight of Prairie and Plains,* foreword by John L. Kessell (1949; reprint, Albuquerque: University of New Mexico Press, 1991).

9. Captain Francisco Chamuscado and a small party of soldiers explored New Mexico in 1581. The following year, Antonio de Espejo led another expedition hoping to find out what happened to the two friars who had refused to return with Chamuscado. In 1590, Gaspar Castaño de Sosa led a group of settlers into New Mexico, where they met with armed resistance at the Indian pueblo of Pecos. Arrested for leading an illegal expedition, Castaño de Sosa returned to New Spain. In 1595, Antonio Gutierrez de Umana and Francisco de Leyba y Bonilla led an expedition into New Mexico, perhaps as far as the plains of Kansas. Marc Simmons, *The Last Conquistador: Juan de Oñate and the Settling of the Far Southwest* (Norman: University of Oklahoma Press, 1991).

10. Ibid., 55.

11. Ibid.

12. Ibid.

13. Ibid., 56–58.

14. Ibid., 68.

15. An *encomienda* was a grant of native tribute. See Robert Himmerich y Valencia, *The Encomenderos of New Spain, 1521–1555* (Austin: University of Texas Press, 1991); Lesley Bird Simpson, *The Encomienda in New Spain: The Beginning of Spanish Mexico* (Berkeley: University of California Press, 1966). The title *hidalgo* was the least impressive of all noble titles, meaning "son of someone." Only those who continued to persevere, remaining in New Mexico for five years and assisting with the betterment of the colony, could be granted the title hidalgo.

16. Virginia became a royal colony by the 1620s and New Mexico by 1610.

17. Oñate to the Count of Monterrey, Río de las Nazas, 13 September 1596, in *Oñate:*

Colonizer of New Mexico, 1595–1628, George P. Hammond and Agapito Rey (Albuquerque: University of New Mexico Press, 1953), part 1, 171.

18. Oakah L. Jones Jr., *Los Paisanos: Spanish Settlers on the Northern Frontier of New Spain* (Norman: University of Oklahoma Press, 1996), 110.

19. Hammond and Rey, *Oñate,* part 1, 183.

20. Simmons, *The Last Conquistador,* 111.

21. Hammond and Rey, *Oñate,* part 1, 323.

22. Ibid., 322–23.

23. Ibid., 320–23, 346.

24. Simmons, *The Last Conquistador,* 114–15.

25. Fray Alonso de Benavides, *The Memorial of Fray Alonso de Benavides, 1630,* ann. Frederick Webb Hodge and Charles Fletcher Lummis, trans. Mrs. Edward E. Ayer (Albuquerque: Horn and Wallace Publishers, 1965), 234. According to Frederick Webb Hodge, Fray Gerónimo de Zárate Salmerón arrived in New Mexico around 1617, serving mostly at Jemez, but also at Zia, Sandia, and Acoma. Zárate Salmerón wrote a book entitled *Relaciónes de Nuevo México* a few years later. Ibid., 200.

26. Ibid., 234.

27. Don Juan de Oñate to the Viceroy of New Spain, 2 March 1599, in Hammond and Rey, *Oñate,* part 1, 481.

28. For a complete transcript of the trial of the Acoma Indians, see Hammond and Rey, *Oñate,* part 1, 428–79; see also Ward Allen Minge, *Acoma: City in the Sky* (Albuquerque: University of New Mexico, 1991); Simmons, *New Mexico;* Simmons, *The Last Conquistador;* Gaspar Pérez de Villagrá, *Historia de la Nueva México, 1610,* trans. and ed. Miguel Encinias, Alfred Rodríguez, and Joseph P. Sánchez (Albuquerque: University of New Mexico Press, 1992), canto XXVII.

29. Hammond and Rey, *Oñate,* part 1, 486.

30. Suzanne Sims Forrest, "A Trail of Tangled Titles: Mining, Land Speculation, and the Dismemberment of the San Antonio de las Huertas Land Grant," *New Mexico Historical Review* 71 (October 1996), 361–93.

31. Eufemia de Sosa, who before 1598 admonished Spaniards who had signed contracts with Oñate not to abandon the expedition, was among those who gave up on New Mexico in 1601. Villagrá, *Historia de la Nueva México;* Simmons, *The Last Conquistador.* More details regarding the Spanish colony's move to Santa Fe would aid modern researchers in establishing accurate priority dates.

32. Simmons, *The Last Conquistador,* 149.

33. John O. Baxter, *Dividing New Mexico's Waters, 1700–1912* (Albuquerque: University of New Mexico Press, 1997), 2; Juan de Torquemada, *Monarquía Indiana,* 3 vols., 5th ed. (Mexico City: Editorial Porrúa, S. A., 1975), I: 678.

34. Ibid.

35. Hammond and Rey, *Oñate,* part 1, 484, part 2, 626, 634.

36. Ibid., part 2, 656.

37. Simmons, *The Last Conquistador,* 160–70. Nine of the friars who Oñate brought

with him into New Mexico in 1598 abandoned the colony three years later. See Benavides, *The Memorial,* 196–200. Hot on their heels was Vicente de Zaldívar, who hoped to set the record straight. Mexico City officials believed the colonists, however, and soon Oñate would be forced to resign his post. Simmons, *The Last Conquistador,* 169–71.

38. Hammond and Rey, *Oñate,* part 2, 693.

39. Ibid.

40. Ibid., 704.

41. Ibid., 729.

42. Benavides, *The Memorial,* 234.

43. That Oñate wrote his resignation letter from San Gabriel and not Santa Fe is another indication that the colony continued to exist in San Gabriel as late as Oñate's resignation in 1607.

44. Hammond and Rey, *Oñate,* part 2, 1044.

45. Simmons, *The Last Conquistador,* 182–85.

46. France V. Scholes, "Civil Government in New Mexico in the Seventeenth Century," *New Mexico Historical Review* 10 (April 1935), 94.

47. Stanley M. Hordes, "Irrigation at the Confluence of the Río Grande and Río Chama: The Acequias de Chamita, Salazar, and Hernández, 1600–1680," draft report for Río de Chama Acequias Association, Española, New Mexico, 31 March 1996 (unpublished manuscript, author's private collection), 4. It was actually Capt. D. Juan Martínez de Montoya, not Capt. Pedro Martínez de Montoya. See "Santa Fe: Est. 1607" *El Palacio* 100, no. 1 (Winter 1994/1995), 15–16.

48. The only Spanish villa in New Mexico prior to the Pueblo Revolt was Santa Fe. See Charles R. Cutter, *The Legal Culture of Northern New Spain* (Albuquerque: University of New Mexico Press, 1995).

49. Scholes, "Civil Government," 93.

50. In this case, the land reverted back to the Crown and would be regranted. Meyer, *Water in the Hispanic Southwest,* 117.

51. Benavides, *The Memorial.*

52. *Estancias* were small ranches whose primary income was derived from livestock. Irrigation was necessary for estancias, pastures, and fields in arid New Mexico.

53. Hordes, "Irrigation at the Confluence," 5. That the son had to ask for a grant to the same piece of land seems instead an indication either that the father did not continually occupy the land, and therefore lost the right to the land, or that his father's petition was not approved. Nevertheless, the title dated before the 1660s indicates that Gómez Robledo did in fact gain title to the land. John O. Baxter strongly disputes this interpretation of Gómez Robledo's continued, yet periodic, occupation of the region. *State of New Mexico v. Ramon Aragón et al.,* U.S. District Court, New Mexico, no. 69-CV 7941 S. C. 1996.

54. Deposition of Stanley M. Hordes, *State of New Mexico v. Ramon Aragón et al.,* 27.

55. Ibid.

56. Andrew L. Knaut, *The Pueblo Revolt of 1680: Conquest and Resistance in Seventeenth-Century New Mexico* (Norman: University of Oklahoma Press, 1995).

57. See John L. Kessell, Rick Hendricks, Meredith D. Dodge, eds., *Blood on the Boulders: The Journals of don Diego de Vargas, New Mexico, 1694–1697*, 2 vols. (Albuquerque: University of New Mexico, 1998); see also Torquemada, *Monarquía Indiana;* Charles Wilson Hackett and Charmion Clair Shelby, *Revolt of the Pueblo Indians of New Mexico and Otermin's Attempted Reconquest, 1680–1682* (Albuquerque: University of New Mexico Press, 1942).

58. Vina Walz, "History of the El Paso Area, 1680–1692," (Ph.D. diss., University of New Mexico, 1951); Sandra K. Mathews-Lamb, "The Nineteenth-Century Cruzate Grants: Pueblos, Peddlers, and the Great Confidence Scam?" (Ph.D. diss., University of New Mexico, 1998).

59. John L. Kessell and Rick Hendricks, *By Force of Arms: The Journals of Don Diego de Vargas, 1691–1693* (Albuquerque: University of New Mexico Press, 1992), 79–80.

60. Ibid., 267.

61. Ibid., 271.

62. Ibid., 276.

63. John L. Kessell, Rick Hendricks, and Meredith Dodge, *To the Royal Crown Restored: The Journals of Don Diego de Vargas, New Mexico, 1692–1694* (Albuquerque: University of New Mexico Press, 1995), 114.

64. Vargas to Viceroy, the Conde de Galve, 11 May 1695, in *The Spanish Archives of New Mexico,* vol. 1, Ralph Emerson Twitchell (Cedar Rapids, Iowa: Torch Press, 1914), 243.

65. Ibid., 251.

66. For example, records indicated that "Just before the Pueblo rebellion of 1680 it contained 300 inhabitants, who were ministered by the missionary at San Ildefonso. . . . [San Juan had a population of] 346 Spaniards and 404 Indians in 1749; in 1782, 500 inhabitants of San Juan and Santa Clara died of pestilence in two months; in 1793 the population was 260 Indians and 2,173 Spaniards; in 1850, 568 Indians. The population in 1910 was 384." Benavides, *The Memorial,* 238; see also Augustín de Vetancurt, *Crónica de la provincia del Santo Evangelio de México: quarta parte del Teatro Mexicano de los sucesos religiosos* (Mexico City: Por Doña Maria de Benavides viuda de Juan de Ribera, 1697); Irving Albert Leonard, *The Mercurio Volante of Don Carlos de Sigüenza y Góngora: An Account of the First Expedition of Don Diego de Vargas into New Mexico in 1692* (Mexico, 1693; reprint, Los Angeles, Calif.: The Quivira Society, 1932).

67. Fred J. Walz, evidentiary hearing, *United States v. Ramon Aragón et al.,* 22–23.

68. One of the interesting debates concerned the translating of a key phrase in a document, *"sacar la acequia."* Dr. Baxter argued that the term meant "to dig or create an acequia," whereas Mr. Walz, attorney for the claimants, argued that the term instead meant "to clean the acequia." *State of New Mexico v. Ramon Aragón et al.,* 440–41, 7. José A. Rivera explains in his glossary that "sacar la acequia" means "to clean out the ditch in the spring, just before the start of the irrigation season," which is sometimes described as *"la limpia de la acequia," "el día de la saca,"* or *"la saca de la acequia."* The phrase, "sacar la acequia" was also used in historical documents to describe the initial digging or construc-

tion of a ditch starting at the point of diversion from the stream source. Rivera, *Acequia Culture*, 230.

69. Rivera also explains that the *saca de agua* is the diversion point from the acequia. Ibid., 228. Michael C. Meyers translates the phrase *"sacadas y corrientes"* when referring to acequias as "ditched and running." Meyers, *Water in the Hispanic Southwest*, 37. *The Collins Diccionario Español-Ingles* (Barcelona: Ediciones Gijalbo, 1979) further supports the definition of sacar as to extract or dig: "to take out, get out; to pull out, draw out, extract; *(chem)* to extract coal etc., to mine." Although this is merely one phrase, the implications of the translation holds key significance to the priority date of the water usage of the Upper Rio Grande. Other cases involving the Rio Chama are currently in litigation, including El Rito.

The Native American Struggle for Water

4

Maggot Creek and Other Tales

Kiowa Identity and Water, 1870–1920

Bonnie Lynn-Sherow

One of the most difficult paths that environmental historians must negotiate these days is the relationship between Indian peoples and their environments. An emerging consensus among environmental scholars holds that people are always active participants in ecological change. Our job as historians requires us to consider the historical relationships between "nature" and culture in geographical and temporal contexts. Very often a region has hosted more than one culturally identifiable group, each competing for the same resources. The arrival of Europeans to the Americas, for example, is this same story writ large. This complex story has been broadly interpreted as the domination of one culture over another and the substitution of one environmental ethos for another. Unbridled Euro-American exploitation of the environment at the expense of Native American resource systems continues to be a stock explanation in the "destruction" of Native American culture.

As a generalization, this conclusion has some legitimacy and will continue to be an area of speculation and revision for a long time to come. Less acceptable, however, are some corollaries to this thesis that have recently gained widespread public acceptance. The most widely accepted is that Indian peoples have an intuitive relationship to "nature" that is both sustainable for human life and beneficial to all life in general. As Daniel Richter has noted, the noncelebration of the Columbian Quincentennial in 1992 suggests that the American public's "unquestioning celebration of European triumph has [moved] to the equally unquestioning worship of Native

American innocence" and their role as "models for democracy and environmental responsibility."[1]

Indian environmental groups have exploited this notion to some advantage in a few highly publicized battles over resources. The leaders of Diné Care's efforts to control mining development on the Navajo Reservation and of Canada's First Nations Crees undeclared war on Hydro Quebec have long intimated a superior Indian environmental stewardship ethic. In these and many other cases, the struggle of native peoples to gain control over the resources their communities have traditionally depended on is crucial to their future as a people. In this sense, Indian environmentalism is a very old, very human story. What is new is the justification some groups have offered for Indians' interest in the future of the environment.

Jewell Praying Wolf James, a lineal descendent of the famous Chief Seattle, recently offered his own ecological interpretation of the history of the United States:

> We knew neither hunger nor disease until contact came in 1492, then our holocaust began and that of the plants, animals and environment. Our oral traditions, teachings and spirituality taught respect for all things. From one generation to the next we were taught the glorification of respecting nature and creation . . . the Earth was our mother, the sky was our father and our prayers brought the two together to form one unified spirit. Our ceremonies and traditions were like the umbilical cord that attached the mother and child and bonded the father to children.[2]

When I came across James's speech, I was immediately reminded of a subway poster for the Native American College Fund I noticed as a graduate student riding the El in downtown Chicago. In stark contrast to the generally garish subway ads, this one featured an airbrushed photograph of a beautiful Indian girl in a pristine wilderness. The caption was a single sentence, "Help save a culture that could save your own." This ad harkened back to the television commercials environmentalists sponsored in the 1970s that made a tearful Iron Eyes Cody, a Hollywood stock player, into a national celebrity. Here was a spokesperson for the environment that need not utter a single word to be effective. Then, as now, Native American empathy for the environment as a racial or cultural trait continues to be a popular truism in American society.

For historians attempting to faithfully re-create Indian resource use, the popularized image of the Indian as environmentalist is not a cause celebre, but a well-intentioned obfuscation. Sympathetic to ideals of environmental justice, many historians find it tempting to disregard evidence of Indian resource use that does not fit the popular stereotype. On a more pragmatic level, historians have also acknowledged that scholars who offer up such evidence risk being branded as Eurocentric or anti-Indian.[3]

Fortunately, there are compelling reasons for environmental historians to document American Indian resource use. The most obvious is that past examples of Indian resource management are still relatively rare. Richard White's *Roots of Dependency* and David Rich Lewis's *Neither Wolf nor Dog* have generated new and important questions about the relationship of Indian peoples to the environment as it relates to cultural change and continuity.[4] These studies are important because contemporary problems of unequal resource allocation, pollution, and cultural maintenance will be more difficult to solve without the depth of past experience that historians provide. This is not advocacy, but only what all historians strive for everyday: to show that the future is never fixed and that every new generation has choices to make.

Far from making Indians "just like us," the study of Indian resource use helps secure Indians' unique place in American society. When we regard all Indian peoples as members of a single conservationist-minded culture, we commit the same error that Christopher Columbus did in labeling an entire hemisphere of different peoples under the single misnomer of "Indian." As the world's economy constricts and American consumer culture is exported to every corner of the globe, it is easy to forget that resource use, like language, religious belief, family structure, art, and politics, is a distinguishing feature of any self-identified society. As Frederick Hoxie has observed, any generalization, well intended or not, of Indian peoples across time and space necessarily "flattens" the identity of tribal groups and their individual histories.[5] Regardless of the different current split between biological and cultural explanations of cultural development, anthropologists and ethnohistorians must continue to ground their conclusions in the everyday material culture that give all people their primary identities. Thus, the study of Indian resource use is an essential component of good ethnohistory.[6]

Kiowas and Prehistory

An analysis of the Kiowas' use of water over time combines the methodologies of environmental history and ethnohistory. Water is an essential and generally scarce resource in the southwestern United States and the Kiowas have long been associated with the history of that region. This chapter attempts to merge the belief systems of the Kiowas into the environmental history of water use in the Southwest. These beliefs changed along with the Kiowas' material circumstances in the years before and after reservation and allotment. This story diverges from the stereotype in that water conservation was not central to Kiowas' religion or philosophy in the years prior to the Indian New Deal (1934). This does not mean that the Kiowas valued water in precisely the same way that Euro-Americans did. The Kiowas' unique world view, although not focused on water, still shaped the ways in which they thought about water and how they used it. Moreover, these views were not static, but changed over time and continue to evolve even today in response to changes in the southwestern Oklahoma environment.

Today, the majority of Kiowas either live or have property in the rolling foothills of the Wichita Mountains between the Washita and Red Rivers. They have not always lived there, but have a long history of migration through the central mountain states. Linguistic anthropologists claim that the proto-Kiowas originated in the present-day American Southwest and migrated northward along the eastern range of the Rocky Mountains to Crow country in present-day Wyoming and Montana. Toward the end of the seventeenth century, the Kiowas moved again, this time migrating eastward with the Crow to the Black Hills between Wyoming and South Dakota. Here the tribes ran headfirst into the better-equipped, horse-mounted Lakota and Pawnee peoples. Quickly acquiring horses of their own, most likely from the Utes, the Kiowas bid farewell to their friends the Crows and moved south, into the mixed-grass regions of Kansas, Oklahoma, and Texas. Kiowa material culture rapidly absorbed the region's abundant resources. The Kiowas adopted, borrowed, and reshaped the cultural trappings of other Plains peoples they encountered. Like the Comanches, who had preceded them, the Kiowas' material well-being improved dramatically and their population numbers soared. Their religious beliefs found new expression and their sense of identity became enmeshed in their surroundings: in the Wichita Mountains; in the flora and fauna of the southern Plains; in the bison herds; in their horses; and in the sun, moon, and stars that guided them in their movements.

Kiowa Legends

The Kiowas' secure sense of place on the southern Plains is clearly revealed in the stories of Saynday (pronounced "Sin-dee"), a skinny, egotistical, and irreverent culture hero. At one time, the telling of the Saynday stories was attached to kinship roles and numerous taboos were associated with their dissemination in the tribe. Old people were the only family members permitted to tell the stories, and then only at night. Their listeners were primarily children who needed to learn right and wrong behavior and to take pride in their identity. However, the Saynday stories also taught the Kiowas' proper relationship to nonhuman nature. It is here that we can begin to outline the Kiowas' relationship to water.

When water is mentioned, it plays a nonanimate role. In one popular story, Saynday interrupts a group of field mice holding a Sundance inside a bison skull. He is so enthralled that he gets right down on the ground to watch and, before he realizes it, has stuck his head all the way into the skull-lodge, where it becomes firmly wedged. The mice are understandably furious with him and leave him there, blinded and stumbling around. After falling down a few times, Saynday decides to feel his way to Cottonwood Creek, which he knows will lead him directly back to his camp. He first gropes his way to an unhappy mesquite tree who tells him to keep going. He next finds a nervous wild plum, so he knows he is going in the right direction. Saynday's next guide is a sturdy elm who directs him over the bank, where he slides into a willow tree, who through sobs announces he has arrived. Saynday then gets into the creek, which carries him over the sand, over the shallows, over the mud, into the deep pools, and finally to his own camp, where he is appropriately ridiculed by all the other animals for his lack of respect and general stupidity. Saynday knows the creek and its nature—sand, mud, shallow, and deep—but of all the elements in the story, including mice, trees, and his animal friends, the creek is strangely silent. Only the creek is without its own personality. It is familiar, but it is only a medium connecting Saynday to the other characters. The watercourse is indifferent to the culture hero's problems.[7]

This lack of relationship between Saynday and the water may have stemmed from the Kiowas' origin story of the half boys. Although another Saynday story recounts how Saynday brought the antlike Kiowas up to the surface of the Earth through a hollow log, most anthropologists consider the half-boys story to have preceded the Saynday tales by several generations, and it may reveal the Kiowas' notions of good and evil even better than the Saynday tales.

In one accepted version of the story, a woman wished to marry Evening Star. He hears her and takes her for a wife and they have a son. Lonely for her people, the woman spies the Earth below the clouds and tries to escape her husband by climbing back down on a buffalo sinew rope. Evening Star ends her escape by striking her with a stone, but their son lands safely on Earth, where he is taken in by a spider woman. His adopted mother makes the boy a web sphere to play with and, against her instructions, he throws it high into the air. As the sphere comes down, it splits him into two half boys. One of the half boys leaves to create trees, mountains, rivers, and animals—the familiar realm of the Kiowas—whereas the other boy walks under a lake, never to be seen again.

This negative association of water with an unproductive underworld parallels the late-nineteenth-century Kiowas' avoidance of ghosts and turtles. These were not the only nonhuman entities the Kiowas feared (owls and bears were also avoided), but it was understood that any medicine man who took one of these "evil" animals as his spirit helper would eventually also be destroyed by its invidious power. This was true of most water creatures, a vestige of the power associated with the half boy who abandoned the world of the Kiowas to live in another, presumably evil, world.

According to a story recorded by ethnologist James Mooney, Ton-ak-a (also known as Notched Tail or Water-Turtle) was a lustful and jealous medicine man who brought attention to himself after stealing another man's wife in 1885. For this, Mooney records, he was whipped and a number of his horses killed by the maligned husband.[8] In interviews done in the early 1930s, tribal elders additionally recalled that Water-Turtle had a long history of mischief and was feared by most of the tribe. Ton-ak-a claimed to have received his power as a child from a water turtle while swimming in a creek. He once demonstrated this power by asking his friends to resuscitate him after death by placing him underwater. Not long afterward, Water-Turtle was found stiff and dead. His friends reluctantly complied with his request by weighing the body down with stones in a nearby creek. They later felt uneasy and sat down to talk about what to do next. Just as they were thinking about leaving, Water-Turtle strode into camp, dripping so profusely that he extinguished their campfire. This only served to increase the Kiowas' fear of the old medicine man, because it was presumed that he frequently misused his power against those who crossed paths with him. As Benjamin Kracht, an anthropologist of the Kiowas' religion, notes, Water-Turtle was not very popular when he eventually died of poisoning and "all the Kiowa were glad."[9]

The source of Water-Turtle's medicine power, or *dwdw,* was closely related to other forms of power in Kiowas' cosmology. Like other Plains peoples, certain animals were associated with death, especially the owl. Owls were thought to be a medium for the voices and spirits (ghosts) of the deceased. Among numerous Plains tribes, ghosts were feared as causing trouble for the living. Jim Whitewolf, a Kiowa-Apache born just before the reservation era, recalled that children were considered extremely vulnerable to ghost medicine. On their excursions for wood, Kiowa-Apache mothers allowed their children to swim in the creeks to keep them busy. When they were through, however, the women would take branches and cover up the children's tracks leading from the water to prevent ghosts from following the children home. In the same way, each mother called her child loudly by name so that the ghosts could not take the children's voices and make them sick later.[10]

The Reservation Era, 1875–1904

This generalized fear of water and its occupants took on new forms during the reservation period. For example, in 1875, several bands of Kiowas were forcibly rounded up and imprisoned at Fort Sill. Government rations at the fort consisted of beef, salt, coffee, flour, rice, and bacon. Some items were appreciated, such as the beef, salt, and coffee, but others were not. The Kiowas did not know what to do with the flour and threw it away along with the rice, which they said was dried maggots. Bacon was summarily rejected—

it was rumored that it was either elephant meat or that it came from the East, where the lakes were filled with giant water snakes.[11]

While supernatural beliefs about water persisted into the reservation period, changes in the Kiowas' resource base necessitated that they pay more attention to their use of water. Long before their confinement on the reservation, the Kiowas were extremely sensitive to the location and quality of the water they needed for themselves and their animals. The names they gave to various streams and creeks reflected these practical considerations, such as Salt Creek, Hog Creek, and Stinking Creek, which denoted their different qualities and uses. Luther Sahmaunt, born in 1860 on the banks of the Arkansas River in Kansas, recalled how as a young man his band was out on a bison hunt together and camped near a stream in present-day Texas. Upstream from the camp, the remains of several dead bison had been left by white hunters to rot in the stream bed. When a sudden cloudburst filled the stream with water, the maggots in the bison carcasses were washed downstream into the Kiowas' camp. The Kiowas promptly renamed the stream Maggot Creek as a reminder of both the location of the camp and the conditions they found there. This form of historical signposting was a well-established Kiowa tradition.

Water quality and availability became even more important after the Kiowas began to farm on a regular basis. Forced to engage in row-crop farming by missionaries and Indian Office personnel, drought consistently destroyed the tribe's crops between 1875 and 1900. Because the Kiowa did not have much faith in the climate to produce crops, their farms were generally small, communally worked affairs. Supported by the local agent, the vast majority of Kiowas were inclined to make their living raising either cattle or horses. This meant that pasturage, sufficient water, and wood for fuel became the Kiowas' primary considerations.

This form of resource use fit in well with the Kiowas' social structure. Previous to their allotment in severalty in 1904, the Kiowas had maintained a semblance of band structure with large, extended family groups camping together in places that offered good wood and water.[12] After allotment, this pattern obtained to a significant degree, with the different bands stringing their individual parcels across the former reservation along several main watercourses.

Lone Wolf II, the plaintiff in the famous Supreme Court Case *Lone Wolf v. Hitchcock* (1901), was a band leader in this fashion. In his correspondence with the Office of Indian Affairs, the leader referred to himself as "Lone Wolf of Elk Creek." Although this identification with a particular band was in keeping with traditional Kiowa social structure, it was also a clear departure from earlier forms of band identification based on leadership and family ties, but one whose geographical location changed with the needs of the people and their animals. By the early 1900s, Kiowas identified themselves as belonging to one of several bands: Elk Creek Kiowa, the Rainy Mountain–Sugar Creek group, Saddle Mountain and Mount Scott People, Hog Creek Kiowa, Mud Creek Kiowa, and Stinking Creek Kiowa.[13]

Allotment

After allotment, access to good water was as essential to Kiowa ranchers as for other raisers of stock in the same area, and the Kiowas' choice of individual allotments generally reflected their intimate knowledge of those watercourses that flowed year-round.[14] However, several things conspired against the Kiowas' success as commercial stockmen. At 160 acres apiece, their allotments were far too small to support more than a few animals at a time. This was noted by the Kiowas, their agents, and even Frederick Newell of the Geological Survey during the Jerome Commission meetings in 1904. Some Kiowas were able to overcome this problem by consolidating their allotments with other family members where possible, but even then they frequently could not obtain enough barbed wire to fence their properties. When Kiowa families attempted farming in addition to raising stock as a way to sustain themselves, they rarely received the needed instruction, seed, and equipment they were promised by the Office of Indian Affairs.[15] However, most damaging to the Kiowas' attempt at western-style farming and ranching was the lack of protection by local law enforcement after 1904. The influx of new white settlers into the previously intact reservation severely mitigated against Kiowa stock raising, as white thieves stole Indian stock at unbelievable rates without any fear of reprisal. Even those Kiowas who did not own stock fell victim to white greed as Texas cattlemen purposely turned their herds out onto Kiowa lands to graze. In years when crops failed and rations were held back by the Indian Office, the Kiowas were forced to eat their own herds. Problems of water quality and access paled in comparison to these other considerations. Water was a moot point with Kiowa stockmen whose grass and herds had a habit of disappearing overnight.

Understandably, most Kiowa allottees concluded that it was better to lease their lands for some guaranteed return than to have their water, grass, and animals systematically stolen from them. By 1920, independent Kiowas who raised stock and farmed had become the exception rather then the rule. Most families continued to live on small parcels of between ten and forty acres on which they had a house, a few animals, and a large garden for the family. The rest of the allotment was then leased to a local white farmer or rancher.

Kiowa children born after 1900 frequently did not live full time on their parents' allotments, but attended reservation and off-reservation boarding schools, where they were taught skills useful in the white society and its industrial economy. Many returned to their communities, but did not take up farming, finding work instead in government service, the missions, or the school system. While all of these adaptations allowed Kiowa "schoolboys and schoolgirls" to maintain their kinship ties and Kiowa traditions, their time away at school also meant they no longer had the same opportunity to intimately explore the nature of the place they called home.

Water Maps: Kiowas Today

Between 1875 and 1920, Kiowas' identity made several shifts in relation to the place they occupied and its available resources. Before 1875, the Kiowas identified themselves according to band affiliation and their attachment to particular leaders. After 1875 and their confinement to the reservation, the bands took up more permanent residence along watercourses or near identifiable mountains. After allotment in 1904, the Kiowas became more scattered and their tribal affiliations were formed around a combination of local places and organizations. These were primarily church membership, school attendance, and township lines. Following the opening of the reservation to whites in 1904 and the Kiowas' subsequent loss of control over their allotments through the lease system, many Kiowas became town dwellers. Ironically, the towns and annual celebrations held on the local fairgrounds were and continue to be instrumental to the Kiowas' maintenance of their traditional band structure.

None of these changes, of course, meant that the Kiowas were unaware of changes to their local environment, including their water sources. A contemporary Kiowa living in Anadarko related that he, his brother, and his sister used to swim in the bend of the Washita River when it was "deep and not so cloudy as now." A natural lake that used to exist near his farm, where he and his playmates hunted swamp rabbits, has completely vanished. Older Kiowas too have noted that the Washita River has changed course in recent memory and this means changes in the landscape, but they also reason (with good evidence) that it is the nature of a river to change its course periodically. They voice concern about the pollution, just as the Kiowas at Maggot Creek did, but theirs is a practical concern, common to rural residents throughout the Southwest.

For the Kiowas, the greatest loss they have experienced as a cultural unit has been the loosening of community bonds as the number of Kiowa living on or near their allotments continues to decrease through inheritance, sale, and migration to towns. As Ray Doyah, a resident of Anadarko and great-grandson of the famous Kiowa leader Ahpeatone, has noted, "Topping a rise near Carnegie in the fifties, it was still possible to identify each Kiowa family by a yard light. Topping that same hill today, I couldn't begin to tell you which yard light belongs to which family."[16] Like Saynday, the most distinguishing features of the landscape the Kiowas inhabited were those that brought them into contact with one another. The watercourses that run through Kiowa country today are not important because they are persons in the same way as turtles, owls, bison, or tornadoes were, but because they connected people in practical ways. Water, especially in the form of rivers and intermittently flowing streams, acted as a road map for Kiowa families to find one another. An intimate knowledge of watercourses also allowed the Kiowas the brief opportunity to make the transition from hunting and raiding to a more sedentary ranching economy, while still very much dependent on their traditional knowledge of the lands' resources. Kiowa beliefs about water were never identical to that of Euro-Americans, but their use of water for everyday needs was not materially different from that of their non-Indian neighbors. Whether in the form of surface water, wells, or seasonal rains, the Kiowas' historical use of water was imminently practical. Water had its own unique place in Kiowa cosmology, but on a day-to-day basis, it existed to meet their needs. As one Kiowa elder recalled from his childhood, even the puddles left behind after a thunderstorm had their purpose, for it was only then that you could get a clear look at yourself.[17]

Notes

1. Daniel Richter, "Whose Indian History?" *William and Mary Quarterly* 2 (April 1993): 382. I would like to thank Ray Doyah of the Kiowa Nation for his candid reminiscences and Jim Sherow, Bunny McBride, and Harald Prins for their careful and critical reading of the manuscript.

2. Donald A. Grinde, Bruce E. Johansen, and Howard Zinn, eds., *The Ecocide of Native America: Environmental Destruction of Indian Lands and Peoples* (Santa Fe, N. M.: Clear Light Publishers, 1995), 250.

3. Richter has noted that "Professional historians who challenge such modern Indian myths as the idea that whites invented scalping or the theory that the Iroquois League provided the model for the United States Constitution can . . . be charged with intellectual elitism at best and racism at worst." Richter, "Whose Indian History?" 383.

4. Richard White, *The Roots of Dependency: Subsistence, Environment and Social Change among the Choctaws, Pawnees and Navajos* (Lincoln: University of Nebraska Press, 1983); David Rich Lewis, *Neither Wolf nor Dog: American Indians, Environment and Agrarian Change* (New York: Oxford University Press, 1994).

5. Frederick Hoxie, "Ethnohistory for a Tribal World," *Ethnohistory* 44 (Fall 1997): 606.

6. Although ethnohistorians acknowledge the elusiveness of claiming to find "the truth" about any "true culture" there remains, according to Fred Hoxie, the primary task of "comprehending and describing as accurately as possible the lives of people and communities whose allegiance is to a variety of cultural traditions other than our own," while resisting the "desire to compress, essentialize and manipulate the people who lie at the heart of [our] inquiries." Frederick Hoxie, "Ethnohistory for a Tribal World," 613. On the relationship between postmodern discourse and ethnohistorical inquiry, see Jennifer Brown and Elizabeth Vibert, eds., *Reading Beyond Words: Contexts for Native History* (Peterborough, Ontario: Broadview Press, 1996).

7. Alice Marriott, *Saynday's People* (Lincoln: University of Nebraska Press, 1947), 44–51.

8. James Mooney, *Calendar History of the Kiowa Indians* (1898; reprint, Washington, D.C.: Smithsonian Institution Press, 1979), 353.

9. Benjamin Kracht, "Kiowa Religion: An Ethnohistorical Analysis of Ritual Symbolism, 1832–1987" (Ph.D. diss., Southern Methodist University, 1989), 118.

10. Jim Whitewolf, *Jim Whitewolf: The Life of a Kiowa Apache Indian* (New York: Dover Press, 1969), 55.

11. Ibid., 54.

12. Lawrie Tatum, *Our Red Brothers* (Lincoln: University of Nebraska Press, 1970), 65; Hugh Corwin, *The Kiowa Indians* (Lawton Okla.: author, 1958), 143.

13. Corwin, *The Kiowa Indians,* 150.

14. For a complete description of the Kiowas' intimate knowledge of forage and medicinal plants, see Paul Anthony Vestal, *The Economic Botany of the Kiowa Indians,* (reprint, New York: AMS Press, 1981).

15. Robert John Stahl, "Farming among the Kiowa, Comanche, Kiowa-Apache and Wichita" (Ph.D. diss., University of Oklahoma, 1978), 129–58.

16. Ray Doyah to author, Manhattan, Kansas, 14 March 1997.

17. "The Life of Rev. Albert Horse," in Corwin, *The Kiowa Indians,* 149.

5

The Dilemmas of Indian Water Policy, 1887–1928

Donald J. Pisani

For decades, historians have explored a paradox in federal Indian policy after the Civil War: It sought to turn Indians into farmers on poor land with inadequate equipment and inadequate instruction in agricultural and marketing techniques. That Native Americans had little understanding of private property and preferred to work the land collectively as tribes, clans, or villages rather than as isolated families compounded the paradox. Yet this explanation of why the policy failed to turn more than a small fraction of Native Americans into farmers misses a fundamental truth. The age of the Jeffersonian yeoman farmer had passed by the time Congress enacted the Dawes Act in 1887, and politicians and bureaucrats in Washington could never decide whether Indians should be separated from or integrated into white society. The family farm represented autonomy and independence, whereas wage labor represented integration into the national economy—although in servile jobs. The Bureau of Indian Affairs (BIA) and politicians in Washington vacillated between agriculture and wage labor, and inconsistencies in Indian policy contributed to failure as much as did devotion to the family farm.[1]

It is misleading to see nineteenth- or twentieth-century Indians as *either* farmers or wage earners; many combined the two. Nor is it accurate to argue that Native Americans became wage workers because, robbed of their land, they had no other choice. Labor off the reservation frequently complemented agricultural income, and Indians found working for wages highly compatible with a traditional economy of hunting and gathering. Also, because work off the reservation often involved "gang labor"—such as harvesting hops or laying railroad tracks—it reinforced social and cultural bonds as it provided much needed income.

Wage labor is a neglected topic in Native American history.[2] Anthropologists have downplayed the interaction of Indians and non-Indians by focussing on "traditional cultures" prior to their "contamination" by Christianity and capitalism. Historians have compounded the problem by concentrating on federal Indian policy rather than on how Indians lived or made a living. Textbooks talk about the disappearance or dispossession of the Indians, not how they fit into the American economy. In addition, historians have fallen prey to the assumption that by the 1880s or 1890s Indians were so thoroughly demoralized that they could not or would not work at "white" jobs—a myth perpetuated by the white farmers who depended heavily on Indian labor.

This assumption flies in the face of the historical record. From the early seventeenth century on, Indians worked for wages in the fur trade, as farm laborers, and as domestic servants. In some of the southern colonies, they were enslaved. Later, as white settlers swept across the continent, Indians became an indispensable source of labor in the American West, particularly in regions where water and other natural resources were scarce, such as the Great Basin. By 1900, in parts of Nevada, for example, Paiutes received more than half of their incomes in wages. When planted in alfalfa, the typical irrigated allotment netted $75 for a season's work. A strong Indian man could earn that amount in a week building railroads off the reservation, and an Indian family could do as well in two weeks picking cantaloupes. "By the turn of the century," the anthropologist Martha Knack has written, "the Walker River Northern Paiutes were clearly in the habit of deciding economic issues on the basis of financial maximization, sometimes to the discomfiture of the agency. They were juggling a complicated economic cycle that included wage labor for both men and women as an essential component. They were very much integrated into the overall economic development of the region, both on the reservation and off."[3]

By 1930, nearly two-thirds of Indian men were still engaged in agriculture, but half worked as farm laborers rather than farm operators. Then, in the 1930s and 1940s, wage labor increased rapidly as the federal government became a major employer in the West. In 1935, wages paid to Indians enrolled in the Civilian Conservation Corps on the Rosebud Reservation exceeded all other sources of income, and when the program was discontinued in 1942, 85 percent of Rosebud men had worked in the corp or other federal work relief projects. After World War II, large numbers of Indians continued to be on the federal payroll, but we should not forget that Native Americans were an important part of the white economy in the nineteenth as well as the twentieth century. They did not become wage laborers solely in response to loss of their land or because they moved in large numbers to cities during and after World War II.[4]

In most of the western half of the nation, agriculture was impossible without irrigation. In 1871, General William Tecumseh Sherman predicted that by irrigating Apache land in the Southwest, the federal government could prevent Indians from raiding white settlements and from leaving reservations in search of water and forage for stock. The cultivation of hay and alfalfa would permit the Apaches simultaneously to supplement the rations issued by the government and to pursue their traditional livestock industry on far less land. In the decades that followed, settled agriculture became synonymous with reducing the size of the BIA. The policy would remove the federal government from the "Indian business," as the champions of fiscal restraint and retrenchment argued. "It has been asserted that with a proper system of irrigation on the Navajo Reservation adjacent to the San Juan River," the Commissioner of Indian Affairs observed in 1902, that "two-thirds of the families occupying that reservation could make homes and become self-supporting." Making the Indians self-sufficient would eventually eliminate the cost of feeding them.[5]

Irrigation became a symbol of agricultural self-sufficiency, autonomy, and independence, but dams, canals, and ditches had an appeal of their own. The construction of public works provided Indians with valuable work experience and a discretionary income that taught the virtues of careful money management. In the last decades of the nineteenth century, most reservations were far-removed from white settlements and the opportunities for manual labor were few. Indian laborers could see direct benefits to their community as well as to themselves, and they took pride in their work. More than one-third of the cost of the Zuni dam at Blackrock, New Mexico, which was begun in 1903 and completed in 1908, was paid out in wages to Indians. As the Commissioner of Indian Affairs proudly noted in 1906: "The lesson taught by the experiment with Indian labor at this dam is unquestionably that if the Indian can be weaned from his habits of irregularity of days and hours, induced to postpone or rearrange his religious festivities so that they shall not interfere with the demands of his employment, and taught the white man's idea of laying something aside for tomorrow instead of spending all today, he can be made into a very valuable industrial factor in our frontier country." In fiscal year 1910, the BIA paid out $145,000 in wages to Indians—more than 10 percent of its annual appropriation.[6]

Annual expenditures for Indian irrigation projects did not begin until the 1890s. The Dawes Act (1887) permitted the president to dissolve tribes and distribute a maximum of 160 acres of former reservation land to each Indian. The legislation provided both the justification for reservation water projects and, through the sale of "surplus" land, the revenue to pay for them. (Congress preferred to spend "Indian money" rather than tap the general treasury.) Allotment would leave the Indians with far less land, but irrigated land would produce much more revenue than dry-farmed land or land used as pasture. Irrigation would pay for itself through higher land values. When a long drought hit large parts of the West at the end of the 1880s, in some places lasting into the mid-1890s, it underscored the need to furnish the Indians with a more dependable water supply than nature delivered.[7]

In 1890, the BIA promised Indians on the Crow Reservation in Montana that the federal government would spend $200,000 irrigating land adjoining the Big Horn and Little Big Horn Rivers if the Indians accepted a smaller reservation. Without irrigation, the allotment of Crow land could not be justified. Therefore, the BIA sent Walter H. Graves, a civil engineer, to the reservation to supervise ditch construction. By 1896, one hundred miles of main ditch, capable of watering about 25,000 acres, had been constructed. The new canal was not designed to introduce the Crows to agriculture, but rather to replace large common farms with individual freeholds to facilitate the production of wheat and hay for sale at Fort Custer and other local markets.[8]

The construction of canals provided steady employment, and 80 percent of the money spent on the Crow project went to Indian workers. Graves informed the Commissioner of Indian Affairs that the Crow were "excellent workmen and in the handling of earth in the construction of channels and embankments . . . some of them can be easily classed as skilled workmen." The Indians did not take to pick-and-shovel work, and they easily became discouraged when harnesses snapped or wagons broke down. Nevertheless, they were adroit at directing horses and manipulating the scrapers used to excavate ditches. Equally important, they appreciated the opportunity to earn sufficient money to buy livestock, wagons, and farm machinery, and a few used their wages to build houses. Unlike the $6 payment each Indian received twice a year for land ceded to the federal government, the wages came in lump sums large enough to matter. "It is remarkable as it is gratifying," Graves reported, "to see how quickly the ideas of individuality, in responsibility, in ownership, in labor, of conservation of their resources, of emulation, etc. are growing and developing among them."[9] By 1910, nearly $1 million—mostly Crow money—had been spent to excavate ditches with the capacity to serve about 70,000 acres. Little more than 27,000 acres were actually irrigated, however, and most of that land was flooded to increase the growth of native grasses fed to stock.[10]

Shortly after work began on the Crow water system, the BIA launched projects on the Fort Peck, Blackfeet, Fort Hall, and Navajo Reservations. Experienced engineers supervised the Crow and Navajo work, but elsewhere Indian agents hired "engineers" who lived near the reservations to design the works. As a result, many canals were poorly conceived and badly constructed and there was no consistency in design or materials from one project to the next. During the 1890s, the BIA and Secretary of Interior repeatedly asked Congress to authorize the bureau to hire a supervising engineer to oversee the construction and maintenance of Indian public works. This was finally done in 1898.[11] Meanwhile, agriculture expanded on many reservations, which prompted one writer to observe that "the plow and the sickle have, to a great extent, driven the tomahawk and the scalping knife from the field."[12]

Appropriations for Indian irrigation projects steadily increased during the first three decades of the twentieth century, from $670,000 in 1907, to $1,065,000 in 1910, to $1,957,700 in 1917, to $2,408,750 in 1924.[13] The Indian Irrigation Service also grew. The 1900 BIA budget provided for two "superintendents of irrigation." By 1911, that number had increased to seven. There was no formal requirement that these men be trained as hydraulic engineers, however, and the Indian Irrigation Service was no match for the

Reclamation Service in numbers or talent. According to historian Douglas Hurt, employees of the Indian Irrigation Service "received the lowest salaries of any in governmental service, and its work suffered from shoddy planning and lack of coordination and standardization among the reservations." Hurt recognized that irrigation projects were designed as much to provide labor as water. "Irrigation projects on reservation land became ends in themselves," Hurt noted, "and their benefits were only of secondary importance to the Irrigation Service."[14] Nevertheless, the Jeffersonian ideal remained powerful. A 1903 editorial in the periodical *Forestry and Irrigation* proclaimed that "the policy of irrigation" was "one of the greatest factors for education and civilization of the American Indian."[15]

When they were created between the 1850s and 1870s, most Indian reservations were far removed from major white settlements. At that time, wage labor focussed on the construction of public works *within* the reservations. By the early decades of the twentieth century, however, the process of allotment turned former reservations into checkerboards of Indian and white holdings. As whites flooded onto Indian land, two things happened: Indians now competed with whites for jobs off the reservation, but the number of jobs expanded, particularly in agriculture. Whites purchased "surplus" Indian land and leased allotments that Indians were unwilling to work or incapable of working. White farmers, the BIA now hoped, would serve both as employers and models of industry to the Indians, thus easing the transition to settled agriculture. This policy was filled with irony. Indian money had been used to pay for irrigating land farmed by whites, now Indians were used to work their own land for the profit of others.

The practice of leasing allotments began in 1891 and grew dramatically in the twentieth century as irrigation made Indian land more attractive. Many native people—young and old—were not capable of cultivating their land; for example, irrigated allotments were often too large for the "head of household" to manage. If Indians could not use the land, it might as well be used by whites. Indian officials were reluctant to discourage leasing, because they believed that native people had to learn to make hard choices and take responsibility for their decisions. Leasing served many purposes, the bureau believed. Whites could provide Indians with an object lesson in how to farm. By encouraging non-Indians to take up reservation land, Indians could model themselves after successful farmers imbued with the ethics of industry and thrift. Equally important, leases provided cash that could pay for farm equipment, livestock, and other agricultural needs. Finally, leasing irrigated allotments insured that there would be water to farm the land if and when the allottee decided to take over cultivation himself. Only occasionally did the BIA grumble about renting the land of Native Americans to whites, as in 1914 when the commissioner called the policy "a poor one at best."[16]

The success of white farmers spelled the doom of Indian farmers. During World War I, Indian people were all but forgotten in the quest to open new acreage to the plow and to provide food for the allies. On the Blackfeet Reservation the number of irrigated farms increased from 18 in 1915 to 328 in 1919. But whereas Indians worked nearly 90 percent of the farms in 1915, by the end of the war white tenants worked more than 70 percent.[17]

Hard times were no kinder to the Indians. Crop and real estate prices plunged during the 1920s, and Congress appropriated very little for Indian irrigation. By the end of the decade conditions on the reservations appeared more bleak and hopeless than ever. The water policy had failed. A 1928 report written by two engineers in the Agriculture Department, Porter Preston and Charles Engle, underscored this conclusion. Their lengthy study represented the most thorough investigation of Indian agriculture compiled to that date.[18]

According to Preston and Engle, there were 150 Indian irrigation projects in 1927, ranging in size from a few acres to the Yakima Reservation's 89,000 acres under ditch. Nearly 700,000 acres within allotted and unallotted reservations had been provided with irrigation at a construction cost of $27 million and an additional expense of $9 million for operation and maintenance. About 30 percent of this land belonged to white farmers. However, whereas Indians watered only 16 percent of their irrigable land, whites cultivated 66 percent. On the Wapato project within the Yakima Reservation, only 6 percent of Indian allotments were actually irrigated by their owners; on the Blackfeet and Flathead Reservations, the figure was roughly 1 percent. In fact, 40 percent of all the land irrigated by Indians in the West was located within the Uintah Reservation in Utah and the Gila River Pima Reservation in Arizona.

As dreary as the statistics seem, actual conditions were worse. According to the Agriculture Department engineers, most Indians who used irrigation simply flooded pasture to stimulate the growth of native grasses for livestock; Indian crops, therefore, returned only half the per-acre income of white farmers. The report concluded: "Many of the so-called Indian irrigation projects are in reality white projects. . . . The continual decrease in the acreage farmed by Indians is the inevitable result of the allotment system."[19]

Nine secretaries of the interior and five commissioners of Indian affairs held office between 1900 and 1929. They promoted irrigation for different reasons: to create work, prevent Indians from straying off the reservations, develop a love of private property, protect water supplies, enhance the value of Indian land, free up more land for white settlers, attract white farmers who would employ Indians and serve as models of industry, increase tribal funds, and reduce the cost of maintaining the reservations. Yet by the 1930s, the dream of turning Indians into farmers had faded; the appeal of wage labor won out. President Herbert Hoover's secretary of the interior, Ray Lyman Wilbur, claimed that Indians were communists "with no incentive to personal endeavor, to accumulation of any sort." They could better develop a sense of private property, Wilbur thought, by "working for hire, under direction" than through farming. "The Indian still seems to lack adjustment to the prevailing civilization," Wilbur observed, "and it is now found that this lack of adjustment is due to the fact that the prevailing civilization works and the Indian does not." Officials in the BIA agreed: The future of the Indian was working for others, not for himself.[20]

However, Wilbur was wrong. Not all Indians were "communists," some worked hard at farming, and at least one historian has concluded that "substantial progress in Indian farming was made before allotment."[21] Nevertheless, on one point Wilbur was right: The federal government's water and agricultural policies had been a disaster.[22] Using agriculture to "transform Indians into whites" had two major pitfalls. First, it ignored the fundamental lesson of frontier farming: With few exceptions, the generation that first plowed "virgin land" rarely enjoyed financial success and seldom put down permanent roots. White pioneers usually moved on after a few years of frustrating, poorly rewarded labor.[23] Given the additional liabilities faced by Native Americans, it was a supreme act of faith to believe that Indians could succeed where so many whites had failed. Second, at the same time that white farmers in the West were beginning to organize and cooperate, Native Americans were asked to give up collectivism in favor of individualism and autonomy. White farmers created such "tribal" organizations as the Grange and Farmers Alliances. During the early decades of the twentieth century the ideal of the yeoman farmer became an anachronism in most parts of the United States. Yet it was deemed a suitable model for Indians well into the 1920s.

The Indians faced many agricultural barriers. Most reservations suffered from poor land and short growing seasons. At the beginning of the twentieth century, officials in the Reclamation Service assumed that virtually any desert soil was productive given sufficient water, but the Indians knew from experience that no amount of water could transform their sterile land into prosperous farms. Furthermore, by the 1920s and 1930s, farming required a far more specialized knowledge of agriculture and far more costly equipment than in the 1870s and 1880s. To buy that equipment and to use land "productively," Native Americans needed capital. Yet they could not mortgage land or borrow money while the title remained with the federal government. Even Indians who owned their allotments outright were reluctant to pledge their land as collateral.

The BIA made the same mistake as the Reclamation Service: It assumed that reclamation was essentially a matter of engineering. As the amount of money spent on dams and canals increased, appropriations for nonreimbursable agricultural purposes and for the purchase of livestock shrank.[24] The $30,000 Congress appropriated for seed and farm equipment in 1888—a modest appropriation considering the goal of the Dawes Act—was cut in half in 1891 and eliminated entirely in 1894. After 1911, tribal funds could be tapped for such purposes, but never to the extent needed by would-be Indian farmers.[25] "We did not give him a penny with which to buy a plow, or a harrow, or a grubbing hoe, or anything at all with which to work the land," Senator Lane lamented on the floor of the U.S. Senate in 1914. "He was left with his bare hands. The white man could not make a success under such circumstances."[26]

Native Americans and reservation agents pleaded with the BIA to free up tribal money for livestock, seed, lumber, farm machinery, and other agricultural necessities. These appeals fell on deaf ears. Congress might have subsidized the growth of certain crops, encouraged the Indians to form cooperatives, helped locate markets, furnished free livestock, or at least paid for irrigation projects out of the general treasury with nonreimbursable funds. Had the reservations been closed to whites; had the federal government provided sufficient money to clear, level, fence, and irrigate the land and purchase farm equipment, seed, and livestock; and had it provided marketing cooperatives and processing facilities (such as canneries), the story *might* have been different.[27]

The social isolation of the individual freehold and the BIA's lack of sympathy for traditional gender roles were equally important barriers to agricultural progress. White farmers complained about the loneliness of farms and ranches in the West. Yet how much worse it must have been for those Native Americans who considered solitude a cruel punishment and defined the quality of their lives in terms of community rather than individualism. In many tribes, farming was women's work; men regarded it as demeaning or degrading. Instead, Indian men took to raising livestock, perhaps because that vocation was consistent with the hunt. The 1928 Meriam Commission recognized that truth and recommended that Indian women be trained to tend gardens, raise poultry, and preserve fruits and vegetables.[28] After 1914, some federal and tribal money was used to purchase herds, but it was too late for many tribes, whose best grazing land had been leased to whites.[29]

Perhaps the biggest obstacle to Indian agriculture was that irrigation encouraged speculation in Indian land. The Indians had far less incentive to farm than not to farm. The only way to protect their land was to do as little as possible to improve it. Without irrigation, most of it was unattractive to whites. To become a good farmer, then, was to invite non-Indians to invade the reservation. Indians could be taught the virtues of private property only if the property they secured was protected—as the law protected the property of whites. The more valuable Indian land became—and irrigation did more to increase its value than any other improvement—the less secure became the Indian's title.

Indian policy reflected the tension between promoting the Jeffersonian freehold and assimilation, between individualism and the cooperation necessary to make one's way in the new corporate order. Was the federal government's paramount goal to help Indians adapt to a harsh desert environment or to produce "imitation white men"? Was its basic purpose to turn the Indians into farmers or to promote individualism? The Indians had to learn that both land and the crops it produced were commodities to be bought and sold. It was not enough to make the Indians self-sufficient, they had to learn the "way to wealth." But was agriculture the best way to teach that lesson? Transforming Indians into farmers was unlikely to break down the reverential view of land at the heart of many Native American cultures.

The Indians had to take responsibility for their own destiny, the BIA argued, but the Indians were rarely consulted in the formulation of federal policies. Nothing better illustrated what John Collier called the "ruthless benevolence" or "benevolent ruthlessness" of the BIA than water policy.[30] Even though they bombarded the BIA and Congress with protests, the Indians got irrigation and paid for it whether they wanted it or not. In 1914, Commissioner Cato Sells categorically denied that the Indians ought to have more say in reservation water projects: "We do not agree that a voice on the part of the Indians as to construction and operation of irrigating systems is a fundamental principle necessary to be observed in handling these matters, as they involve questions of great economic importance and require the consideration of trained minds."[31] In the following year, the Board of Indian Commissioners observed that whites who lived on or near reservations had far more influence within the BIA than did Indians. The Indians could not be blamed for concluding that reclamation was just another scheme to escape treaty obligations and transfer control over their land to whites. How, the board asked, could Native Americans be expected to take interest in a policy imposed upon them by force?[32]

In the end, it may well have been the Indian view of law that defeated federal land and water policies on the Indian reservations, as historian Barbara Leibhardt has shown. Many features of American law seemed irrational to Native Americans. What sense did it make, for example, to separate rights to land and water when nature did so much to unite the two? Many Indians clung to the idea that treaties between the federal government and Native Americans expressed an immutable higher law and, despite repeated violations and abuses, they assumed that treaties could not be rescinded or abridged. In speaking of how the Yakima Indians viewed the 1855 treaty that created their reservation, Leibhardt notes: "The treaty became the unchanging standard against which they evaluated the integrity and legitimacy of all future federal laws and policies. In that view, law was a static, nearly divinely sanctioned agreement among individuals."[33]

Subsequent attempts to redefine treaty rights by allotment and the diversion of tribal funds to irrigation could not alter timeless responsibilities and obligations. Perceiving the world in these terms, the Indians hoped that one day the federal government would honor its covenants. In contrast, whites took a much more flexible and pragmatic view of law. They revered the Declaration of Independence and Constitution—both expressions of Higher Law—but they saw Indian treaties as agreements that could be altered to suit new circumstances and conditions, just as statutes could. To whites the law was less a moral obligation than a tool to promote the rapid development of natural resources and the creation of new wealth. Its basic goal was to reward individual enterprise and only incidentally to promote a just society. Indians and whites looked at the land and agriculture in profoundly different ways, but, equally important, they differed over the nature and moral purpose of the law.

There is no simple explanation for the failure to turn Native Americans into family farmers; water policy is only part of a complicated story. Reduced to essentials, the family farm was inconsistent with modern America. Long after rural gave way to urban America and long after factories eclipsed farms, agriculture was perceived widely as the foundation of civilization. Most Americans believed that human societies evolved through certain predictable stages. That was why Frederick Jackson Turner's frontier thesis had such wide appeal. Just as human beings had to walk before they ran, Indians had to become farmers before they could be absorbed into the larger society.

Yet to become a farmer required the same sort of separation and isolation that Ole Rölvaag, Willa Cather, Hamlin Garland, and so many other American writers considered the curse of rural life. If Euro-Americans found such a life stultifying, what chance did Native Americans have? In the West, irrigation agriculture also required vast amounts of capital, which is one reason the family farm gave way to agribusiness. Wage labor offered an alternative: It promoted thrift, acquisitiveness, punctuality, and many other values consistent with urban-industrial America. Native Americans did not regard agriculture and wage labor as inconsistent, and, originally, Indian policy found room for both. By the end of the 1920s, however, agriculture faced a bleak future on the reservations. It should come as no surprise that during and after World War II, wage labor did far more to "Americanize" the Indian than agriculture had in the decades from 1887 to 1940.

Notes

1. The best survey of Indian agriculture is R. Douglas Hurt, *Indian Agriculture in America: Prehistory to the Present* (Lawrence: University Press of Kansas, 1987). This chapter concerns federal water policy. It does not distinguish between Indians who were traditionally agrarian or nomadic, individualistic or communal. Nor does it explore social and cultural differences within and among Native American communities. My focus is on the Bureau of Indian Affairs' efforts to amalgamate Indians into American society. Future historians will need to pay much greater attention to how the Indians themselves attempted to reconcile federal policy with their aspirations and traditional ways of life.

2. The best overview of Indian wage labor is Alice Littlefield and Martha C. Knack, eds., *Native Americans and Wage Labor: Ethnohistorical Perspectives* (Norman: University of Oklahoma Press, 1996). For a sampling of the basic literature, see Kathryn MacKay, "Warrior into Welder: A History of Federal Employment Programs for American Indians, 1879–1972" (Ph.D. diss., University of Utah, 1985); John H. Moore, "The Myth of the Lazy Indian: Native American Contributions to the U.S. Economy," *Nature, Society, and Thought* 2 (1989): 195–215; Alice Littlefield, "Native American Labor and Public Policy in the United States," in *Marxist Approaches in Economic Anthropology*, ed. Alice Littlefield and Hill Gates (Lanham, Md.: University Press of America, 1991), 219–32; J. S. Holt, "Introduction to the Seasonal Farm Labor Problem," in *Seasonal Agricultural Labor Markets in the United States*, ed. Robert D. Emerson (Ames: Iowa State University Press, 1984), 3–32; Alan Sorkin, "Trends in Employment and Earnings of American Indians," in *Toward Economic Development for Native American Communities: A Compendium of Papers Submitted to the Subcommittee on Economy in Government of the Joint Economic Committee, Congress of the United States* (New York: Arno Press, 1970), 107–18; Donald L. Parman, "The Indian and the Civilian Conservation Corps," *Pacific Historical Review* 40 (February 1971): 39–56; Albert L. Hurtado, "California Indians and the Workaday West," *California History* 69 (Spring 1990): 2–11; Rolf Knight, *Indians at Work: An Informal History of Native American Labour in British Columbia, 1858–1930* (Vancouver, B.C.: New Star Books, 1978); and William Y. Adams, "The Development of San Carlos Apache Wage Labor to 1954," in *Apachean Culture, History, and Ethnology*, ed. Keith H. Basso and Morris E. Opler (Tucson: University of Arizona Press, 1971), 116–28.

3. Littlefield and Knack, *Native Americans and Wage Labor*, 173–74.

4. Ibid., 18–19.

5. W. T. Sherman to C. Delano, 7 November 1871, in RG 75, Records of the Bureau of Indian Affairs, Irrigation Division, General Correspondence, Entry no. 653, Box 52, "Ft. Apache, 1871–1918"; L. W. Cooke to Commissioner of Indian Affairs, 20 October 1893, RG 75, Special Cases, Entry no. 190, "Blackfeet" Box; *Report of the Commissioner of Indian Affairs, 1902* (Washington, D.C.: Government Printing Office, 1903), 63–64; *Report of the Commissioner of Indian Affairs, 1909* (Washington, D.C.: Government Printing Office, 1910), 10.

6. *Report of the Commissioner of Indian Affairs, 1906* (Washington, D.C.: Government Printing Office, 1907), 13–15; *Report of the Commissioner of Indian Affairs, 1908* (Washington, D.C.: Government Printing Office, 1908), 58. Also see *Report of the Commissioner of Indian Affairs, 1897* (Washington, D.C.: Government Printing Office, 1897), 29; *Report of the Commissioner of Indian Affairs, 1911* (Washington, D.C.: Government Printing Office, 1912), 14; *Report of the Commissioner of Indian Affairs, 1916* (Washington, D.C.: Government Printing Office, 1917), 41, 44.

7. On allotment and water policy, see Donald J. Pisani, "Irrigation, Water Rights, and the Betrayal of Indian Allotment," *Environmental History Review* 10 (Fall 1986): 157–76; Leonard A. Carlson, *Indians, Bureaucrats, and Land: The Dawes Act and the Decline of Indian Farming* (Westport, Conn.: Greenwood Press, 1981); Janet McDonnell, *The Dispossession of the American Indian, 1887–1934* (Bloomington: Indiana University Press, 1991).

8. Frederick E. Hoxie, *Parading through History: The Making of the Crow Nation in America, 1805–1935* (New York: Cambridge University Press, 1995), 228, 231, 273–78; *Report of the Commissioner of Indian Affairs, 1891* (Washington, D.C.: Government Printing Office, 1892), 51; *Report of the Commissioner of Indian Affairs, 1896* (Washington, D.C.: Government Printing Office, 1897), 29; *Report of the Commissioner of Indian Affairs, 1897* (Washington, D.C.: Government Printing Office, 1897), 33; *Report of the Secretary of the Interior, 1892* (Washington, D.C.: Government Printing Office, 1893), LIV; *Report of the Secretary of the Interior, 1896* (Washington, D.C.: Government Printing Office, 1896), XLVI; *Report of the Secretary of the Interior, 1898* (Washington, D.C.: Government Printing Office, 1898), XLVII; *Board of Irrigation, Executive Departments*, 54th Cong., 1st sess., S. Doc. 36 (Washington, D.C.: Government Printing Office, 1896), serial 3349, 15–16.

9. W. H. Graves to the Commissioner of Indian Affairs, 25 October 1892, 14 October 1894, 9 August 1895, 10 March 1898, RG 75, Records of the Bureau of Indian Affairs, Special Cases, Entry no. 190, "Crow"; *Report of the Commissioner of Indian Affairs, 1898* (Washington, D.C.: Government Printing Office, 1898), 49. Reservation agents had good reason to exaggerate Indian enthusiasm and progress. Their jobs depended on demonstrating success, and many agents expected to profit in various ways from the irrigation of reservation land.

10. *Report of the Commissioner of Indian Affairs, 1910* (Washington, D.C.: Government Printing Office, 1911), 22.

11. *Report of the Secretary of the Interior, 1892* (Washington, D.C.: Government Print-

ing Office, 1893), LIV; *Report of the Commissioner of Indian Affairs, 1892* (Washington, D.C.: Government Printing Office, 1892), 92–93; *Report of the Commissioner of Indian Affairs, 1893* (Washington, D.C.: Government Printing Office, 1893), 47–50; *Report of the Commissioner of Indian Affairs, 1894* (Washington, D.C.: Government Printing Office, 1894), 24–26; *Report of the Commissioner of Indian Affairs, 1895* (Washington, D.C.: Government Printing Office, 1896), 28.

12. Hiram Price, "The Government and the Indians," *Forum* 10 (February 1891): 715; Hurt, *Indian Agriculture in America*, 118.

13. Laurence F. Schmeckebier, *The Office of Indian Affairs: Its History, Activities, and Organization* (Baltimore, Md.: Johns Hopkins University Press, 1927), 239.

14. Hurt, *Indian Agriculture in America*, 170–71.

15. Anonymous, "The American Indian and Irrigation," *Forestry and Irrigation* 9 (May 1903): 236.

16. *Report of the Commissioner of Indian Affairs, 1914* (Washington, D.C.: Government Printing Office, 1914), 29. For good overviews of the leasing system, but with little attention to irrigation and water rights, see Delos S. Otis, *The Dawes Act and the Allotment of Indian Lands* (Norman: University of Oklahoma Press, 1973), 98–123; J. Kinney, *A Continent Lost—A Civilization Won: Indian Land Tenure in America* (Baltimore, Md.: Johns Hopkins University Press, 1937), 214–48. The states, not the federal government, regulated water rights, and state officials insisted that water had to be used continuously to perfect claims. In most western states, water rights did not attach directly to particular parcels of land. They could be acquired by tenants as well as landowners. Indian rights were no different. Because most Indians did not irrigate, the Bureau of Indian Affairs relied on whites to "reserve" water for the time that Indians would be ready to use it.

17. Department of the Interior, *Nineteenth Annual Report of the Reclamation Service, 1919–1920* (Washington, D.C.: Government Printing Office, 1920), 439–40. Agricultural development on the Flathead and Fort Peck Reservations followed a similar course. See the statistics on pp. 447 and 456 of the same report.

18. Porter J. Preston and Charles A. Engle, "Report of Advisors on Irrigation on Indian Reservations," in U.S. Senate, *Hearings before a Subcommittee of the Committee on Indian Affairs*, 8, 10, 11, 12, 17 July 1929, congressional print S545-2-B (Washington, D.C.: Government Printing Office, 1930), 2210–61. Preston and Engle were chosen to conduct the study presumably because they would take a more detached view than engineers in the Department of the Interior. For more general surveys of reservation conditions, see "Survey of Conditions of Indians in the United States," in U.S. Senate, *Hearing before the Committee on Indian Affairs*, 70th Cong., 1st sess., 10 and 13 January 1928 (Washington, D.C.: Government Printing Office, 1928); *Hearings before a Subcommittee of the Committee on Indian Affairs*, U.S. Senate, 12–14 September 1932 (Washington, D.C., Government Printing Office, 1934).

19. Preston and Engle, "Report of Advisors on Irrigation on Indian Reservations," 2217–20.

20. Ray Lyman Wilbur, *Conservation in the Department of the Interior* (Washington, D.C.: Government Printing Office, 1931), 118.

21. Carlson, *Indians, Bureaucrats, and Land,* 116, 120, 130.

22. Indians did not suffer equally from federal policies. For a case study of a tribe that maintained traditional patterns of life and enjoyed considerable agricultural success in the twentieth century, see Thomas R. Wessel, "Phantom Experiment Station: Government Agriculture on the Zuni Reservation," *Agricultural History* 61 (Fall 1987): 1–12; Thomas R. Wessel, "Agriculture on the Reservations: The Case of the Blackfeet, 1885–1935," *Journal of the West* 18 (1979): 17–24.

23. Douglas Hurt recounts that as early as 1881 the agent at the Pine Ridge Reservation in Dakota Territory had observed: "White men well trained in farming, have tried to till the soil in this vicinity in Northern Nebraska and have lost all the money invested, and have not produced enough to pay for the seed. I can confidently venture to state that, if the experiment were tried of placing 7,000 white people on this land, with seed, agricultural implements, and one year's subsistence, at the end of that time they would die of starvation, if they had to depend on their crops for their sustenance." Hurt, *Indian Agriculture in America,* 132.

24. *Report of the Secretary of the Interior, 1882* (Washington, D.C.: Government Printing Office, 1882), IX; *Report of the Commissioner of Indian Affairs, 1895* (Washington, D.C.: Government Printing Office, 1896), 26; Hurt, *Indian Agriculture in America,* 109.

25. Schmeckebier, *Office of Indian Affairs,* 238; McDonnell, *The Dispossession of the American Indian,* 26.

26. *Congressional Record,* 63rd Cong., 2d sess., U.S. Senate, 17 June 1914, 105–90.

27. A few examples speak volumes. At the Oto Agency in Nebraska, Indians used knives, tin cups, and even their hands to dig post holes on the agency farm. At the Ponca Agency in Dakota Territory, Indians harvested wheat with butcher knives. An agent for the Kiowa, Comanche, and Wichita in Oklahoma complained that there was no way to divide 115 plows among the 300 Indians who wanted to use them. And all too often those lucky enough to receive a plow found that Indian ponies were too weak to pull them. Hurt, *Indian Agriculture in America,* 125, 147.

28. Hurt, *Indian Agriculture in America,* 110–11, 134, 169.

29. Schmeckebier, *Office of Indian Affairs,* 247. Thomas Wessel has pointed out the irony of the Bureau of Indian Affairs' position. Lewis Henry Morgan, one of the leading ethnographers of the late nineteenth century, argued that stock raising was an intermediate stage on the evolutionary ladder—a rung up from savage hunters and a rung down from civilized farmers. As Morgan wrote in 1878: "We have overlooked the fact that the principal Indian tribes have passed by natural development out of the condition of savages into that of barbarians. In relative progress they are now precisely where our own barbarous ancestors were when by the domestication of animals they passed from a similar into a higher condition of barbarism, through still two ethnical periods below civilization." See Thomas R. Wessel, "Agent of Acculturation: Farming on the Northern Plains Reservations, 1880–

1910," *Agricultural History* 60 (Spring 1986): 235. In short, the myth that Indians could not "progress" unless they practiced settled agriculture was questioned even at the time.

30. The Collier phrases are from the annual report of the Commissioner of Indian Affairs for 1940, as published in the *Report of the Secretary of the Interior, 1940* (Washington, D.C.: Government Printing Office, 1940), 357.

31. Cato Sells to Secretary of the Interior, 13 March 1914, RG 75, Irrigation Division, General Correspondence, Entry no. 653, Box 3.

32. Board of Indian Commissioners, "Brief on Indian Irrigation, 1915," RG 75, Irrigation Division, General Correspondence, Entry no. 653, Box 3.

33. Barbara G. Leibhardt, "Law, Environment, and Social Change in the Columbia River Basin: The Yakima Indian Nation as a Case Study, 1840–1933" (Ph.D. diss., University of California, Berkeley, 1990), 176.

6

First in Time

Tribal Reserved Water Rights and General Adjudications in New Mexico

Alan S. Newell

A judicial determination of local custom was the common method for resolving water disputes in New Mexico at the turn of the twentieth century. Although promoters of large- and small-scale irrigation systems favored a more centralized and expedient system for governing this scarce natural commodity, the territory's legislative assembly expressed little interest in revamping the system of local use and custom that had evolved in New Mexico over the previous two hundred years. It was not until 1905 that New Mexico's legislators made their first foray into comprehensive statutory water administration with the designation of a territorial irrigation engineer, who had responsibility to evaluate water resources and to authorize the construction of dams and dikes. The acknowledged limitations of the engineer's power quickly led to an expanded statute in 1907 whereby the territorial engineer acquired the authority to supervise all New Mexican waters. The territorial engineer's principal tool was the hydrographic survey, which was intended to provide a simple and speedy method for adjudicating and allocating the territory's water resources.[1]

It is likely that few, if any, of the territorial legislators at work on the new water statute in New Mexico conceived of how long or how difficult the process, which began in 1907, would be. It also is doubtful that they were aware of how a case moving swiftly through the federal courts in Montana would affect future adjudications in New Mexico and other western states. In February 1906, news reached the citizens of the small farming town of

Chinook, Montana, that the 9th Circuit Court of Appeals, sitting in San Francisco, had affirmed Montana District Judge William Hunt's 8 August 1905 decision that the Assiniboine and Gros Ventres Tribes of the Fort Belknap Indian Reservation had a prior right to use water from the Milk River. Judge Hunt's decision and injunction in *United States v. Moses Anderson et al.* permanently barred upstream irrigators from depriving tribal members of the five thousand miners inches that had been reserved for them by the United States when Congress created the reservation in 1888. Few participants in this unfolding drama on Montana's high plains or in other states and territories in the arid West anticipated the far-reaching effects of what became known as the Winters doctrine.[2]

U.S. Attorney for Montana Carl Rasch had offered a panoply of arguments defending the Indians' water right to the lower court, including the notion that the reservation had a riparian right to the use of water that coexisted with those claimed under Montana's statutory scheme of prior appropriation. Fort Belknap Superintendent William R. Logan clearly hoped to keep the case within the confines of the state's prior appropriation doctrine and had chosen the defendants on the basis of what he believed to be their junior dates.[3] Both advocates were surprised perhaps when Judge Hunt and later the 9th Circuit Court and U.S. Supreme Court based their favorable decisions on the implied reservation of water by the federal government pursuant to its agreement with the Assiniboine and Gros Ventres Tribes that established the reservation in 1888.[4]

The Supreme Court's 1908 decision in *Winters v. United States* created a judicially recognized water right that effectively inserted the United States into state water proceedings. Since 1866, Congress had recognized the paramount right of states to adjudicate water rights on the basis of local, customary practices, and federal officials generally acknowledged that the way to provide water to their facilities was to file a notice under state law.[5] The Winters doctrine offered the federal government a different, and perhaps paramount, means to assert its right to water for federal reservations.[6] The doctrine also imposed upon the United States, as trustee for tribes, the responsibility for claiming federally reserved water rights for the nation's western tribes, who relied on scant water supplies available on their arid reservations. As a result, since 1908, the United States has brought suit against nonfederal irrigators in support of Indian water rights.

Since the mid-1970s, this "trust responsibility" has required the federal government to become a party to state efforts to recognize and to quantify water rights throughout the West in what are known as "general stream-wide adjudications." As an advocate for tribes, the U.S. Department of Justice has recently been engaged in as many as forty-eight ongoing water disputes in twelve western states, most of these involving litigation.[7] Virtually all of these cases require the United States to both assert the reserved water right of the tribe or tribes involved in the adjudication and to review critically and, if necessary, to challenge the rights of nonfederal water users.

Evolving legal doctrines frame these often-contentious debates. The Winters doctrine has not remained static during the last ninety years and neither have other areas of Indian law. To properly view how the Winters doctrine is applied to tribal water rights, one must first understand how the Winters doctrine has developed and how it intersects with what has been termed the "canon of construction" governing agreements between the United States and Native Americans. An example of such an intersection is apparent in a recent state water case involving the Mescalero Apache Tribe of southeastern New Mexico. The work undertaken in that case readily demonstrates the problems associated with reconciling a state's water law with an evolving judicial concept.[8]

The Legal Context for Historical Research

The Winters doctrine's judicial past provides the context for current historical inquiry. The Supreme Court's decision in January 1908 confirmed Judge Hunt's opinion and analysis made two years earlier by acknowledging a reserved water right. However, the high court's ruling did not delineate either the scope or the quantity of that water right. Lacking a statutory mandate from Congress, it has been left to the courts to define the parameters of the "Winters right" on a case-by-case basis.

Fort Belknap Superintendent William Logan recognized the limitations and fragility of his judicial victory. Announcing the appellate court's decision to Commissioner of Indian Affairs Francis Leupp in February 1906, Logan speculated that "outside of the borders of the reservation in this immediate neighborhood, it is fair to say that I am about the most unpopular Indian in this country."[9] Hoping to assuage local concerns, the superintendent voluntarily closed headgates on the reservation once the Indian lands had been sufficiently irrigated, allowing non-Indian farmers "an opportunity to irrigate and thereby save their crops."[10] Logan also enlisted the service of the local Reclamation Service engineer Cyrus Babb to fashion an agreement with the farmers of Chinook to provide water to both interests. Although not adverse to challenging his non-Indian neighbors when necessary, the Fort Belknap superintendent recognized that cooperation was preferable to contest.[11]

Logan and other reservation superintendents also believed that Congress or the courts might eventually revisit the decision in *Winters* to the detriment of the tribes. Consequently, they placed their hopes on developing Indian agriculture as a way of perfecting the reserved right. For example, Logan encouraged the leasing of 10,000 acres of reservation land to a sugar-beet producer in 1907. Despite congressional funding and the clearing of thousands of acres of land by tribal members, the venture failed in 1910.[12] On the adjacent Blackfeet Reservation, Special Indian Agent Yvan Pike also moved aggressively to develop an irrigation project that would provide water to both Blackfeet allottees and possible non-Indian entrymen. Similarly, on other reservations, particularly those in the Southwest, Indian irrigation engineers viewed as imperative the application of water to beneficial use to perfect a water right under state law, despite the prima facie evidence of an earlier federally reserved water right.[13]

This concern over the ambiguity of the Winters doctrine was warranted. As the debate proceeded, decisions in state and federal courts appeared to narrow the scope of the right. Courts in Arizona and Oregon found a means to both accept and to limit the impact of *Winters*. In *United States v. Wrightman*, the United States sought an injunction to prohibit a non-Indian from diverting water from a spring located on the San Carlos Apache Indian Reservation in Arizona. The waters from the spring had apparently been used by the settlers since 1896. In 1913, an Apache woman received an allotment of land encompassing the spring with the understanding that she would use spring water to develop a farm. In its 1916 ruling, the district court acknowledged that the decision in *Winters* could apply to the San Carlos Reservation, but only if the reservation of the spring was "necessary to the objects for which the reservation was created." The court found that the flow of the spring in question was negligible and, thus, by adding this limitation to the Winters doctrine, was able to find for the defendant and against the United States.[14]

One year later, in 1917, the Oregon Supreme Court entered a more sweeping and damaging ruling for the United States in the case of *Byers v. Wa-Wa-Ne et al.* The issue involved the use of water from the Umatilla River, which had been diverted across the Umatilla Indian Reservation since 1874 for the purpose of powering a flour mill in the town of Pendleton. The right-of-way for the ditch had been approved by the reservation superintendent and had been confirmed by Congress in 1885. Seeking to use this water to irrigate the only irrigable land on the reservation, the United States asserted the paramount right of the tribe pursuant to the 1855 Treaty of Walla Walla that established the reservation. The lower court found in favor of the tribe, but on appeal, the Oregon Supreme Court had a different view. The appellate court found that Congress's 1885 affirmation of the private right took precedence over the treaty and, more striking, denied "that the mere creation of an Indian reservation by treaty impliedly secures to the Indians all water in streams which touch the reservation which they may at any time desire to put to a useful purpose." The high court in Oregon found that few lands on the Umatilla Reservation had been irrigated and, absent this history of use, the treaty could not be read to imply a reserved water right.[15]

Both of these cases circumscribed the reserved right found in *Winters* and put the burden on the United States to show that the purpose of establishing a reservation during treaty negotiations, as demonstrated by subsequent reservation history, included an express or implied reliance on irrigation water to support agriculture. Of course, both of these cases did not carry the authority of the U.S. Supreme Court in the 1908 *Winters* decision. It was left to that court to once again address the Winters doctrine in its 1939 decision in *United States v. Powers et al.*

The case of Thomas Powers, a non-Indian settler on the Crow Reservation in southeastern Montana, involved a long-standing dispute between irrigators and the Indian Irrigation Service. The reservation had been allotted and opened to non-Indian entry in 1908 and, as a result, much of the arable land had fallen into nontribal ownership. Some of these lands were within the Crow Indian Irrigation Project, which had been constructed between 1890 and 1915. Many of these non-Indian landowners used water from the same sources as irrigation project lands, albeit through a series of private, rather than project, ditches. Beginning in 1919, a series of periodic droughts on the reservation prompted white farmers to seek judicial recognition and protection of their water rights.[16] Although the local Montana court recognized these appropriative rights, it did not pass judgment on the Crow's reserved water right. As the dry years continued, federal officials on the reservation began asserting a Winters right and notifying private irrigators that project lands would be served before they were allowed to take water. In 1921, District Engineer William Hanna believed that the situation was sufficiently desperate to warrant the closing of private headgates on the Little Big Horn River and the stationing of Indian police at key locations to ensure that the headgates were not reopened.[17]

Although wet years through the 1920s ameliorated the situation somewhat, the return of drought to the Great Plains beginning in 1931 rekindled the dispute and forced the United States in 1934 to file suit against Thomas Powers and twenty-one other nonproject irrigators on the reservation. In *United States v. Powers et al.* federal attorneys argued for a Winters right predicated on an 1868 treaty with the Crow that established the reservation. They also claimed that congressional authorization of the Crow Indian Irrigation Project gave the Interior Department control of the reservation's water—a power that was paramount to private water rights, whether Indian allottee or non-Indian. The lower courts' decisions and, eventually, the Supreme Court's ruling in *Powers* surprised Indian officials. The court held that, even though the Crow had a reserved water right pursuant to *Winters,* each tribal member had a beneficial interest in that right. Consequently, when reservation lands were allotted to individual Crow for agricultural purposes, a portion of the water also was assigned to that tribal member. This right was then transferable when the land was sold into non-Indian ownership. The secretary of the interior's authority to administer the reservation irrigation project did not carry with it the right to deny private irrigators the water that they needed for their lands.[18]

The Supreme Court's decision in *Powers* introduced new uncertainty into the scope and value of a Winters right. Was the right a tribal right or an individual one? If the right could be sold to non-Indians, then was it severable? If the right could be put to use by other than tribal members, then how was it to be quantified in terms of the original purposes of the reservation? The *Powers* court seemed to imply that the reserved right was intended principally for irrigation and thus was appurtenant to the land. This last point was particularly important when the high court next forayed into water litigation in the interstate dispute of *Arizona v. California*.

This lawsuit began when Arizona filed an action against California under the original jurisdiction of the Supreme Court in 1952. The principal issue in this long and intricate case (aspects of which continue to the present day) was how to apportion the waters of the Colorado River. Eventually, Nevada, New Mexico, Utah, and the United States joined the litigation. Five Indian reservations situated in Arizona and California had a claim to the apportionment based on the Winters doctrine. After conducting more than two years of trial, hearing 340 witnesses, and compiling more than 25,000 pages of transcript, Special Master Simon H. Rifkind recommended to the Supreme Court an equitable apportionment of the river in January 1961. The court heard oral argument on the report a year later and rendered its decision in June 1963.[19]

With respect to the reserved water right claimed by the United States for the tribes, Justice Hugo Black, writing for the majority, found that the right had been established on the date that Congress or the president had created the various reservations. In affirming the right, Justice Black consciously sought to connect the present decision to the court's earlier rulings in *Winters* and *Powers*.[20] But Justice Black and the majority of the Supreme Court went further in their decision than had previous courts. Because of the need to apportion the Colorado River, the justices had to devise a standard by which to quantify the reserved water right for these tribes. The standard that they accepted was one that would "satisfy the future as well as the present needs of the Indian Reservations." The federal government's efforts to irrigate those reservations suggested to Rifkind and to the court that those needs could be met by delineating the "practicably irrigable acreage" (PIA) on the reservations.[21]

The Supreme Court's decision in *Arizona v. California* both confirmed the existence of a reserved water right and served to clarify how such a right might serve present and future needs and how it might be quantified.[22] Since 1963, the "PIA standard" has been a departure point for many investigations undertaken by public historians, who are retained to evaluate Indian reserved water rights. Developing a "PIA claim" involves research into the aboriginal use of water by various tribes as well as a detailed analysis of the historical use of irrigation water on the reservation. Given the dismal record of BIA-sponsored irrigation projects on most reservations, the historian seeks to understand not only what happened, but what planners hoped would happen.

The focus on determining the present and future "practicably irrigable acreage" of a reservation (or for that matter determining what is "practicable") can document a critical economic need of the tribe. But does this inquiry really capture Congress's purpose in creating an Indian reservation? If a reservation was eventually determined to be more valuable for grazing, for minerals, or for its timber, does that mean that the tribe is deprived of a reserved water right? Since 1963, lawyers have debated whether the PIA standard enunciated in *Arizona v. California* is specific to the reservations involved in that case or whether it has universal application. Some Indian advocates assert that if the United States limits its defense of Winters rights to a reservation's PIA, then it is breaching its fiduciary responsibility to tribes.[23] For this and other reasons, the United States frequently broadens its historical research into the purposes of the reservation—an inquiry that leads to methods of water right quantification not covered by the PIA standard.

The federal emphasis on developing Indian irrigation during the late 1890s and early 1900s is undisputed and supports the applicability of the PIA standard to many arid reservations in the West.[24] However, to overemphasize the agricultural motive behind federal reservation policy ignores a crucial principle of federal Indian law. The "domestic dependent nation" status of native tribes first articulated by Justice John Marshall in *Cherokee Nation v. Georgia* and *Worcester v. Georgia* in the 1830s laid the cornerstone for a unique "canon of construction" for treaties between tribes and the United States.[25] This interpretive approach is fundamental to federal Indian law and requires the court to first identify any ambiguity in Indian treaties and agreements and, if found to be ambiguous, to interpret them in a manner most favorable to the tribe. The canon of construction places a burden on the court to elicit expert historical testimony concerning the intent of the parties at the time that the treaties and agreements were negotiated.

The 9th Circuit Court of Appeals clarified the application of these principles to water rights cases in a 1921 lawsuit involving the Bannock and Shoshone Indians on the Fort Hall Reservation in Idaho. In *Skeem v. United States* the court held that the reserved right must be viewed as a reservation by the tribes of what they possessed at treaty time, rather than a grant to them from the United States. This decision was decidedly more consistent with the Supreme Court's rulings in the pivotal treaty interpretation cases of *Winters v. United States* and *United States v. Winans* than were the decisions in *Wrightman* and *Byers.*[26] In *Skeem* the court implied that the water right was really an *Indian* reserved right, rather than a *federal* reserved right. Although the 9th Circuit Court justices did not offer a historical account of Indian understanding of the applicable treaties and agreements at issue in *Skeem,* they nonetheless suggested the importance of this analysis.[27]

In attempting to understand more than just federal interest in the treaty process, it is possible to discover a broader range of perceptions on the purpose of the reservation. Recent scholarship in the history of Native Americans, particularly focused upon the reservation period, has sought to portray a multifaceted native response to federal policy.[28] It is not accurate to view the treaty process as one entirely dictated by U.S. officials, adhering to a prescribed policy emanating from Washington. Treaty commissioners usually negotiated with a large measure of discretion—often because of the expressed or perceived needs of one or more factions of the tribe. For example, Gov. Isaac Stevens recognized the need to adjust federal policy to the conditions of Washington Territory when he negotiated nine agreements with Pacific Northwest tribes between 1854 and 1855. Longtime Indian Office Clerk and Acting Commissioner of Indian Affairs Charles Mix instructed Stevens to employ sufficient discretion to secure the necessary treaties with a disparate group of tribes ranging from Puget Sound to the Missouri River. For most of these tribes, the liberality of Stevens's instructions resulted in the tribe retaining broad fishing and hunting rights both within the reservation and on ceded lands.[29] These rights could implicitly affect the water rights of the various tribes.[30]

Similarly, many tribes, such as the Nez Perce of Washington and Idaho, had multiple and often contradictory objectives in agreeing to negotiate. In the 1855 Walla Walla Council, some Nez Perce acknowledged the changes occurring around them and accepted the initial terms of a treaty. Others, more wedded to a traditional, extensive hunting territory, staunchly opposed a large land cession but recognized the value in the United States' efforts to secure a peaceful continuance of on- and off-reservation hunting and fishing activities.[31] This factional split continued with the Nez Perce after 1855, as non-Indian settlement forced the tribe into negotiations of supplemental treaties and agreements. Eventually, Nez Perce were recognized as either "treaty" and "nontreaty" Indians—a division that contributed to the much heralded flight of some Nez Perce across Idaho and Montana in 1877. Although not always ending in armed conflict, the rapid pace of settlement in the intermountain and plains regions of the West after 1870 contributed to intertribal tensions, which are apparent in numerous formal and informal agreements made between tribes and the United States.

One common element that did bind various tribes together in their negotiations with the United States was the desire to secure a firm foothold in a territory that was under assault from impending or actual white settlement. A recurring theme in the records of negotiations for treaties after 1850 is the expectation that the reservation would provide a sanctuary for tribal communities. Of course, this notion fit squarely into federal plans to isolate Native Americans from non-Indians and to provide time for them to adopt new habits. American negotiators routinely emphasized this benefit of a reservation when trying to convince tribal leaders to cede the bulk of their aboriginal territory to the United States.[32]

If tribes believed that the reservation was to provide them with a secure homeland, whether held in common by the tribe or allotted to individuals, then the "purpose of the reservation" may be broader than the narrow agricultural objectives of the United States. The reserved water right may be necessary to support other tribal activities on a reservation, such as the need for water for ceremonial use or for the maintenance of a fishing right.[33] A wider focus of reservation purposes and adherence to the canon of construction require detailed historical examination of contemporaneous understandings of tribal as well as federal negotiators. Only through this analysis can the court render a sound judgment of the breadth and quantity of the Indian reserved water right.

The court also must consider the relationship of the reserved right to other non-Indian water rights in a general streamwide adjudication. Large and complex general adjudications are presently underway in four western states (Montana, Idaho, Arizona, and New Mexico) that seek to determine the priority and quantity of hundreds of thousands of individual rights claimed under state law.[34] The United States is a party to these proceedings pursuant to legislation introduced by Nevada Sen. Pat McCarran in 1952.[35]

The broad scope of these adjudications requires the United States to look at the tribal water right's priority relative to other rights in the basin. Where the right is premised on the negotiation of a treaty that extinguished aboriginal Indian title and established a reservation, the tribe's use of water is generally senior to other appropriators. However, for tribes whose dealings with the United States have only been through executive or secretarial orders, the priority date for the reservation may be later then other non-Indian water users. In such cases, courts must evaluate the legitimacy of those non-Indian claims under the requirements of state law.

Historical research into the priority, purpose, and quantity of both the Indian reserved right and the non-Indian state water right thus occurs concurrently in many general adjudications—albeit under different legal principles of state and federal law. Historians who enter the fray must immerse themselves in the sources and history of Indian and non-Indian land and water use in a specific geographical area. By recognizing the legal context for proposed research, historians have a unique opportunity not only to influence court decisions, but to add a considerable body of research to the scholarly library.

Historical Inquiry and the Mescalero Apache Reserved Water Right

Most of the contemporary questions surrounding reserved water rights for Indian tribes were at issue in a recent case involving the Mescalero Apache Tribe in southeastern New Mexico. Measured by many water cases, *State of New Mexico v. L. T. Lewis et al.* was not one of the longest running or better known cases, having been filed in 1973 and finally decided in the appellate court in 1993. However, it is one of the more interesting cases in its application of federal Indian law to a state court's determination of a reserved tribal water right. The *Lewis* case not only illustrates the historical trends in the evolution of the Winters doctrine, but also demonstrates the role of historians and the research potential for historical inquiry in water rights cases.

New Mexico's State Engineer Stephen Reynolds filed his motion to adjudicate the waters of southeastern New Mexico's Rio Hondo system on 5 December 1973 in Chaves County district court in Roswell.[36] Reynolds quickly followed this motion with another against the United States as trustee for the Mescalero Apache Tribe seeking to enjoin the tribe from diverting water from the Rio Ruidoso, a tributary to the Rio Hondo, to supplement the supply needed for a new resort (The Inn of the Mountain Gods). Although the United States sought to dismiss the suit, the Supreme Court's recent ruling in *Akin v. United States* provided grounds for keeping the Rio Hondo litigation in state court. With great reluctance, the Mescalero Apache Tribe and the United States prepared for a lengthy and complicated general water adjudication.[37] Determining the priority date and quantities of water used by both Indians and non-Indians in the Rio Hondo Basin required an understanding of the earliest history of Hispanic and Anglo settlement in southeastern New Mexico.

That settlement began in the mid-1850s with the establishment of Fort Stanton on the Rio Bonito and the location of a number of small clusters of non-Indian communities along the Rio Bonito and upper Rio Ruidoso. Trader Paul Dowlin located a saw mill at Ruidoso in 1853 to provide lumber to settlers and the military. The community of Missouri Plaza, located in 1866 on the Rio Hondo twelve miles upstream from the river's mouth, was the first town to be located on that stream.[38]

Water was the life blood of settlement in the basin and contributed to many of the antagonisms between settlers. Upstream diversions along the Rio Ruidoso prompted early abandonment of Missouri Plaza, whose residents found relocation preferable to conflict. Rights to use water also played a prominent role in the land disputes throughout the Pecos Valley during the 1860s and 1870s and contributed to the notorious Lincoln County War.[39] It is no surprise then that the Rio Hondo Basin received early attention from the newly created State Engineer's Office soon after passage of New Mexico's foundation water statutes in 1905 and 1907.

One of the first hydrographic surveys conducted by the State Engineer's Office under the new statutes was in response to litigation involving the Rio Hondo. This 1908 survey also was prompted, in part, by interest in developing the Rio Hondo Project as one of the territory's first federally supported reclamation projects. Although completed in 1907, the project never officially opened. The United States eventually lost interest in the project and the 1913 litigation that it spawned, but a subsequent controversy involving one earlier litigant, the El Paso and Rock Island Railway Company, prompted the state engineer to undertake a new hydrographic review in 1930.[40]

All of this interest in the Rio Hondo Basin had little involvement with or by the Mescalero Apache Tribe, despite the federal government's prominence in these disputes. Although the tribe's reservation encompassed the headwaters of the Rio Hondo system, neither the Mescalero Apache Tribe nor the United States acting as a trustee were parties to the earlier adjudications. The Mescalero did not become embroiled in basin water disputes until they were named as a party in the 1973 filing by State Engineer Reynolds. Their entrance into the lawsuit introduced the history of the tribe's use of the region prior to and after their signing the treaty of 1 July 1852.

The Mescalero Apache had been present in the Southwest since the fourteenth century. Prior to 1850, they ranged widely from the Sandia Mountains to the north to the present-day Mexican states of Chihuahua and Coahuila to the south. After becoming established in the White and Sacramento Mountains in the late 1770s, the Mescalero generally confined themselves to a territory between the Pecos River and the Rio Grande, with Sierra Blanca to the north and the Davis Mountains to the south. Within this area, they subsisted by hunting and gathering. The Mescalero were also notorious raiders of other Indian communities as well as the few Hispanic settlers bold enough to venture into the region.[41]

Mescalero raiding increased after American occupation of New Mexico in 1846, thus prompting federal interest in negotiating a treaty. Establishing peace in New Mexico Territory and bringing the Mescalero Apache within the purview of U.S. reservation policy led Gov. James Calhoun to instruct his secretary, John Ward, to meet with tribal leaders in late June 1852. Shortly thereafter, Indian Agent John Greiner concluded a treaty with the tribe on 1 July 1852. The treaty called for peace and amity between the United States and the Mescalero and between the Apache and other tribes. It also placed the Mescalero Apache under the exclusive jurisdiction of U.S. officials, who agreed to at their "earliest convenience, designate, settle, and adjust their [Mescalero] territorial boundaries." Within these boundaries, the United States sought to encourage Mescalero agriculture.[42]

The Apache peace deteriorated soon after ratification of the July 1852 treaty. Much of the blame for the subsequent resumption of Apache raiding rested on political instability in the territory as well as the failure of the United States to provide adequate funding for the expected and difficult transition from raiding to farming. Some of the Mescalero Apache leaders remained at peace despite these unfulfilled promises. However, it was not until establishment of Fort Stanton on the Rio Bonito (one of the provisions of the 1852 treaty) that a permanent peace seemed achievable.

Some Mescalero bands established small agricultural settlements in the valleys of their White Mountain homeland during the late 1850s. However, these efforts were minimal and tribal members had to be supported by rations supplied from Fort Stanton.[43] One principal reason for the inattention paid to the Mescalero immediately before and after the Civil War was the failure of the United States to establish a permanent reservation boundary for the tribe. Despite efforts to locate the reservation in the White Mountains during the mid-1850s, it was not until 1873 that President Grant issued the first of five executive orders establishing and later adjusting the reservation boundary.[44] Between 1873 and 1883, the United States unilaterally adjusted the reservation boundary four times to exclude non-Indian farms, mills, and mining interests and to correct surveying errors.

The Mescalero Apache Tribe's long tenure in the White and Sacramento Mountains of southeastern New Mexico was the subject of numerous historical and anthropological reports and testimony during the course of the state engineer's adjudication of the Rio Hondo system in *State of New Mexico v. L. T. Lewis et al.* Of most interest to the court was the intent of the parties in negotiating the treaty of 1 July 1852 and in the later definition of reservation boundaries. The absence of a treaty journal made the former analysis amenable to a variety of interpretations.

The United States and their experts argued that the treaty had to be viewed in the context of a changing federal Indian policy in the 1850s. In this first and only ratified treaty with the Mescalero Apache, the United States hoped to introduce the Mescalero to sedentary agriculture within the valleys of their White Mountain homeland. The experts presented evidence that the United States and the Mescalero intended to locate the reservation in the traditional Mescalero homeland, but that various historical factors intervened to delay this survey until 1873. While acknowledging the overly ambitious plans of federal officials and the Mescalero resistance to change, the United States argued that the canon of construction required the court to date the tribe's reserved water right from the signing of 1 July 1852.[45]

The state of New Mexico and the nonfederal irrigators, organized as the Water Defense Association, took a different and opposing view of how the law should apply to history. Arguing that the 1852 treaty had not established a reservation within the context of the Winters doctrine, the state also pointed to their experts' testimony that the Mescalero Apache undertook little if any agricultural development until after the boundary was located in 1873. That boundary had changed so often between 1873 and 1883 that the state believed a reserved water right could only have been created at the time of each executive order. The effect of the state's position was that the Mescalero Apache Tribe had five distinct reserved dates, rather than one Winters' date of 1 July 1852. According to the state, the Winters' right was truly a *federal* rather than an *Indian* right and had no relationship to the parties' intent in 1852.[46]

Lafel Oman, a retired New Mexican jurist, was the trial judge in *Lewis*. Judge Oman's 1989 decision affirmed a federal reservation of water for the Mescalero Apache Tribe and ignored the application of the canon of construction. Oman found that the 1852 treaty had no effect on the reservation's creation or the intent of the parties. Rather, it was the five separate executive orders between 1873 and 1883 that created the Mescalero Apache Reservation. Judge Oman's strict adherence to the notion that the establishment of boundaries defined the reservation of water led him to the impractical conclusion that the Mescalero had not one, but five separate priority dates for the various portions of the reservations.[47]

On appeal in 1993, the New Mexico State Court of Appeals reversed Judge Oman's ruling on the reservation's priority date and found instead that both parties to the treaty of 1 July 1852 anticipated that water would be reserved for the Mescalero Apache. Two of the three appellate judges hearing arguments in *Lewis* found that the district judge had inappropriately applied the law to his determination of the historical facts. The appellate judges, Benny E. Flores and Lynn Pickard, viewed the historical intent of the parties and the application of the canon of construction as integral to a decision on priority date. Using this standard, the court had to consider what the parties believed was the purpose of the 1852 treaty as well as the history of the Mescalero Apache from that date through the final executive order in 1883. Flores and Pickard concluded that Oman ignored entirely "the treaty and the history leading up to it."[48]

The decision by the New Mexico State Court of Appeals in *Lewis* suggests the importance of viewing the Winters right as both an Indian and a federal water right. It confirms the evolutionary course of the reserved right doctrine since the Winters decision and places that history squarely within the twentieth-century development of Indian law. The Court of Appeals' review of relevant cases, some dating to the early twentieth century, suggests the significance that the court placed on viewing all documents that reflect promises made by one party to another (e.g., treaties, executive orders, secretarial orders) within the purview of canon.[49]

The significance of *State of New Mexico v. L. T. Lewis et al.* transcended the ultimate ruling on the nature of the reserved water right. The case also represented the first time that the United States systematically examined the effect of nonfederal claims to water, particularly those of private landowners, on the tribal reserved right. The nature of the general adjudication offered the United States, as trustees for the tribe, an opportunity to challenge both the priority date and the quantity of water used by individuals under state law. The research required for this challenge uncovered new opportunities to examine the local and regional use of land and water in southeastern New Mexico and contradicted established notions of early water use in the region.

The nature of early New Mexican Hispanic agriculture and its foundation in the community acequia is well documented in historical literature.[50] Ira Clark, in what may be the most comprehensive work on the evolution of New Mexico's water law, describes the community acequia as "rooted in both the indigenous and the imported systems."[51] The legal foundation of this early system differed from those that followed American sovereignty in 1846 by vesting ownership and control of the ditch system in the town leaders, who appointed a *mayordomo* with broad power over water users in the community. Given its prominence in New Mexico, particularly along the northern reaches of the Rio Grande and Pecos River systems, it is understandable that the legal and social principles of the community acequia were adopted in the first laws established for the new American territory in

1851. Although these principles remained and even strengthened during the subsequent decades, they also proved anachronistic to both Hispanic and Anglo settlers, ditch companies, and federal hydrographers with plans for reclaiming the arid regions of southern and eastern New Mexico after 1870. The increasing complexity of water development in New Mexico prompted a recodification of water law in the territory in 1895 and a general weakening of the community acequia as an institution.[52]

Settlement in the Rio Hondo Basin illustrates the evolution of New Mexican water use and the deterioration of the traditional community acequia as an institution, which had been so prevalent in the northern regions of the state. As noted earlier, small and sometimes ephemeral Hispanic communities arose at La Placitas on the Rio Bonito, Dowlin's mill on the Upper Rio Ruidoso, and Missouri Plaza on the Rio Hondo after the establishment of Fort Stanton in 1855. Yet not until the end of the Civil War in 1865 and the settlement of the Mescalero Apache Reservation boundaries in 1873 did emigrants feel comfortable venturing into these river valleys. Unlike older areas of the territory that were subject to Spanish and Mexican land grants, settlers on the Rio Hondo took advantage of the various federal homestead laws enacted by Congress during the second half of the nineteenth century.[53]

This later era of settlement under federal land law affected the nature of water use in the Rio Hondo Basin. Although there may have been a few small, community acequias associated with the earliest Hispanic settlement on the Rio Bonito, this institution quickly succumbed to the construction of private ditches and corporate ditch ventures. Evidence of land and water use available in General Land Office patent files, cadastral surveys, and Surveyor-General reports creates a picture of rapid development along the three rivers in the 1870s and 1880s and provides a glimpse of earlier activity in the region. These sources, when compared to later testimony from water litigation, probate records, and the state engineer's hydrographic surveys in 1908 and 1930, show a pattern of ditch extension, land consolidation, and, eventually, ineffective federal reclamation efforts that characterized the Rio Hondo Basin. They also show that many of the *Lewis* court's initial historical impressions of water use in the basin were wrong.

For example, a special master in the *Lewis* case had originally assigned an 1855 priority date to the Lincoln Ditch, located on the southern side of the Rio Bonito near the town of Lincoln. This date corresponded with the establishment of Fort Stanton and the small Hispanic community of La Placita. However, a more detailed examination of the land settlement in the vicinity of the Lincoln Ditch revealed a different story of the land use around La Placita and a later priority date for the acres irrigated by the Lincoln Ditch. General Land Office cartographic data and survey notes from 1867 documented the existence of only one ditch south of the Rio Bonito, which was located very close to the river. This canal, later known as the Titsworth Ditch, apparently irrigated river bottoms, perhaps in the manner described by John Wilson for lands along the Rio Grande during the same era.[54] In 1867, the General Land Office's contract surveyor also noted this limited irrigation when he described existing cultivation as confined to the "River 'Bottom' proper" and that it did not "embrace the 'Second Bottom.' "[55] Land records and testimony from defendants in a 1913 water litigation involving the El Paso and Rock Island Railway Company confirmed that this "second bottom" was not placed under irrigation until the early to mid-1870s and only after construction of the private Lincoln Ditch in 1870.[56]

This specific and detailed history of water and land use is critical to establishing priority dates in a general water adjudication. Vague and imprecise characterizations of adaptation of the "traditional" community acequia to all regions of the state is both inaccurate and of little use to a judge charged with determining when water was first applied beneficially to a particular tract of land. Both the State of New Mexico and the intervening association of private water users sought to portray water development along the Rios Ruidoso, Bonito, and Hondo as a network of community acequias in an effort to defend the special master's award of early priority dates—some as early as 1853. The United States' tract-by-tract analysis of settlement clearly demonstrated the imprecision of earlier research. Although some scholars might argue that such an effort to understand land and water use by researching specific tracts is inaccurate, impracticable, and disruptive of traditional community values, it is clearly research that provides a more accurate and useful means of understanding the historical nuances in water- and land-use patterns. Moreover, such analysis is required by well-established New Mexican law.[57]

Conclusion

The Winters doctrine has engendered substantial controversy in legal circles since its initial confirmation by the Supreme Court in 1908. At a time when federal policy encouraged the alienation of tribal lands through allotment, reservation openings to non-Indian settlement, and competency commissions, the enunciation of a tribal right to reserved water may appear to be an aberration. If viewed as a right to provide sufficient water to support irrigation on the reservation, however, the Winters right seems clearly to fit within the agricultural paradigm that has been a constant in federal Indian policy since the early 1800s. The truly remarkable feature of the decision in *Winters v. United States* is apparent only when one moves beyond the specific facts at issue on arid western reservations at the turn of the twentieth century.

Viewed as a right derived from a negotiation between unequal parties, the reserved right must be construed as intending to further both federal and Indian purposes. The maintenance of economically viable Indian communities, whether based solely on irrigated agriculture, stock raising, or a mixture of traditional and adapted subsistence strategies, was the common element that brought both parties together. Despite the legal meandering of the Winters doctrine in some state and federal opinions since 1908, the initial Hunt decision in 1905 and the two appellate affirmations clearly place the doctrine within the mainstream of contemporary Indian law.[58]

The New Mexico State Court of Appeals' decision in *State of New Mexico v. L. T. Lewis et al.* followed this precedent and construed the history of Mescalero Apache and U.S. relations in light of the intentions of the parties and the canon of construction. In so doing, it found that the Mescalero had reserved a water right in 1852, despite the fact that federal officials did not delineate a specific reservation until 1873. Hispanic and Anglo settlers to the Rio Hondo Basin after 1852 appropriated water subject to the paramount rights of the tribe.

Notes

1. Ira G. Clark, *Water in New Mexico* (Albuquerque: University of New Mexico Press, 1987), 115–22.

2. *United States v. Moses Anderson et al.* was renamed *Winters v. United States* on appeal to the U.S. Supreme Court when Moses Anderson failed to join in the appeal. The name Henry Winter was next on the list of defendants.

3. "Bill of Complaint in Equity," 26 June 1905, *United States v. Moses Anderson et al.*, Case 747, U.S. Circuit Court, 9th Circuit, District of Montana, Records of the District Courts, RG 21, National Archives and Records Center, Seattle; Carl Rasch to Attorney General, 8 August 1905, Folder 1, Box 221, Straight Number File 587.30, CF, General Records of the Department of Justice, RG 60, National Archives; Carl Rasch to Morris Bien, 25 November 1905, Bureau of Reclamation, Project Files 1902–1919, Crow Project 548 Milk River, Records of the Bureau of Reclamation, RG 115, National Archives; Attorney General to President Roosevelt, 18 December 1905, and W. R. Logan to Commissioner of Indian Affairs, 3 June 1905, both in Folder 1, Box 221, Straight Number File 587.30, CF, General Records of the Department of Justice, RG 60, National Archives.

4. *Winters v. United States,* 148 Federal Reporter 684, 207 U.S. 564.

5. 26 July 1866, 39th Cong., 1st sess., U.S. Statutes at Large, vol. 14, 86.

6. Harold A. Ranquist, "The Winters Doctrine and How It Grew: Federal Reservation of Rights to the Use of Water," *Brigham Young University Law Review* (1975): 648.

7. Lloyd Burton, *American Indian Water Rights and the Limits of Law* (Lawrence: University of Kansas Press, 1991), 50–57.

8. Norris Hundley Jr. has been one of the few historians to attempt to understand the historical development of the Winters doctrine. In two articles, "The Dark and Bloody Ground of Indian Water Rights: Confusion Elevated to Principle," *The Western Historical Quarterly* 11 (October 1978): 455–82 and "The 'Winters Decision' and Indian Water Rights: A Mystery Reexamined," *The Western Historical Quarterly* 13 (January 1982): 17–42, Hundley explores the events precipitating the doctrine and its evolution in subsequent years. In the 1978 article, Hundley concludes that the Supreme Court in *Winters* and subsequent decisions has been ambiguous about whether the right was reserved by the tribes or by the United States for the tribes. This ambiguity has affected how various parties characterize the scope and extent of the right today. In his second, more in-depth look at the events surrounding the *Winters* decision, Hundley analyzes both the reasons for the litigation and local reaction to it and concludes that the decision was less ambiguous than he had initially thought. Hundley's pioneer work in this area is sound and suggestive of the value of such focused research. In this chapter, however, I differ with Hundley in a few fundamental areas. In his 1982 essay, Hundley suggests that the *Winters* decision was predicated on both concepts that (1) the lands reserved by the United States had a riparian right to water; and (2) the Indians reserved water through their "right of occupancy." Although he correctly places the foundation for aboriginal title in the early rulings of Justice John Marshall, Hundley fails to distinguish between a water right derived from aboriginal title and one reserved through a treaty. This confusion leads Hundley to conclude incorrectly that legal scholars at the time believed that *both* Indians and the United States could reserve land. What Norris Hundley is alluding to is the application of the "canon of construction" to negotiated treaties and agreements between the United States and tribes. As discussed in this

paper, the 9th Circuit Court of Appeals applied this concept in the 1921 case of *Skeem v. United States.*

9. William Logan to Commissioner of Indian Affairs, 17 February 1906, SC 190, Fort Belknap Agency, RG 75, National Archives.

10. William Logan to Commissioner of Indian Affairs, 12 June 1906, SC 190, Fort Belknap Agency, RG 75, National Archives.

11. Logan praised Babb's diplomatic skills and admitted that "with my combative disposition I could not have come within a mile of making as good an agreement." William Logan to C. C. Babb, 19 April 1906, Misc. Letter Book 1, 1893–1905, Box 57, Fort Belknap Agency, RG 75, National Archives–Rocky Mountain Region (NA-RMR), Denver.

12. The failure of the venture coupled with drought in 1910 caused Logan to comment that he had to "thank the great spirit" that the Indians had not actually planted sugar beets. William Logan to Henry H. Rolapp, 24 January 1906, Misc. Book 1, Box 57, Fort Belknap Agency, RG 75, NA-RMR, Denver; "Annual Report of the Commissioner of Indian Affairs for 1908," 60th Cong., 2d sess., H. Doc. 1046; William Logan to M. J. Costello, 2 July 1910, Misc. Book 4, Box 60, Fort Belknap Agency, RG 75, NA-RMR, Denver.

13. During the time that *United States v. Moses Anderson (Winters)* was proceeding through the courts, a comparable case was underway involving the Blackfeet Tribe. The ruling in *Conrad Investment Co. v. United States* (1908) followed the principles established in *Winters.* Despite the ruling in both cases, a congressional commission appointed to review the Blackfeet Irrigation Project in 1914 urged the beneficial use of water "in order to perfect title." "Report of a Commission to Investigate Irrigation Projects on Indian Lands," 63rd Cong., 3rd sess., H. Doc. 1215, 13–21. See also R. Douglas Hurd, *Indian Agriculture in America: Prehistory to the Present* (Lawrence: University Press of Kansas, 1987), 171.

14. *United States v. Wrightman,* 230 Federal Reporter, 282.

15. *Byers v. Wa-Wa-Ne et al.,* 169 Pacific Reporter, 127. See also Rebecca Wardlaw, "The Irrigable Acres Doctrine," *Natural Resources Journal* 15 (1975): 377.

16. For example, see *Daniel Sullivan v. O. T. Souder et al.,* Bill of Complaint, Montana District Court, 13th Judicial District, File W-1 Water Rights, Box 22749, Billings Area Office, RG 75, NA-RMR, Denver.

17. William Hanna to J. L. Slattery, U.S. Attorney, 7 July 1921, File W-1 Water Rights 1921, Box 22750, Billings Area Office, RG 75, NA-RMR, Denver.

18. *United States v. Powers et al.,* 305 U.S. 527.

19. *Arizona v. California,* 373 U.S., 548–49. Arizona had filed suit against California twice before in the 1930s when it refused to accept the Colorado River Compact. See Norris Hundley Jr., *Water and the West: The Colorado River Compact and the Politics of Water in the American West* (Berkeley: University of California Press, 1975), 282–306.

20. *Arizona v. California,* 600. See also Ranquist, "The Winters Doctrine and How It Grew," 648.

21. Hundley, *Water and the West,* 600–601.

22. The 9th Circuit Court of Appeals was the most active and aggressive court in de-

fining Indian reserved water rights. The court in *Conrad Investment v. United States* was first to expand on the Winters doctrine by suggesting the right was intended for present and *future* uses. See *Conrad Investment Co. v. United States,* 161 Federal Reporter, 832. The court also was particularly active in reviewing these issues in a series of cases that came before it in the 1930s, 1940s, and 1950s. See Wardlaw, "The Irrigable Acres Doctrine," 380–81.

23. Ibid., 382.

24. Hurd, *Indian Agriculture in America,* 136–73.

25. Francis Paul Prucha, *American Indian Treaties: The History of a Political Anomaly* (Berkeley: University of California Press, 1994), 165–67; Rennard Strickland, ed., *Felix S. Cohen's Handbook of Federal Indian Law: 1982 Edition* (Charlottesville, Va.: Michie Bobbs-Merrill, 1982), 221–25.

26. *United States v. Winans* was a 1905 case involving the Yakama Tribes access to a "usual and accustomed" fishery in the Columbia River. The court ruled that the Indian understanding of the 1855 treaty by which they ceded their aboriginal territory to the United States was that they reserved access to the fishery. The right was reserved by them and not granted to them by the United States. *United States v. Winans,* 198 U.S. 371 (1904) 371.

27. *Skeem v. United States,* Federal Reporter, vol. 273, 95–96.

28. For an example of such a study, see Melissa L. Meyer, *The White Earth Tragedy: Ethnicity and Dispossession at a Minnesota Anishinabe Reservation, 1889–1920* (Lincoln: University of Nebraska Press, 1994).

29. Acting Commissioner of Indian Affairs Charles E. Mix to Gov. Isaac I. Stevens, 30 August 1854, Letters Sent from the Office of Indian Affairs, 1824–1881, M21, Roll 50, Bureau of Indian Affairs, RG 75, National Archives. Mix's instructions echoed those of Commissioner Manypenny a year earlier. Commissioner of Indian Affairs George Manypenny to Gov. Isaac Stevens, 9 May 1853, "Annual Report of the Commissioner of Indian Affairs, 1853," 33rd Cong., 1st sess., H. Exec. Doc. 1, 453–57.

30. For example, see *United States v. Adair,* 723 F. 2d. 1394 (9th Circuit, 1983).

31. "A true copy of the Record of the [O]fficial [P]roceedings at the Council in the Walla Walla Valley, held jointly by Isaac I. Stevens Gov. & Supt. W. T. and Joel Palmer Supt. Indian Affairs O. T. on the part of the United States With the Tribes of Indians named in the Treaties made at the Council June 11th and 9th, 1955," T494, Roll 5, Bureau of Indian Affairs, RG 75, National Archives.

32. This argument usually was manifested in the treaty by a statement that the reservation was for the tribes "exclusive use," as in the Walla Walla Treaty of 1855, or, in the case of the Navajo Treaty of 1868, the agreement that the reservation was a "permanent home" from which tribal members could not leave. See "Treaty with the Walla Walla, Cayuse, etc, 1855" and "Treaty with the Navaho, 1868," in Charles J. Kappler, *Laws and Treaties,* vol. II (Washington, D.C.: Government Printing Office, 1904), 694–98, 1015–20.

33. The 9th Circuit Court of Appeals ruled in the 1983 case of *United States v. Adair* that the Klamath Indian Tribe of Oregon had reserved water for the support of a fishery

when it entered into a treaty with the United States in 1864. This case also is noteworthy for the fact that the court held that this reserved water right continued despite the reservation's dissolution pursuant to the Klamath Termination Act of 1954. *United States v. Adair* 723 F 2d. 1394 (cert. denied).

34. Other, specific stream adjudications are also being litigated in state courts in California, Washington, Oregon, and Nevada. Many of these cases also involve one or more claims for Indian reserved water rights.

35. The McCarran bill was inserted in the 1952 general appropriation bill for the Departments of Justice, State, and Commerce. The Department of Justice resisted state efforts to join the United States as party to these streamwide adjudications until the Supreme Court's 1976 decision in *Akin v. United States,* which confirmed that the United States could be joined as a party to state general water adjudications. See "McCarran Amendment," 66 Stat 560. See also Elizabeth McCallister, "Water Rights: The McCarren Amendment and Indian Tribes' Reserved Water Rights," *American Indian Law Review* 4 (1976): 303–10.

36. The Rio Hondo is formed by the convergence of the Rio Ruidoso and Rio Bonito, which flows east-southeast from the White and Sacramento Mountains. The Rio Hondo joins the Pecos River approximately twelve miles above the town of Roswell. Although all of the rivers are considered perennials, they frequently experience low or no flow during the summer months.

37. Ira G. Clark, *Water in New Mexico,* 663. Clark's book is undoubtedly the best single volume on the history of New Mexico's water law.

38. Frederick Nolan, *The Lincoln County War: A Documentary History* (Norman: University of Oklahoma Press, 1992), 37–38, 68.

39. Clark, *Water in New Mexico,* 49–50.

40. "Bill of Complaint, *United States of America v. El Paso and Rock Island Railway Company, a corporation, et al.,*" Law and Equity Case Files, 1913, Box 24, no. 238, U.S. District Court Records, District of New Mexico, RG 21, National Archives, Denver; "Report on the Hondo Hydrographic Survey by the Territorial Engineer to the Court of the Sixth Judicial District of the Territory of New Mexico," n.d., State Engineers Papers, State Records and Archives, Santa Fe. Maps, Bonito River Hydrographic Survey 1930, State Engineers Office, Santa Fe.

41. Albert Schroeder, *Apache Indians I: A Study of the Apache Indians,* vol. III, ed. David Agee Horr (New York: Garland Publishing, 1974). See also Ralph A. Smith, "Apache Plunder Trails Southward, 1831–1840," *New Mexico Historical Review* 37 (1962): 20–42.

42. "Treaty with the Apache, July 1, 1852" 10 *Stat.* 978. See also Annie Heloise Abel, "The Journal of John Greiner," *Old Santa Fe: A Magazine of History, Archaeology, Genealogy, and Biography* 3 (July 1916): 219–21.

43. Michael Steck to Superintendent James Collins, 20 April 1858, Box 2, Folder 6; Major Charles Ruff to Michael Steck, 6 July 1859, Box 3, Folder 1, both in Steck Papers, Zim-

merman Library, University of New Mexico, Albuquerque; Steck to Collins in "Annual Report of the Commissioner of Indian Affairs for 1858," 35th Cong., 2d sess., H. Exec. Doc. 2, 547–51.

44. There were a number of reasons for the United States to begin addressing the Mescalero Apache problem in the 1870s. The Mescalero's escape from incarceration at the Bosque Redondo during the Civil War as well as increased settlement along the Rios Hondo, Ruidoso, and Bonito after 1865 prompted the military, local Indian agents, and settlers to argue for a defined reservation. For example, see "Petition to Commissioner of Indian Affairs," 22 January 1862, Letters Received, New Mexico Superintendency (M-234, Roll 554), RG 75, National Archives.

45. Historical Research Associates, "A Historical Study of the Mescalero Apache Indian Tribe: 1848–1873," U.S. Exhibit 28, *State of New Mexico v. L. T. Lewis et al.*, nos. 20294 and 22600 consolidated, District Court of Chaves County, Roswell, New Mexico.

46. See "Post-Trial Brief of the State of New Mexico," *State of New Mexico v. L. T. Lewis et al.*, 23–36.

47. State of New Mexico, Court of Appeals, Opinion, *State of New Mexico v. L. T. Lewis et al.*, 12 May 1993, *New Mexico Reports*, vol. 116, 198.

48. Ibid., 202.

49. Ibid., 202–3. Despite the Appeals Court's favorable ruling on the priority date for the Mescalero Apache and its acceptance of the United States's argument that the Mescalero right should include water for recreation, religious, and other purposes, the court refused to quantify based on a PIA standard. Consequently, the water given to the Mescalero was substantially less than they had claimed.

50. In his *Water in New Mexico*, Ira G. Clark provides a valuable bibliographic essay on the history of water development in New Mexico, including community acequias.

51. Clark notes that there were two forms of community acequias in New Mexico. One was the "municipally owned ditch" and the other was a more informal, cooperative ditch owned by private irrigators. Ibid. 15.

52. Ibid., 25–30. Clark explains how this shift in the importance of the community acequia resulted in New Mexico's adoption of a "dual administrative system" by 1903 (Ibid., 100–101).

53. Settlement pressure in the Rio Hondo Basin was so light that the surveyor general did not even begin establishing township boundaries until 1867. See "Annual Report of the Surveyor General's Office, 1866," 39th Cong., 2d sess., H. Exec. Doc. 1, 470–71.

54. John P. Wilson, "How the Settlers Farmed: Hispanic Villages and Irrigation Systems in Early Sierra County, 1850–1900," *New Mexico Historical Review* 63 (October 1988): 333–56.

55. General Land Office, "Survey Notes," Township 9 South, Range 16 East, Surveyed by Robert B. Willison, 5–15 October 1867, vol. S-976, Bureau of Land Management, Santa Fe, New Mexico, 65.

56. "Answer of Cruz de Jara Community Acequia et al., *United States of America v. El*

Paso and Rock Island Railway Company, a corporation, et al.," Law and Equity Case Files, 1913, Box 24, no. 238, 7.

57. Francis Levine, "Dividing the Water: The Impact of Water Rights Adjudication on New Mexican Communities," *Journal of the Southwest* 32 (Autumn 1990): 269–74; Ira Clark, *Water in New Mexico*, 103–4.

58. Hundley, "The Winters Decision and Indian Water Rights," 34.

7

Winters Comes Home to Roost

Daniel McCool

When Henry Winter diverted a small amount of water onto his parched fields in northern Montana's Milk River Valley, the fledgling twentieth century held a fragile promise. With enough water, some luck with the weather, and a great deal of backbreaking work, Mr. Winter and his fellow settlers figured they could scratch out an existence in the rolling prairie. Aside from the fellow settlers, Winter's only other neighbors were the Assiniboin and Gros Ventre Indians living on the Fort Belknap Indian Reservation just downstream.

History did not record Winter's thoughts as he diverted that first gallon of water, but we can surmise that he had no idea his action would spark one of the West's longest and most bitter conflicts—a conflict that would pit governments against governments, bureaucrats against bureaucrats, Indians against Anglos, and ultimately lead to a convoluted and complex set of court decisions that came to be called the Winters doctrine.[1]

In the American West, a unique water law developed called the Prior Appropriation doctrine. Its principal focus was on time: The first to divert and apply water was awarded a full water right; all subsequent diverters held an inferior water right. It is ironic that a water doctrine based on who arrived first would work such a disadvantage to Indian people, but the realm of water law often defies logic. The "first in time, first in right" principle of the Prior Appropriation doctrine narrowly construed the concept of use. In the Anglo ethos, use meant dams and diversions; it was assumed that water that did not create a product was wasted. The great challenge to these water pioneers was to defy nature and allow no river to reach the sea.

Indian people had, of course, used the waters of America for centuries, but not in a manner that satisfied the Prior Appropriation doctrine. That would have to wait until the Bureau of Indian Affairs (BIA) fulfilled its promise of irrigation projects so that Indians might feed themselves in the absence of their traditional forms of sustenance. But the dilatory BIA irrigation program always lagged behind the better-financed Anglo projects; it built Indian irrigation projects so slowly that the initial stages of the projects became dilapidated before the latter stages were begun; the agency's construction schedule was slower than the rotting wood and rusting metal that characterized Indian irrigation projects. In addition, many tribes were hesitant to pick up the plow. The culture-bending shift from the freedom of the hunt to the toil of the farm was too much to ask of a people so recently at liberty.

While Indian people were attempting to adjust to their new confinement on reservations, the landscape around them was being transformed. As settlers moved west they attempted to bring the East with them; they wanted green fields where there was only the brown-hued tint of deserts; they wanted trees where only grass could be seen to the limitless horizon. It took a singular kind of courage—and no little imagination—to look upon the parched lands of the West and assume that providence and the plow would transform it into a breadbasket. And it took a great leap of reasoning to justify stealing it all from the Indians. But the early settlers were as self-righteous as they were hard working. It was this unbending belief that God was on their side that permitted the settlers to view Indian reservations as temporary holdings that could, with clear conscience, be whittled down to small waterless tracts. The Dawes Act would do the whittling; the Prior Appropriation doctrine would make the tracts waterless.

Or so the settlers thought. At a time when Indians were routinely robbed of everything they possessed, it is easy to imagine that a group of settlers, confronted with a claim that Indians actually had water rights, would blithely assert that, yes, some land was reserved for the Indians, but the treaties made no mention of water. Thus, Indian land did not have any Indian water. It was Henry Winter and his neighbors who made this claim when the attorney for the Fort Belknap Reservation filed his case. Think of the surprise he must have felt when the judge ruled in favor of the Indians:

> Prior to 1888 nearly the whole of northern Montana north of the Missouri River, and eastward from the main chain of the Rocky Mountains was recog-

> nized as Indian country, occupied in part by the tribes of Indians now living upon the Fort Belknap Reservation. By treaty of May 1, 1888, the Indians "ceded and relinquished to the United States" their title and rights to lands not embraced within the reservation then established as their permanent homes. The purposes of the treaty were that means might be had to enable the Indians to become "self-supporting as a pastoral and agricultural people, and to educate their children in the paths of civilization."

The judge then held that the Indians were entitled to sufficient water "to an extent reasonably necessary to irrigate their lands."[2]

The settlers, stunned at being subjected to sensitivity to Indian needs that was wholly uncharacteristic of the times, quickly appealed. They argued, with considerable justification, that their claims were congruent with the "customs of the country."[3] They lost again in appellate court, and then went to the Supreme Court. The government's case on behalf of the Fort Belknap Indians offered a useful insight into prevailing Indian policy. BIA bureaucrats argued that the Indians should be awarded a water right, not because it was the right thing to do, but because it would allow the Indians to become self-sufficient and thus make them "independent of future Government support," hastening the day when Indian people, following the "paths of civilization," could be absorbed into Euro-American society.[4]

The Supreme Court, ignoring many of the arguments presented by both sides, based its decision on treaty rights and held that the government implicitly reserved for the Indians water "for a use which would be necessarily continued through the years."[5] It was a rare victory for American Indians, but it did not signal a fundamental shift in either BIA policy or the attitude of Anglos. Rather, it went largely unheeded outside the courtroom. Inside the courtroom it gave rise to an unending legacy of conflict.

Water in Court

The Winters doctrine was initially viewed as a minor irritant by officials in western states and their allies in the Bureau of Reclamation. The bureau continued to build projects for Anglos as though the Winters doctrine did not exist. In the meantime, however, tribes won a host of victories in the courtroom, giving rise to the saying that "Anglos got wet water and Indians got paper water." In effect, the Bureau of Reclamation and other federal and state agencies had built a host of water projects that did not hold a clear title to the water.

Much of this short-sighted behavior can be attributed to the great contrast between the Prior Appropriation doctrine and the Winters doctrine. Each developed independently of the other; and they are a study in contrasts. Prior Appropriation requires a diversion and use; Winters is contingent upon future development of the reservation. Prior Appropriation was adopted by western states as they attempted to grapple with an aridity that made eastern water law impractical. In contrast, the Winters doctrine is federal case law and, unlike Prior Appropriation, it has never been recognized by the U.S. Congress. The purpose of Prior Appropriation was to create an advantage for those who first diverted water, thus lending economic stability to newly developing regions of the West. The purpose of the Winters doctrine was to allow Indian people sufficient time to transform themselves into farmers in the European tradition, and thus make reservations viable economic entities that would not require continued government assistance.[6]

After some hesitation, the Bureau of Reclamation placed its faith in the Prior Appropriation doctrine as a way of insuring a water supply for its projects. This permitted the bureau to lavish money on Anglo projects while the BIA irrigation program languished, the result of public indifference, hostility from western interests, and the Bureau of Reclamation's zeal to build more projects for Anglos. Consequently, water projects were built all over the West that depended on water resources that flowed through, or bordered, Indian reservations.[7]

But the threat of Winters lay on the horizon like an approaching storm. Over time the accumulated court cases upholding the Winters doctrine eventually gave it weight. A major victory for tribes occurred in 1963 with the landmark case of *Arizona v. California*. The Supreme Court determined that Indian reservation land held a reserved water right to all the water necessary to irrigate all the "practicable irrigable acreage." The court also applied the doctrine to all federally reserved lands, not just Indian reservations.[8] The western water establishment began to take notice. After seventy years of litigation, Anglos began to fear that one day the entire western water regime could become unhinged if the government ever acted to give substance to Indian rights. Their fear of a suddenly energized reserved rights doctrine gave them an incentive to attempt a negotiated solution to the problem.

Indian tribes, in contrast, initially had little faith in a politically charged negotiation where they would face a formidable array of adversaries across the table. However, their attitude began to change in the mid-1970s, when a series of court cases shifted water litigation from federal courts to state courts.[9] Indian victories had been won in federal courts; the states promised to be much less receptive to Indian claims. In addition, even the federal courts, including the high court, were becoming more hostile to Indian claims. Indian tribes lost a series of water cases before the high court in 1983.[10]

The most recent court test of the Winters doctrine occurred when the Eastern Shoshone and Northern Arapaho Tribes of the Wind River Reservation in Wyoming claimed reserved water rights in the Big Horn River watershed. Because of the McCarran Amendment, the case went to state district court, was appealed to the Wyoming Supreme Court, and then appealed to the U.S. Supreme Court. This series of legal contests resulted in twenty years of acrimonious litigation—*Big Horn I, II, and III*—and cost upward of $20 million in legal fees.[11] What the case did not provide was a clear and definitive contemporary concept of reserved rights. Perhaps the clearest lesson from these cases was that it was becoming increasingly risky for tribes to take their reserved water claims to court.

By the late 1970s and early 1980s, non-Indians were afraid the Winters doctrine would never just disappear, as they had hoped. But Indian tribes were equally fearful that the doctrine would be whittled down by hostile state courts and less receptive federal courts. Thus, after years of conflict, both sides became amenable to the possibility of negotiating settlements to Indian water claims.

Water Treaties

Beginning in 1978 with the Ak Chin settlement, the federal government espoused an official policy of supporting negotiated Indian water settlements as an alternative to litigation. In the ensuing years, thirteen additional settlements have been signed and at least two dozen are currently in negotiation or awaiting a negotiation team.[12] This new policy is best characterized as a second treaty era. The first treaty era, in the nineteenth century, concerned Indian land; the second treaty era concerns Indian water.

There are six significant similarities between the first treaty era and the second. First, these negotiations have taken place between unequal parties, often following bitter conflict. In the first treaty era, the power differential was military; in the second era, the inequality concerns money, organization, and access to water. This forced Indians to enter into treaty negotiations from a disadvantaged position, but with the hope that they could salvage at least some of their rights to land and water.

Second, the written agreements resulting from the negotiations were intended to be permanent settlements to long-standing conflicts. The hope was that a formal, congressionally recognized agreement would provide lasting legitimacy. In the first treaty era the commitments made were seldom honored; although it is too early to make conclusive generalizations, the second treaty era has initially met with mixed success in meeting agreed-upon commitments.

A third similarity is that both treaty eras involved government-to-government agreements between semisovereign entities. The nature of that relationship has changed over time, but the basic approach is still the same: If either side refused to recognize the legitimacy of the other, there would be no recognized need to negotiate. In both treaty eras all parties have had to grudgingly recognize that the other side held at least some of the cards. Thus, the formal treaty process has helped to establish the legitimacy of each side's claims.

A fourth similarity is that the options available to Indian people were extremely limited; the choice of the lesser of two evils is a common theme in both eras. Thus, it has seldom been the case that tribes argued for what they wanted; rather, they argued for what they thought they could get, considering the overpowering dominance of the Anglos.

Fifth, nearly every treaty from both eras has involved a direct trade-off of Indian resources in exchange for money. In the first treaty era, this often meant that Indian people relinquished large tracts of land in exchange for a promise of goods and cash. In the second era, Indians have given up future claims of reserved water, which were potentially enormous, in exchange for economic development funds, funding for water projects, and other financial assistance.

Finally, nearly all settlements from both eras are based on the assumption that Indians were willing to give up large claims in exchange for a steadfast guarantee that their remaining resources would remain forever in their possession. The idea of sacrificing hope for reality is a common theme in the Indians' discussions over whether they should negotiate.

Contemporary negotiations have the advantage of a better understanding of how to conduct a conflict-resolution process. Current literature on the subject often characterizes negotiation as a "win-win" situation where all parties gain from the agreement.[13] This is a valid characteristic for negotiations between parties of roughly equal power, but it breaks down quickly when one side in the negotiation has a long tradition of overpowering the other. In regard to Indian water settlements, it is more accurate to characterize the negotiations as a highly politicized process of give and take. Anglo water users are not interested in negotiating with Indian tribes unless they get something significant out of the process; the most common item on their agenda is a strict limit to the water that Indians can claim. Indian tribes have been willing to accept such limitations in exchange for a recognized right to their remaining water and some form of financial assistance. There is a definite zero-sum quality to these exchanges of water rights and financial resources.

Why Settlements?

There is an extensive literature on negotiation, conflict resolution, and cooperative management, nearly all of it produced within the last fifteen years. This literature reflects an effort to move beyond lawsuits and focus on substantive solutions rather than winning and losing. The appeal of negotiation as an alternative to litigation quickly became apparent in regard to Indian water rights, where court cases have become legendary for their duration, bitterness, and inconclusiveness.

The federal government began espousing negotiated settlements during the Carter administration, but the new policy did not fully develop until the Reagan administration made it a priority. Secretary of Interior James Watt claimed it was part of the administration's "cooperative good neighbor policy."[14] The administration's commitment to negotiated settlements was reflected in its budget priorities; suddenly funding was shifted from litigation to negotiation. This shift in priorities was supported by most Indian and non-Indian parties to the water conflicts. Many practitioners as well as academics began to extol the advantages of negotiation:

> The costs of litigation can be tens of millions of dollars for each side and can perpetuate bitterness with non-Indian communities. Thus, parties to many of these Indian water conflicts are opting for the negotiating table, as they grow acutely aware of the immense financial burdens and lost opportunities caused by lengthy litigation. Negotiated solutions also have the capacity to deal with practical problems of using and developing common water resources to the advantage of both Indians and non-Indians.[15]

Much of the support for negotiated settlements focuses on the practical economic advantages that are much more likely to develop in a cooperative relationship: "The most valuable result of a negotiated settlement is the establishment of commercial and governmental relationships which remain after the negotiations are concluded. In most cases, water rights negotiations are the first substantive opportunity for the local non-Indian community to begin to understand and appreciate the needs and capabilities of their Indian neighbors."[16]

These advantages appealed to the veterans of the Indian water wars; tribes began lining up to negotiate. So many parties were interested in negotiation that the Interior Department had difficulty fielding a sufficient number of negotiation teams. Following the 1978 settlement for the Ak Chin community, settlements were signed with increasing frequency.[17] In 1982, a large, complex settlement involving thousands of southern Arizona water users and the Tohono O'odham Tribe (formerly the Papago) was signed with much fanfare (P. L. 97–293). By this time, Montana had formed a special commission to negotiate reserved water rights; it bore fruit in 1985 when the state signed a compact for the Fort Peck Reservation.[18] The settlement era was in full-swing. In 1988, three major settlements were signed: San Luis Rey (the Mission tribes) in California (P. L. 100–675); Colorado-Ute (P. L. 100–585); and Salt River Pima–Maricopa in Arizona (P. L. 100–512).

The emphasis on negotiation continued under the Bush administration. Three more settlements were signed in 1990: Fort Hall in Idaho (P. L. 101–602); Fort McDowell in Arizona (P. L. 101–628); and the very complicated settlement that involved two states—California and Nevada—and two Indian reservations—Fallon Paiute–Shoshone and Pyramid Lake (P. L. 101–618). Two years later, four more settlements became law: Jicarilla Apache in New Mexico (P. L. 102–441); Northern Cheyenne in Wyoming (P. L. 102–374); the Northern Ute in Utah (P. L. 102–575); and the San Carlos Apache in Arizona (P. L. 102–575). Another settlement, signed in 1994, resolved a dispute between the city of Prescott, Arizona, and the local Yavapai Tribe (P. L. 103–434).[19] The most recent settlement, signed 10 December 1999, was for the Chippewa Cree Tribe of the Rocky Boys Reservation.[20]

The success of the process slowed considerably under the Clinton administration, especially after the Republicans gained control of Congress. This occurred despite a continued commitment from Congress and the president to favor settlements over litigation. But settlements cost money, and this policy, like so many others, became a victim of the great effort to cut spending and balance the federal budget. An unofficial yet undeniable new aspect of the policy was to settle these conflicts without a significant outlay of federal expenditures—"settlements on the cheap," as one congressional staffer called it.[21]

At this point it is too early to tell if the settlement policy, as executed thus far in the form of fourteen settlements, has been a success. Each settlement is unique, and none of them have met everyone's expectations. Indeed, the negotiating table is where the wish list is left behind and the *realpolitik* of water and power determines the allocation of acre-feet and funding. However, certain trends indicate how successful the policy has been thus far. Relying on the literature on settlements, four advantages are routinely claimed by proponents: settlements are cheaper, they achieve finality and certainty, they can allocate water and fund the projects to deliver that water, and they enhance comity and good relations among Indians and their non-Indian neighbors. Each of these will be examined by identifying the nature of the claimed advantage and then briefly surveying the existing settlements to assess the extent to which the expected benefits have been achieved.

Saving Money

There are two distinct philosophies regarding funding for water settlements. One approach, favored by successive administrations and reflected in Office of Management and Budget policies, is that the cost of a settlement should never exceed legal liability. In other words, the accumulated costs of damages, court costs, and attorneys fees should be used as a ceiling in deciding how much to spend on a specific settlement. The other view is that water settlements are an integral part of the federal government's trust responsibility. Thus, the cost of settlements should reflect the dimensions of that responsibility and should not be limited to speculative concepts of legal exposure. In 1990, the Bush administration attempted to standardize the negotiation process by developing official "criteria and procedures" for negotiations. That document reflects both of these perspectives; first, negotiators are instructed to limit funding to "calculable legal exposure—litigation cost and judgment obligations if the case is lost," but another section of the document approves of "costs related to Federal trust or programmatic responsibilities."[22] Merely identifying both approaches did nothing to resolve the conflict between the two.

If we use the first criteria for cost—legal exposure—it is fairly clear that settlements have exceeded the government's probable costs if it had chosen to litigate rather than negotiate. The first thirteen settlements have cost approximately $495.7 million.[23] That is an average of $38 million per settlement. For sake of comparison, the Big Horn litigation involving the Wind River Reservation and Wyoming is often touted as a glaring example of how much it costs to litigate; in that litigation total costs are estimated at $20 million. But most settlements have exceeded that figure by millions of dollars. The high costs of settlements are not necessarily due to an unprecedented level of generosity to Indian tribes. Quite the reverse, many settlements require a large payoff to non-Indians to grease the political wheels. An Interior Department official made this point clear: "Most of the cost [in settlements] is to mitigate impacts on Anglos, that is a big part of the cost." This is a new version of a strategy that old hands in the BIA call the "Indian blanket," where programs for Indians are wrapped in favored pet projects for western Anglos.[24]

If the second criteria for cost is used—the trust responsibility—it becomes much more difficult to ascertain a relative sense of cost. Would it be less expensive to honestly meet the trust responsibility by investing in some other kind of Indian program? Probably not. The real problem with the trust aspect is the troubling possibility that funding for Indian water settlements has come out of other Indian programs. In the contemporary politics of budgeting, all funding is zero-sum: Money for one program can only be obtained by making a cut of similar magnitude in other programs. Many Indian leaders have claimed that this is the case, as one Senate staffer explained: "Overall, Indian country does not gain from these settlements. . . . I can assure you that the money comes out of other Indian programs." Other commentators, especially those in the Clinton administration, claim that this problem has been solved.[25]

The settlements have resulted in an unprecedented investment in Indian water, but they have not magically transformed the government's traditionally poor performance in meeting its commitment to Indian people. It is fair to conclude that the settlement policy has not been an effective money-saving strategy. However, that does not mean that the funding is not money well spent.

Finality and Certainty

The first treaty era was notorious for broken promises. Although there were actually very few treaties that promised to last "as long as the rivers flow and grass grows," the clear assumption among Indian signatories was that each treaty resolved their land disputes with the White Man and left them with a permanent title to their remaining lands. History tells us their expectations were not fulfilled.

In the second treaty era, the hope of a permanent settlement has played a prominent role in the negotiations. For Anglos, this took the form of quantification—an empirical measure of just how much water Indians are entitled to, and no more. The great fear of Anglo water users was the infinitely expandable nature of reserved rights; the amount of water claimed could grow with the needs of the reservations. Anglos wanted to cap that growth at a specified level, leaving all remaining water at their disposal.

To achieve this, a careful water inventory has to be completed and the settlement must contain language stating that no further reserved claims can be filed. This took the form of a fairly standard disclaimer statement that is found in most of the settlements. For example, the Tohono O'odham settlement (the southern Arizona–Papago settlement) states that the tribe "executes a waiver and release" of "any and all claims of water rights or injuries to water rights . . . [and] any and all future claims of water rights."[26] Similarly, the Jicarilla settlement "is intended to provide for the full, fair, and final resolution of the water rights claims of the Tribe."[27]

In other words, tribes have forsaken forever any claims to reserved water rights, which is a fairly significant sacrifice in return for the expected benefits of the settlements. But have settlements truly achieved a permanent peace with this formula? Thus far, the track record is decidedly mixed. Ak Chin had to be renegotiated, although now it is on track. Amendments had to be made to the Tohono O'odham and San Carlos settlements, yet the tribes are still experiencing problems. The Colorado Ute, Northern Ute, and San Luis Rey settlements have run into major political obstacles; all three of these may require renegotiation and significant modification. In addition, the Fort Peck compact is now in search of a congressional approval for a marketing provision. The hoped-for finality has only been achieved by some of the settlements. Thus, even the sacrifice of all future Winters claims has not created the kind of finality and certainty that was expected.

Indeed, it may be folly to expect certainty regarding a natural resource that is best known for its fickleness; we can adequately predict neither nature's allocation nor the wildly varying human uses of water. It may be more practical to view settlements as permanent only for a specified period of time. Who can predict the West's water needs in a hundred years? The only certainty is that those uses will look quite different from those of today; should we expect today's water laws and policies to be adequate for such a different setting? Given the vagaries of both water and water policy, is it reasonable to expect tribes to relinquish all reserved rights claims *forever?*

Wet Water

One of the great constraints on litigation is its inability to move beyond court-granted recognition of rights. In the years since the Winters case, judges declared in favor of millions of acre-feet of water on behalf of Indian reservations. But the water allocation regime in the West remained virtually unchanged; the Bureau of Reclamation and the Army Corps of Engineers continued to construct massive projects for Anglos, many of which diverted water away from reservations. In the meantime, the BIA's irrigation program wallowed in failure. This gave rise to the commonly used phrase that makes a distinction between "paper water," which is the result of a court victory, and "wet water," which is actually delivered and used. Tribes had a surfeit of the former and an acute shortage of the latter.

The great hope of the settlement policy was that, finally, tribes could negotiate for wet water. But this represents a difficult challenge. In many areas of the West, all practical sources of water are already allocated, necessitating an expensive and controversial re-allocation of this precious commodity. The search for wet water has also been hampered by an implicit policy of avoiding appreciable water losses among Anglos.[28] This "no-harm-to-Anglos" rule is especially constrictive in southern California and the arid mountain-states region. The only way to meet both of these goals—more water for Indians and no appreciable loss to non-Indians—is to rely on large federally funded water projects. Some settlements, such as the Northern Cheyenne settlement and most of the Arizona settlements, use water storage from existing projects. These settlements were not opposed by environmental groups. In contrast, the Colorado Ute settlement attempted to resolve its water supply problems by authorizing a large federal water project, the Animas–La Plata. Environmentalists and downstream water users have vociferously attacked this project, and thus far have prevented its construction. The lesson is fairly clear: Building large new water diversion projects is not a politically viable method of solving the "wet water" problem.

These constraints have limited the ability of settlements to achieve the goal of wet water. On some reservations, such as Fort McDowell, Ak Chin, and Salt River Pima–Maricopa, new sources of water have been successfully put to use as a direct result of their settlements. On other reservations, notably Northern Cheyenne, Pyramid Lake, and Fort Hall, increased tribal use of water appears likely in the future. But numerous settlements have yet to provide additional water, and little or no progress is being made toward that goal; specifically, the San Luis Rey, Tohono O'odham, Colorado Ute, San Carlos Apache, and Jicarilla Apache settlements are stalled due to a variety of water supply limitations. At best the promise of wet water has been only partially fulfilled by the settlement strategy.

The problem of securing wet water via settlements is just one of the many difficulties tribes have encountered when trying to implement settlements. In some cases settlements have been left purposefully vague to achieve agreement; this has later caused numerous problems. In the highly conflictual world of water law and politics, every word can become a point of contention as parties vie for every possible advantage.

Comity

Nearly all parties agree that settlements have improved the relationship between tribes and non-Indian water users. In contrast to litigation, negotiation places the stress on solutions rather than winning or losing. This has helped all parties to achieve at least some of their goals through the settlement process. Negotiations also provide a forum for communication, an opportunity to meet face-to-face, and a working environment that gives all stakeholders a common set of goals. These advantages are often cited in the literature supporting negotiation as a conflict-resolution strategy.

But there is another reason why settlements have enhanced comity; federal dollars, supplied liberally, have lubricated the process. Most settlement have not focused on re-allocating benefits, but expanding the pie via federal financing of water projects, development funds, buy-outs, subsidies, and loan forgiveness. It is easy to be good neighbors when a third party pays the costs of making everyone happy, or at least satisfied.

Still, comity is one of those benefits that is difficult to measure in economic terms, but is nevertheless precious. An amiable agreement over water can create partnerships that reach into other areas of life. An Interior Department official explained: "If settlements are done correctly, you can have a stable community, and you have the kind of benefit you can't put on a balance sheet. You have harmony in the community. These are intangible benefits that clearly outweigh the costs of a settlement." When Congressman Kolbe of Arizona rose on the House floor to speak on behalf of the San Carlos settlement, he spoke of comity: "Many formerly contentious issues have been resolved. The progress on this bill has been nothing short of miraculous, making clear that parties with varied interests can come together to reach agreement without resorting to litigation."[29]

However, not all issues are resolved in settlements, and in some cases, the settlement has actually increased tensions. Nearly all of this new conflict has occurred, not between Indians and non-Indians, but either among Indians of the same tribe or between tribes. Thus, many settlements have simply transformed water conflict from a white-Indian issue to an Indian-Indian issue. For example, on both the Southern Ute Reservation in Colorado and the Tohono O'odham Reservation in Arizona, the tribes are bitterly divided over their water settlements. There are still on-going conflicts at Pyramid Lake and Uintah and Ouray (Northern Utes of Utah) as to how to respond to their settlements.

Settlements have also incited conflict between tribes. The Navajo Nation opposed diversions from the San Juan River that are critical to the success of the Colorado Ute settlement. The Nevada settlement brought the Fallon Paiute–Shoshones and the Pyramid Lake Tribe into conflict. Also, the Northern Cheyenne settlement sparked animosities between that tribe and the neighboring Crow regarding the use of the Big Horn River. Comity is a laudable goal, but it is increasingly difficult to find in the highly contentious arena of western water, even with the added incentive of federal funding.

Conclusion: Sellout or Fulfillment?

To some, the water settlements are a fulfillment of the promise of Winters; for them the settlements offer the possibility of wet water flowing to tribal homelands. To others, each settlement is a sellout, an abandonment of all future reserved rights claims. This brief examination of existing settlements provides support for each of these perspectives; the record thus far is mixed. The settlement era has produced some notable successes as well as some enormously problematic situations.

If the settlement policy continues—and nearly everyone expects that to be the case—then it is worthwhile to identify factors that may have a significant influence on future settlements.

Perhaps the most obvious factor is the federal budget. Continued budget-cutting may well deny to negotiators their greatest ally—the promise of federal funding. In recent years there have been several attempts to dramatically reduce funding for the BIA. It is conceivable that the budget for settlements could face the same challenge, but less likely due to the popularity of settlements among western Anglos, who are particularly well represented in the Senate. If settlements only benefited Indians, they never would have generated sufficient political support. But much of the funding and benefits of settlements go to non-Indians who have considerable political clout. Indeed, a cynical view of settlements would argue they are simply a new form of pork barrel for traditionally powerful Westerners, with some additional benefits to Indians.

Another factor is the dramatically changing water-use patterns in the West. Today about 85 percent of the water diverted in the West is used by agriculture, mostly for hay and other low-value crops. But economics, population growth, and political opposition to irrigation subsidies are all placing pressure on the old water regime. It is probable that much of the water currently used by agriculture will be re-allocated. If, for example, the federal government stops subsidizing hay farming, millions of acre-feet of water would become available. It would then be much easier for cities, recreationists, and, yes, Indian tribes to stake a claim for more water. In a very real sense the American West does not have a water problem; it has an irrigation subsidy problem.

The third trend that will affect all water users, including Indian tribes, is water marketing. An open market in water would result in a significant reallocation. It could also make Indian reservations big players in the water market, *if* they have not signed away their marketing rights. Unfortunately, nearly every Indian water settlement contains a prohibition on any marketing outside the local area. This was a massive sacrifice of Indian marketing potential, one that is not yet apparent in its scope. But water marketing is inevitable—all the economic forces are in that direction—and the tribes that have signed settlements will be at a significant disadvantage.

The final trend that may affect future negotiations is the increasing awareness that the first treaty era and the second treaty era are intimately related; the land affects the water and vice versa. In the future, tribes may expand the scope of their negotiations to include land as well as wet water. In exchange for the water concessions demanded by Anglos, tribes may parley their reserved water rights into a demand for the return of lands lost through allotment, theft, and sale.

The mixed blessings of the settlement era have, at the least, provided a sense of hope that progress on Indian water issues can be made. After nearly a century of litigation, failure, and frustration, the glimmer of a better future is now a possibility. Settlements have a great potential, but they are not created equal. A good settlement can guarantee the future; a bad one looks a great deal like the settlements negotiated in the first treaty era.

Notes

1. A court clerk mistakenly added an "s" to Mr. Winter's name.

2. *U.S. v. Moses Anderson et al.*, Memorandum Order (December 1905). Anderson dropped out of the group of petitioners when the case was appealed, thus moving Mr. Winter into position as the named petitioner.

3. Appeal from the Circuit Court, District of Montana, to the 9th Circuit Court of Appeals, *Henry Winters et al., v. the United States of America*, no. 1243, 5 February 1906.

4. Letter from W. B. Hill, U.S. Indian Inspector, Fort Belknap Agency, Montana, to W. H. Code, Chief Engineer, Indian Service, Los Angeles, California, 29 August 1907, 20, RG 75, Fort Belknap file, National Archives.

5. *U.S. v. Winters*, 207 U.S. 564 (1908), 568.

6. For a succinct explanation of the Winters doctrine and its legal implications, see David Getches, "Indian Water Rights Conflicts in Perspective" in *Indian Water in the New West*, ed. Thomas McGuire, William Lord, and Mary Wallace (Tucson: University of Arizona Press, 1993,) 7–26; Jana Walker and Susan Williams, "Indian Reserved Water Rights," *Natural Resources and the Environment* 5, no. 4 (1992): 50.

7. For a discussion of the conflict between the Bureau of Reclamation and Indian tribes, see Daniel McCool, *Command of the Waters: Iron Triangles, Federal Water Development, and Indian Water* (Berkeley: University of California Press, 1987; reprint, Tucson: University of Arizona Press, 1994), ch. 6; Monique Shay, "Promises of a Viable Homeland, Reality of Selective Reclamation: A Study of the Relationship between the Winters Doctrine and Federal Water Development in the Western United States," *Ecology Law Quarterly* 19 (1992): 547–91.

8. *Arizona v. California*, 373 U.S. 546 (1963).

9. For a discussion of the McCarran Amendment cases, see Mary Wallace, "The Supreme Court and Indian Water Rights," in *American Indian Policy in the Twentieth Century*, ed. Vine Deloria Jr. (Norman: University of Oklahoma Press, 1985), 197–220.

10. *Nevada v. U.S.*, 103 Sup. Ct. 2906 (1983); *Arizona v. California*, 103 Sup. Ct. 1382 (1983); *Arizona v. San Carlos Apache Tribe of Arizona et al.*, 103 Sup. Ct. 1382 (1983).

11. In *The General Adjudication of All Rights to Use Water in the Big Horn River System and All Other Sources*, 753 P. 2d 61 (Wyoming 1988); *Wyoming v. U.S.*, 492 U.S. 406

(1989). See Wes Williams, "Changing Water Use for Federally Reserved Indian Water Rights: Wind River Indian Reservation," *University of California Law Review* 27 (1994): 501–32.

12. The settlements are: Ak Chin, 1978, renegotiated 1984; Papago (Tohono O'Odham), 1982; Fort Peck, 1985; Colorado Ute, 1988; Salt River Pima–Maricopa, 1988; San Luis Rey, 1988; Fallon Paiute–Shoshone/Truckee Carson–Pyramid Lake, 1990; Fort Hall, 1990; Fort McDowell, 1990; Jicarilla Apache, 1992; Northern Cheyenne, 1992; San Carlos Apache, 1992; Ute Indian (Utah), 1992; Yavapai-Prescott, 1994. For a complete listing of the settlements and their various features, see Jon Hare, *Indian Water Rights: An Analysis of Current and Pending Indian Water Rights Settlements* (Washington, D.C.: The Confederated Tribes of the Chehalis Reservation and the Bureau of Indian Affairs, 1996); Elizabeth Checchio and Bonnie Colby, *Indian Water Rights: Negotiating the Future* (Tucson: Water Resource Research Center, University of Arizona, 1993); Daniel McCool, "Intergovernmental Conflict and Indian Water Rights: An Assessment of Negotiated Settlements," *Publius* 23 (Winter 1993): 85–101; Daniel McCool, "Indian Water Settlements: The Prerequisites of Successful Negotiation," *Policy Studies Journal* 21, no. 2 (1993): 227–42.

13. The best known is Roger Fisher and William Ury, *Getting to Yes* (New York: Penguin Press, 1981). Also see Howard Raiffa, *The Art and Science of Negotiation* (Cambridge, Mass.: Harvard University Press, 1982).

14. James Watt, Secretary of the Interior, Department of Interior news release, 8 December 1982. While talking of being "good neighbors," Watt called reservations examples of failed socialism.

15. David Getches, *Forward to the Tribal Water Management Handbook* (Oakland, Calif.: American Indian Resources Institute/American Indian Lawyer Training Program, 1988), xv.

16. Christopher Kenney, "The Legacy and the Promise of the Settlement of Indian Reserved Right Water Claims," *Water Resources Update* 107 (Spring 1997): 22.

17. This settlement had to be renegotiated in 1984 (P. L. 98–530) and was amended again in 1992 (P. L. 102–497).

18. Fort Peck–Montana Compact, 15 May 1985.

19. An amendment to the San Carlos settlement was also signed in 1994 (P. L. 103–435).

20. Department of the Interior, press release, 10 December 1999.

21. Anonymous interview by the author. Much of the negotiation and implementation of Indian water settlements remains ongoing, and participants in these discussions often prefer that their comments remain anonymous so as not to affect their participation in subsequent negotiations.

22. *Federal Register* (12 March 1990) 55, no. 48, 9223.

23. Office of Management and Budget, accounts 34020 and 34420; Budget Office of the Department of the Interior, 10 June 1996.

24. Anonymous interview by the author. See note 21.

25. Anonymous interview by the author. See note 21.

26. P. L. 97–293.

27. P. L. 102–441.

28. Reid Peyton Chambers and John Echohawk, "Implementing the Winters Doctrine of Indian Reserved Water Rights: Producing Indian Water and Economic Development without Injuring Non-Indian Water Users?" *Gonzaga Law Review* 27, no. 3 (1991/1992): 449.

29. *Congressional Record—House,* 22 October 1990, H11498.

Agricultural Conundrums

8

Water, Sun, and Cattle

The Chisholm Trail as an Ephemeral Ecosystem

James E. Sherow

Try to imagine the Chisholm Trail as an ecosystem rather than simply as a pathway for Texas cattle herds.[1] Think of ecosystems as dynamic communities of life and physical forces. In this light, the Chisholm Trail was a short-lived system, an ephemeral ecosystem bridging a previous one largely shaped by the presence of Indian peoples and a later ecosystem formed when farmers dominated the landscape. How life and physical forces came together to shape the Chisholm Trail ecosystem and how the trail served as a bridge between the former and latter ecosystems are the subjects of this story.

Daniel Botkin once astutely wrote, "Life and the environment are one thing, not two, and people, as all life, are immersed in the one system."[2] From 1860 to 1885, Texas drovers endeavored to control the water and solar energy reserves of the Chisholm Trail in culturally shaped pursuits. Doing so, they altered the dynamic properties of the Chisholm Trail environment. These changes affected more than just the plants, animals, and water resources of the trail; they also shaped the culture and lives of the people who occupied the region.

The ecosystem of the trail encompassed vastly more territory than a narrow cattle path leading from central Texas to the railheads in Kansas. The southern portion was a huge expanse of rangeland in central Texas that nourished millions of cattle and horses. In the middle portion were lush grasslands and ample water sources throughout Indian Territory, or present-day Oklahoma, both essential elements in the fueling of the herds driven north.

In many places, the north-south corridor itself embraced a swath more than fifty miles across. In some spots, millions of cattle and horse hooves beat the trail as bare and as hard as a modern concrete highway. To graze cattle meant leading herds well away from where the routes had become barren rangelands. Under such conditions, prairies, once luxuriant with nutritious little bluestem, the mainstay fuel for cattle and horses, quickly became sparse pastures of buffalo grass. Although buffalo grass and little bluestem possess roughly equal nutritional values, the former never produces in the quantity of the latter.[3] Of course, quantity of forage was an overriding consideration to drovers when hundreds of thousands of animals were relying upon the same sources of grass.

In the summer of 1880, Joseph McCoy commented in his diary on the broad extent of the trail. He noted how drovers in Indian Territory, while awaiting orders from their Texas bosses, routed herds far away from the trails to secure good grass and water. The valleys and flats near the mainline of the trail showed the effects of overgrazing, and this forced drovers to take their herds to the uplands in search of "sagegrass," or little bluestem.[4] On these grounds the cattlemen ran the risk of poor water supplies, but they lacked alternatives. There was simply not enough grass and water to supply simultaneously all of the herds moving through the same vicinity.

At the northern end, the Chisholm Trail ecosystem braced the grasslands and water courses near and west of the major cattle towns in Kansas. These open ranges sustained herds until the cattle were loaded onto the transportation systems leading to eastern markets. The way in which drovers used the broad prairie flats near the Great Bend of the Arkansas River to graze their herds before taking them to Wichita, Kansas, illustrates the point. In 1874, drovers wintered more than 40,000 cattle between Great Bend and Wichita in anticipation of spring markets.[5]

In a similar manner, the wide Smoky Hill River Valley provided water and grass for herds loading onto shipping points at Abilene, Hays, and Ellsworth, Kansas. The Kansas Pacific Railway Company published a pamphlet devoted to extolling the grass conditions in the valleys. The company spent considerable sums of money distributing the publication among Texas drovers.[6]

On the mixed-grass prairies, water, the solvent of life, was always limited and localized and varied in quality. Streams' courses flowed intermittently and in variable volumes depending on chaotic weather patterns. Moreover, great stretches of the prairies had little surface water, and the life thriving on free-flowing water naturally congregated in or near those sources. Indian peoples, cattlemen, and farmers always sought a reliable source of water to sustain their respective material cultures.

Whenever possible, drovers avoided going more than a day without watering their stock. Richard Withers, an experienced cowboy, knew the great difficulty in herding thirsty cattle. He likened driving a water-deprived herd that had caught the scent of alkali springs to handling "a bunch of mixed turkeys." The best any cowboy could hope for in such a situation was to catch up with the herd and mill the animals once they had quenched their thirsts.[7]

Summer herding in the proximity of railheads always meant locating a "living" water source. In the summer of 1873, San Antonian C. W. Ackermann kept his father's herd on the Ninnescah (an Osage word meaning clear water) River for more than three months before loading the cattle some thirty miles to the north in Wichita, Kansas. Of course, no one else could use the same water and grass with someone like Ackermann and his longhorns there, and often Indian peoples and farmers were placing overlapping demands on these same resources.[8]

The ecodynamics of trail driving affected scores of interrelated systems—biotic, physical, and cultural. On the mixed-grass prairies, the plants acted as one vast solar collector and animals consumed the stored energy in the stems and leaves. Essentially this was a biosystem fed by a renewable solar energy supply collected through plant production. However, like any fuel, the consumptive demands placed on it could exceed its productive and reproductive abilities. Whenever this happened, certain consumers flourished at the expense of others.

Indian peoples, cattlemen, and farmers all understood how water and solar energy resources in any given area of the grasslands could support their individual needs, but not their collective ones. In the early 1860s, Indian peoples dominated the use of solar energy and water sources along and near the trail; by the 1880s farmers had taken over the neighborhood. In between, the great cattle drives along the Chisholm Trail created an ephemeral and transitional ecosystem, one temporarily spanning the prairie environments shaped by various Indian peoples and those later shaped by farmers.

The characteristics of this ephemeral ecosystem can be illustrated by depicting the forces leading to its inception and to those shaping its historical life. The Chisholm Trail biome, or life community, was a grouping of plants, animals (especially longhorns, horses, protozoa), and humans, including most notably Indian peoples, drovers, merchants, and farmers. This community operated within a web of other physical and cultural forces such as rain, wind, fire, law, markets, and cultural worldviews. The dynamism of the Chisholm Trail ecosystem is illustrated by the plight of Indian peoples attempting to preserve their energy supplies; the conflicting claims to water and stored solar energy; the perilous intersection of markets and winter grazing in 1871; and the cultural collision between farmers, cattlemen, and Indian peoples as expressed through the management of prairie fires and "Texas fever."

Creating of the Chisholm Trail

In 1861, Col. William Emory found himself in a bind. He needed to escape northern Texas. His small command of Union soldiers, who had manned frontier forts, could not march east through hostile Confederate lands. Emory had to ride north to reach friendly Union territory, yet he had no clue where he could find good grazing and water for his mounts. He persuaded a noted Delaware chief, Capt. Black Beaver, to guide him and his troops north. Black Beaver knew an unimpeded way, one plentiful in grass and water, and he led Emory's command north out of Texas to an intersection with the Santa Fe Trail in present-day Kansas.[9]

In 1865, Jesse Chisholm, a man of Anglo and Cherokee parentage, built an Indian trading post in the shady grove of trees just above the juncture of the Little and Big Arkansas Rivers. The Wichitas and other Caddoan-speaking peoples, who had fled Confederate forces to the south, had already settled there. By the end of the Civil War, Chisholm had decided to use this location to launch a freighting and trading business with the Indian peoples living in present-day Oklahoma. His first trek south followed the still visible trace left by Emory's escape. Chisholm traveled during a rainy season and his wagons left deep, easily followed ruts along those of Emory's command. By the time Chisholm's mule skinners had driven their lumbering wagons to his trading post near present-day El Reno, Oklahoma, an easily pursued path cut north and south through mixed-grass prairies.[10]

Chisholm and Black Beaver both traveled this route because they knew its resources provided the water and solar energy, or grass, needed for their draft and pack animals. The route also provided a nearly level route with generally safe fords north and south. The way was also free of mettlesome woods. In a short time, these same features would attract the cattlemen of Texas.

Joseph McCoy understood the positive features of the trail quite clearly: "[The Chisholm Trail] is more direct, has more prairie, less timber, more small streams and less large ones, and altogether better grass and fewer flies—no civilized Indian tax or wild Indian disturbances—than any other route yet driven over, and is also much shorter in distance because [it is] direct from Red river to Kansas."[11] These energy-rich resources and geographic features would sustain the great cattle drives on the Chisholm Trail.

The territorial surveys of Kansas clearly illustrate the advantages of cattle driving along McCoy's route. In 1871, A. J. and E. L. Argell mapped Downs Township, Sumner County, which is in south-central Kansas. The map of this area shows features common to the entire length of the trail. The surveyors traced the main route of the Chisholm Trail on a rise between two creeks in a treeless grassland. No trees hindered the movement of cattle, pasture and water were on either side of the route, and traversing the rise gave drovers a clear view of the countryside for miles around.[12] Driving their herds ten to fifteen miles a day through such resource-rich grasslands allowed drovers to fatten their cattle while on the trail.[13]

Before 1867, Texas cattlemen had driven their stock north along what they called the Shawnee Trail. North of central Texas the trail led through Indian reservations to the southeastern border of Kansas, crossed the southeastern quarter of Kansas, and terminated in Sadilia, Missouri. However, farmers in Kansas, Missouri, and Illinois (where Texas cattle were fattened before heading to the Chicago stockyards), greatly feared Texas cattle would bring the dreaded "Texas fever" to their domestic herds.

Several times before this disease had decimated domestic cattle herds throughout the Midwest. A microscopic protozoa *(Pyrosoma bigeminum)* lived inside ticks that infested Texas longhorns. Texas cattle had acquired immunity to the lethal effects of these microorganisms, but short-horn domestic cattle were at great risk whenever a Texas tick found a home on their hides. Midwestern farmers, like everyone else at the time, misunderstood the cause of their cattle's deaths. They did understand, however, that their cattle died in large numbers whenever Texas cattle were in the vicinity.[14] Predictably, Midwestern farmers and ranchers violently resisted the presence of Texas herds.

The cattle drives of 1866 graphically illustrate the harsh situation faced by Texas drovers when they encountered too much water. While driving their herds across the swollen Red River, several cowboys drowned. After fording the river, one outfit encountered three days of driving thunderstorms and pounding hailstones. The cattle stampeded, and the cowboys toiled to keep the terrified animals rounded. Some hands found the work too onerous, and they simply turned their horses south and headed home to Texas.[15]

Those drovers who continued on to Missouri faced vigilantes on the southern border of Kansas. With hickory switches in hand, these farmers tied the Texas drovers to trees and whipped their backs into bloody pulps. These same farmers, along with Indian peoples to the south, burned the grass along the Shawnee Trail. In essence, they destroyed the solar fuel supply for any herds that could have followed. The drovers who finally made it to Missouri or to Illinois lost heavily in terms of animals and money. One Texan, W. W. Sugg, discussed his situation with an Illinoisan stock buyer, Joseph McCoy, who undertook an alternative for these Texans by developing the Chisholm Trail to Abilene, Kansas.[16]

Markets, a derivative of culture, and the geography of Black Beaver's and Jesse Chisholm's route had joined to shape the historical Chisholm Trail ecosystem. Market forces combined with rain, wind, and sun as powers mixing with grass, water, and animals. Markets and geography lured McCoy to Abilene, and the Texas cattle followed.

The intersection of markets, animals, geography, and climate would make some Texas ranchers wealthy. Col. Dillard R. Fant of San Antonio was one such fellow. During fifteen years, he drove more than 200,000 cattle to markets. Notice the word "markets," which means more than just

Chicago. Texas drovers sent herds to California; Indian reservations throughout the West; port cities like New Orleans; and they even supplied markets in Latin America, like the one in Havana, Cuba. Fant drove longhorns to Kansas, Nebraska, and Wyoming; supplied agencies at Yankton and Standing Rock; and filled military contracts at Fort Reno and Fort Sill. In 1884, he employed more than two hundred men with more than 1,200 saddle horses to drive 42,000 longhorns to Wyoming.[17]

Certainly, Fant had profited very nicely from this trade. But cattle driving had more than economic or social effects. Fant's herds and those of other drovers became intertwined in the ecosystem dynamics of the grasslands, and this vibrancy worked as a feed-back loop that influenced culture and economics.

Markets and Winter Grazing

The confluence of economics, cattle, and weather in the grasslands did not always work out as well for others as it did for Fant. The largest cattle drive to Abilene and its surrounding markets occurred in 1871. In Kansas, Texas drovers had somewhere between 600,000 and 700,000 beeves to place in the market. However, several forces worked against the shipment of these animals. Heavy rains fell throughout the spring and into the summer months with the exception of June, which was a little drier than it had been in the previous few years. As a result, the grass grew "coarse, washy, and spongy," and its nutritional value for cattle plummeted. Worse, herds frequently stampeded during these spring storms. Consequently, these animals failed to fatten and fetched low prices for their owners.[18]

By late fall, stock prices were still flat. Some drovers and local farmers considered wintering approximately 350,000 beeves in the anticipation of better prices in the spring. Previous experience indicated such a strategy could prove profitable. In the *Abilene Chronicle,* one writer using the pseudonym "Ibex" encouraged farmers to buy unmarketed cattle in the fall and to winter these animals on prairie grasses. By early spring, Ibex predicted, farmers could double or triple their investment. Better yet, Ibex cheerily proclaimed, this strategy required "little labor and care."[19] But the winter of 1871–1872 proved otherwise.

Wintering a large cattle herd required hard work. Drovers had to pick a suitable location, one with reliable water, good grass, and protection for workhands during inclement weather. Often, drovers picked riparian ecosystems in the middle and western extent of the Solomon, Smoky, Saline, and, later, Arkansas River Valleys. Cowboys carved dugouts in the sides of stream banks where they would while away long winter nights. They cut and cured riparian grasses for winter feed to supplement cattle grazing on the short buffalo grass. Typically, cowboys stacked about a ton of hay per animal. The hands fire-guarded a large circular area where they could drive and mill their herds in case of a prairie fire. Finally, they stockpiled their own food supplies and corn and oats for their horses.[20]

In the fall of 1871 fire swept across the bluestem prairies. This forced drovers farther west onto the short-grass plains than they would have preferred. Moreover, many of these same drovers, lacking contracts for their cattle, had taken the risk of grazing their animals on their own accounts. Poor grazing conditions and weak markets left men and cattle in a precarious situation. Hardly had the drovers finished their dugouts when a freezing storm swept the short-grass prairies. Sleet fell and collected as a solid sheet of ice two or three inches thick, effectively denying the cattle any fodder. Strong gales made it impossible for cowboys to bring their wind-driven herds to where they had hay stored.[21]

This blizzard lasted for a full three days and killed thousands of cattle, hundreds of horses, and several cowboys. Herd owners suffered staggering losses. McCoy estimated that more than 250,000 cattle perished. One firm working on the Republican River began with more than 3,900 head; by the spring, cowboys found only 110 head remaining. Many drovers had stripped the hides from their dead cattle, and that spring they shipped more than 100,000 hides to eastern markets.[22]

D. S. Combs of San Antonio took a herd of steers from San Marcos, Texas, to Red Cloud, Nebraska. Even bypassing the glutted Abilene market did not help, and he tried to winter graze his herd. The November blizzard destroyed him financially. Of course, his cattle did not do well either. The ice had deprived his cattle of water and grass. "Our cattle," he later recalled, "literally starved to death, snow covered the grass and the water froze so they could not drink. I left in the spring, a busted and disgusted cowman."[23] A winter remembered by McCoy as one in which many cattlemen "met reverse, loss, and financial ruin" actually encouraged others to fill the market void created by their misfortune. The storms of November 1871 should have served as powerful warning against wintering cattle on the open short-grass range, but cattlemen had weak collective memories, as the destructive, open-range blizzards of 1886–1887 would later show.

Indian Peoples and Energy Stores

In his promotion of the Chisholm Trail, McCoy had underestimated Indian peoples' resistance to the passage of Texas cattle over their lands. Comanches, Kiowas, Southern Cheyennes, and Osages, for example, maintained large horse and cattle herds. Some, like the Kiowas and Southern Cheyennes, engaged in profitable freighting businesses and used the Chisholm Trail as their major avenue of commerce. This meant that they relied upon the same water and fuel sources as did Texas drovers.

Sometimes drovers could reach amicable terms with Indian peoples while crossing Indian Territory. Quannah Parker, a renowned Comanche and a fairly prosperous rancher in his own right, often met drovers entering Comanche territory. He told them that his wife (actually Quannah had several wives) was hungry, and Quannah often received one or more beeves from Texas drovers. This was Quannah's way of exacting a toll on the drover's herds crossing the Comanche Reservation.[24] In fact, Quannah Parker and George Bent of the Southern Cheyennes often worked alongside drovers to arrange rights of passage for longhorns or to negotiate grazing leases with Texas ranchers.[25]

All went well for drovers and Indians alike when the tolls were paid. A trail boss seldom had to give more than a few beeves from his herd to meet the tax. Such was the experience of George Saunders while negotiating with several Comanches in the summer of 1884. In gratitude, the Comanches even helped Saunders and his cowboys swim the herd across the Canadian River after Saunders had paid them a horse and "some provisions."

In fact, the drovers and Indians were all having such a good time that they invented and played a fantastic game. Several Comanches would ride past the cowboys and dare the Texans to lasso them. The speeding warriors continued this for some time until Saunders managed to catch one fellow and throw him to the ground. Most of the cowboys feared a fight next, but Saunders and a few others rushed to the Comanche, loosened the rope, and prayed for the warrior to regain his breath. All heaved a sigh of relief when the man burst out laughing for being caught, and Comanches and Texans reveled in the hilarity of the moment.[26]

At other times considerable friction arose between drovers and Indian peoples. This same outfit of Texans later greeted around two hundred Kiowa men and women. T. T. Hawkins, a cowboy with Saunders, remembered the Kiowas as being in an "ugly mood." Unbeknownst to the Texans, the Kiowas had many reasons to be sullen. They watched longhorns mow flat some of their prime grazing lands, which deprived them of the collected solar energy needed for their extensive horse herds. A succession of agents had been suppressing their Sun Dance, a ceremony crucially important to the cultural well-being of the Kiowas. Bison, an important sacred symbol as well as a source of subsistence, were a rarity. In part rightly so, the Kiowas blamed the Texans for the disappearance of the bison. All together, these factors made an already difficult transition to reservation life more painful for the Kiowas.[27] More than two hundred warriors led by Bacon Rind and Sundown unsuccessfully demanded beeves from Saunders. An armed standoff left both sides jittery. The Kiowas left Saunders and killed ten beeves in a nearby herd for retribution.[28]

In 1869, cowboy L. D. Taylor of San Antonio remembered a time that "four hundred" Comanches simply helped themselves to twenty-five beeves in the herd while he and other cowboys nervously watched.[29] In 1868, while following the Arkansas River toward Wichita, Kansas, M. A. Withers encountered thirty Osages who liberally helped themselves to the cowboy's barbecued meat and mess provisions. In addition, the Osages demanded and received one head of cattle.[30] Numerous accounts like these attest to the efforts of Indian peoples to exact compensation for the destruction or use of their energy supplies.

Prairie Fires

In consideration of their respective solar energy needs, the people of the Chisholm Trail ecosystem calculated the effects of prairie fires. They took a varied approach toward prairie fires depending upon their cultural views. For example, Indian peoples used fires to maintain their herds of mules and horses and to promote rich grasslands for the herbivores they hunted.

Farmers both set and fought prairie fires to protect their crops. Drovers fought prairie fires to meet the grazing requirements of their cattle herds. All of these people had differing views of prairie fires, and the gradual extinguishing of prairie fires was, again, the powerful force of culture working to modify geography. In other words, the landscape came to reflect the values of those people who occupied it.

In the 1860s and 1870s, many people in the Chisholm Trail ecosystem clearly understood the connection between Indian peoples, prairie fires, and grasslands. "Indian countries are *clean* countries," so one Junction City writer noted in 1870. This correspondent made the connection between Indian hunting and grass-fire practices. However, he misunderstood how burning contributed to a lush grassland biosystem, one that was rich in stored solar energy. Instead, the writer thought of these fires as a cause of "sterility" in grasslands.[31]

The reporter was certainly correct about one thing: Indian hunting and burning practices complemented each other. Kaws, Pawnees, Osages, and many other nations had their material culture mortally weakened in the wake of Anglo fire-suppression designs. Solar energy could then be employed in the interests of either farmers or ranchers.

Indians, ranchers, and farmers all needed access to solar energy. The ranchers used it in the form of grasses for forage. Farmers harvested it in the form of their crops, which displaced native grasses. Consequently, all these groups wanted a way to tap the solar energy basking the Plains, but they could not simultaneously collect this source in the same area given the different ways in which they harvested it. The eventual dominance of farmers meant the eradication of grassland biosystems, the elimination of Indian hunting practices, and the end of open-range grazing practices of Texas drovers.

Consider the way in which fire was managed by one group or the other to accomplish their respective goals. In the spring of 1866, Indian peoples burned the grass behind the first herds of Texas cattle. Two purposes were undoubtedly served: The fires would eventually enhance the grasses for bison in the summer, and the fire put a quick stop to drives through Indian Territory in the immediate future. In the fall of 1867, after the first herds reached Abilene, drovers reported the grassland was "black" back to the Texas border at the Red River. They blamed this burning on Indian peoples and settlers.[32] Again, these fires could put a quick, temporary stop to trail driving.

In 1871, grass fires contributed to the disaster that Texas drovers encountered that winter. For miles around Abilene, fires had burned off the mixed-grass prairies. This forced drovers to take their herds westward onto the short-grass prairies, upon which fires had made little progress. There, on the exposed High Plains, thousands of longhorns met their deaths in a frigid, icy embrace.

As their numbers swelled around the Abilene, Wichita, and other cattle towns, farmers burned the prairie grasses to protect their homesteads and to deprive Texas drovers of forage for their herds. Many farmers believed Texas cattlemen to be notorious for running their herds roughshod over crops. What better way to prevent this than to deprive drovers of the energy sources needed to fuel their herds? By 1873, drovers arriving in Wichita faced the unpleasant prospect of taking their cattle west as far as Great Bend to avoid confrontation with farmers.

The Ephemeral Ecosystem

Cattlemen, by moving their herds through the grasslands toward the shipping points at Abilene, Wichita, and Ellsworth, had altered the grasslands shaped by Indian peoples burning practices. This, in turn, had a significant effect on prairie biomes. As farmers entered the area, through their use or suppression of fire, they too shaped an entirely new biome, one that was neither conducive to open-range cattle grazing nor to Indian hunting practices. The long-standing biomes of Indian peoples, prairie grasses, and animals disappeared. The grassland biome, necessary to Texas drovers, gave way to agricultural biomes, which completely displaced the grasslands. In effect, the sort of short-lived ecosystems shaped by Texas drovers' and their longhorns intersected with two incompatible biomes, the prior one of Indian peoples and the latter of farmers. In the end, Texas drovers were denied their energy sources just as assuredly as were Indian peoples. For both, the results were the same: Indian peoples and drovers alike could not maintain their material culture.

The emergence of the farming ecosystem out of the previous ones is clearly depicted by a writer for the *Wichita Eagle* in 1872. The correspondent had counted more than 70,000 Texas cattle in a twenty-five-mile radius around Wichita. Herders were fattening these animals while waiting for the first opportunity to secure a good price at the stockyards. The columnist also simultaneously noted maturing corn fields in the region intermixed among luxuriant wild-grass pastures.[33] Within fifteen years, only one of these elements dominated the region: domestic grain fields.

After 1900, J. J. Roberts of Del Rio, Texas, traveled north and visited Wichita, Newton, and other former cattle shipping points on the Chisholm Trail. "[W]here the old Trail passed through in those early days . . . the change that meets your eyes is but little short of marvelous. Where saloons and dance halls stood are now substantial school buildings and magnificent churches and the merry prattle of happy children. And it was a deep feeling of pride that came to me, to know that I had had a humble part in bringing about this wonderful change."[34]

The change Roberts observed was the result of a farming culture, one, as he had correctly noted, that had its way paved in part by the cattle drovers of Texas. This old cowboy poem beautifully describes the end of this era:

> Cowboy rest, thy labor o'er.
> Sleep the sleep that knows no breaking;
> Dream of cattle drives no more,
> Days of toil and nights of waking.[35]

The Chisholm Trail ecosystem, a bridge between an Indian and farming landscape, was no more.

Notes

1. There are other ways of depicting the Chisholm Trail. For example, William Cronon, in *Nature's Metropolis: Chicago and the Great West* (New York: W. W. Norton and Company, 1991), 218–24, sees the great cattle drives as the creation of a "second nature." I think this is an erroneous reading and misapplication of Marxist ideology. See James Sherow, "An Evening on Konza Prairie," in *A Sense of the American West: An Anthology of Environmental History*, ed. James Sherow (Albuquerque: University of New Mexico Press, 1998), 15–16. Donald Worster simply contends that ranching has "had a degrading effect on the environment of the American West." Worster tends to see ecosystems as self-regulating entities of nature that more often than not are fouled by human activities, which are apart from nature. See Donald Worster, "Cowboy Ecology," in *Under Western Skies:*

Nature and History in the American West (New York: Oxford University Press, 1992), 34–52. To me, a more interesting story lies in the depiction of how forces blend to create vibrant ecosystems or of how these powers tear each other apart and debilitate life. There is no steady-state grassland ecosystem against which the environmental effects of culture can be measured. Rather, the effects of culture must be understood as culture intertwines with other forces to shape historically dynamic ecosystems.

2. Daniel Botkin, *Discordant Harmonies: An Ecology for the Twenty-First Century* (New York: Oxford University Press, 1990), 188.

3. B. W. Allred, "Bluestem on the Longhorn Trails," *Journal of Soil and Water Conservation* 5 (October 1950): 150–57, 198.

4. Joseph Geating McCoy, Diary, 1880–1881, MS 406, entries for 21–22 June 1880, Kansas State Historical Society, Topeka.

5. *Sumner (Kansas) County Press,* 3 July 1874, 3.

6. Kansas Pacific Railway Company, *Guide Map of the Great Texas Cattle Trail from the Red River Crossing to the Old Reliable Kansas Pacific Railway* (Kansas Pacific Railway Company, 1874).

7. Richard Withers, "The Experience of an Old Trail Driver," in *The Trail Drivers of Texas,* 2d. ed., ed. and comp. J. Marvin Hunter (Nashville, Tenn.: Cokesbury Press, 1925), 314.

8. Joseph G. McCoy, *Historic Sketches of the Cattle Trade of the West and the Southwest* (1874; reprint, Washington, D.C.: The Rare Book Shop, 1932), 121; C. W. Ackermann, "Exciting Experiences on the Frontier and on the Trail," in Hunter, *Trail Drivers of Texas,* 157.

9. John Rossel, "The Chisholm Trail," *Kansas Historical Quarterly* 5 (February 1936): 5–7; *Wichita (Kansas) Eagle,* 1 March 1890; "Beginning and End of the Old Chisholm Cattle Trail," *Kansas City Times,* 9 December 1924.

10. Ibid.

11. McCoy, *Historic Sketches of the Cattle Trade,* 23.

12. Territorial Survey Map, Township 32 South, Range 2 West, February and March 1871, Kansas State Historical Society, Topeka.

13. Mrs. A. Burks, "A Woman Trail Driver," in Hunter, *Trail Drivers of Texas,* 296.

14. Cecil Kirk Hutson, "Texas Fever in Kansas, 1866–1930," *Agricultural History* 68 (Winter 1994): 74–104.

15. McCoy, *Historic Sketches of the Cattle Trade,* 122–23; see also Texas drover J. M. Daugherty's similar account, "Harrowing Experience with Jayhawkers," in Hunter, *Trail Drivers of Texas,* 696–99.

16. McCoy, *Historic Sketches of the Cattle Trade,* 122–23; see also the contribution of Texas rancher Col. J. J. Myers to the creation of the Chisholm Trail drives to Abilene in Hunter, *Trail Drivers of Texas,* 637–42.

17. Anonymous, "Colonel Dillard R. Fant: Sketch of One of the Most Prominent of All Trail Drivers," in Hunter, *Trail Drivers of Texas,* 515–18.

18. McCoy, *Historic Sketches of the Cattle Trade,* 226.

19. *Abilene (Kansas) Chronicle*, 2 February 1871; similar advice was given in the Topeka, Kansas, *State Record*, 14 June 1871.

20. McCoy, *Historic Sketches of the Cattle Trade*, 217.

21. Ibid., 226–28.

22. Ibid.

23. D. S. Combs, "Spent a Hard Winter near Red Cloud," in Hunter, *Trail Drivers of Texas*, 467–70.

24. John Wells, "Met Quanah Parker on the Trail," in Hunter, *Trail Drivers of Texas*, 162–68.

25. William T. Hagan, *Quanah Parker, Comanche Chief* (Norman: University of Oklahoma Press, 1993), 28–39; Donald J. Berthrong, *The Cheyenne and Arapaho Ordeal: Reservation and Agency Life in the Indian Territory, 1875–1907* (Norman: University of Oklahoma Press, 1976), 91–117.

26. T. T. Hawkens, "When George Saunders Made a Bluff Stick," in Hunter, *Trail Drivers of Texas*, 390–96.

27. James Mooney, *Calendar History of the Kiowa Indians*, intro. John C. Ewers (Washington, D.C.: Smithsonian Institution Press, 1979), 352–54.

28. Hawkens, "When George Saunders Made a Bluff Stick," 390–96.

29. L. D. Taylor, "Some Thrilling Experiences of an Old Trailer," in Hunter, *Trail Drivers of Texas*, 501.

30. M. A. Withers, "Killing and Capturing Buffalo in Kansas," in Hunter, *Trail Drivers of Texas*, 97.

31. *Junction City (Kansas) Union*, 19 February 1870.

32. *Junction City (Kansas) Union*, 2 November 1867.

33. *Wichita (Kansas) Eagle*, 6 September 1872.

34. J. J. Roberts, "Fifty Years Ago," in Hunter, *Trail Drivers of Texas*, 785–86.

35. C. S. Brodbent, "Lost Many Thousands of Dollars," in Hunter, *Trail Drivers of Texas*, 594.

9

Private Irrigation in Colorado's Grand Valley

Brad F. Raley

Historians often depict the American West as a colony for eastern capital. Walter Prescott Webb, William G. Robbins, and others argue that nineteenth-century western communities were controlled, not by local farmers and ranchers, but by investment bankers in Chicago and London.[1] Clearly, eastern capital financed many western enterprises and natural resources did flow eastward. Although these theories may apply to such extractive industries as mining and timber, irrigated agriculture offers a different twist to this story, suggesting that historians cannot ignore the role local communities played in their economic destiny. Instead of corporate managers exploiting small towns eager for population growth, sometimes it was the other way around.

Few enterprises required outside intervention more than did irrigation development in the arid West. Private irrigation companies routinely failed. Often connected to land-speculation ventures, private canal companies in Colorado and California succeeded in watering land adjacent to rivers, but found more extensive projects nearly impossible to construct at a profit.[2] Private ditch companies often served struggling farmers who lacked capital, experience, and income to fund irrigation projects. Early settlers to Colorado's Grand Valley also encountered this problem when they found agriculture virtually impossible without irrigation. When local capital failed to construct expensive canals, local boosters recruited and even exploited outside investors. The private canal companies in Colorado's Grand Valley failed to return a profit, however, primarily because promoters built the projects ahead of need, rather than responding to population pressures. Their major focus on constructing a canal opened more farmlands than the valley had farmers. Low population pressure depressed land prices and reduced potential revenue for the irrigation company, both in land and water sales. In the end, the local farmers gained ownership of a completed ditch built for them by outsiders and eastern capital.

Many western communities were highly risky ventures, and Grand Junction was no exception. The remoteness and extreme aridity of the Grand Valley emphasized its speculative nature. The rest of Colorado was extensively explored, mapped, and recorded by everyone from the early Spanish travelers to American government topographers. Almost all these travelers skirted the Grand Valley—the valley was more barren and arid than the rest of the Rocky Mountains and far more remote than the accessible and widely farmed eastern slope. Yet, when the government opened west-central Colorado to white settlement, several enterprising individuals chose to establish a community in the Grand Valley. But what made Grand Junction unique among its neighbors was that, whereas the others could exist as cattle or mining towns, Grand Junction could not survive without irrigation development. Community boosters needed irrigation canals to attract the stabilizing population they so desired.

For years following statehood in 1876, west-central Colorado remained the domain of several Ute Indian tribes. In September 1881, government officials opened the former reservation to white settlement. White Coloradans had complained loudly about the "troublesome Utes" after the so-called Meeker Massacre in 1879, when a small group of Utes killed an inept Indian agent and his family. In response, Indian agents deemed the Grand Valley unsuitable for a reserve and relocated the Utes to lands in Utah and southwestern Colorado. Impatient white settlers moved in without permission, and within days staked out a townsite at the confluence of the Grand and Gunnison Rivers. During that first year, the town added a newspaper, bank, store, and post office; what had begun as a company of six shareholders, by 1885 had grown to 850 people. The town was remote, even by western standards, but was fortunate to gain a narrow gauge connection with the Denver & Rio Grande by 1882.

The community's most valuable natural resource was the seemingly unlimited water supply in the Grand River. That water did little good within the river's banks, however, and soon after settling the small community,

Grand Valley residents turned their attention to constructing irrigation canals. The town founders, like many nineteenth-century Americans, believed progress meant attracting a stable population and developing an economic base. To do so, they firmly believed the valley's future lay in agriculture. The valley was capable of great production, but early visitors all understood that this would only happen with the creation of a series of irrigation ditches. Yet, building private irrigation projects in the Grand Valley proved much more difficult and expensive than initially expected.

The valley's first white visitors compared the Grand Valley to a desert, but they hoped irrigation could transform it into a productive garden. The inaugural issue of the *Grand Junction News* noted that the valley's agricultural potential was limitless. Edwin Price, the paper's editor, argued that the area could easily produce everything from wheat to melons. A Gunnison pioneer said of the Grand Valley: "Though a wilderness now, in five years it will be the nicest and best part of Colorado to live in."[3] George Crawford, one of the initial founders, reported that a friend had brought in a stalk of corn ten feet high, grown in unplowed land for only a month.[4] The *Greeley Tribune* reported that forty-pound watermelons and "two to three pound" potatoes grew in the Grand Valley with "no special care or attention given to them."[5] The valley's growing potential seemed clear. The only barrier was getting water on this fertile soil.

The volume of water was no problem. The Grand and Gunnison Rivers supplied a large amount of water to the valley. All visitors noted the plentiful supply of water available for potential irrigation. D. S. Grimes, a Denver fruit promoter, wrote that "the supply of water for irrigation, which is already causing serious apprehension in the minds of farmers living along the streams that flow east from the mountains, is in these western rivers inexhaustible."[6] A reporter from the *Denver Republican* noted that Grand Junction had many advantages that Denver lacked, including a river "several times bigger than the Platte."[7] *Grand Junction News* editor Price continually referred to the valley's "never-failing" supply of water which "largely exceeds the tillable land."[8]

For Grand Valley residents the difficulty was not finding water, but transferring it to the land—a problem that would complicate each of the four stages that defined the Grand Valley's development of private irrigation. During the first, farmers tried small, local, cooperative efforts aimed at irrigating land close to the river. When that failed to open enough acres to irrigation, a local entrepreneur named Matt Arch raised $200,000 to finance a more extensive canal. Arch succeeded in constructing part of the ditch, but had to recruit outside capital to continue construction. He did so when he brought in an agricultural journalist and promoter named William Pabor, who gave Arch connections to Denver capitalists. One of these investors, T. C. Henry, then bought Arch out and finished the ditch. Even though Henry and Pabor would complete the ditch, their company failed to make a profit and was eventually sold to the local farmers as a nonprofit corporation.

Early Cooperative Canal Building

All of the initial irrigation attempts were small cooperative projects designed to divert waters from the Grand River onto adjacent land. In October 1881, before the U.S. government opened the land to legal settlement, four men began digging the Grand Valley Ditch with picks and shovels.[9] They soon found this small ditch beyond their limited capital and engineering skills. However, not all early projects failed. In March 1882, a group of settlers led by J. P. Harlow and Patrick Fitzpatrick formed another company of twenty-one stockholders to build the Pioneer Ditch, which was completed that April.[10] It was a very small venture, six miles in length and able only to water a few acres close to the river, but it proved that irrigation was possible in the region. This was the easiest canal to build and the most accessible land to irrigate. The initial canals were limited in size due to limited equipment and engineering expertise. This attempt also proved how expensive such projects could be. The original incorporated value was $50,000, a steep price for such a small canal.[11] Irrigation had begun, but from this point forward it would be much more costly and difficult to water the remaining lands. The remaining project, the Pacific Slope Ditch, was constructed to provide domestic water to the town.

Canal construction took place haphazardly during the region's early history. The state created water districts in the area in 1887, but until then the farmers tapped the Grand River without any state supervision and with little local planning.[12] Area ranchers desperately needed irrigation canals to make their lands productive, but few had the resources to construct anything of size. To fund their fledgling operations, many worked as ranch hands in the Ouray area or for the railroad.[13] As a result, these early canals were substandard products, to say the least. Spring floods washed out unreliable headgates and flooded farmlands.[14] The Pacific Slope Ditch, built to supply domestic water, often flooded parts of the town and was rarely clean. The *Grand Junction News* reported that the "irrigating ditches on Main street and Colorado avenue are literally reeking with filth. The question of health and decency aside, we ought to have pride enough in our city to be certain that we produce a good impression on visitors."[15] As was often the case with initial community development, boosters often exchanged planning and quality for expediency. Valley residents desperately needed a more stable and secure source of irrigation water.

Grand Junction's leaders believed that a viable irrigation infrastructure would attract outside investment in their community, as well as future settlers. The land farther from the river appeared rich, but had very little value without water. They understood that the community had to attract attention or forgo the life-giving capital necessary for economic growth. The *Grand Junction News* editorialized "all that the Grand Valley wants now is the capital to build one or two large ditches, and then we can show to the people of the state that the reservation is worth redeeming from solitude. With the water on our lands, Grand Valley will make a showing in the agricultural yields of the state that will surprise the people."[16] If they did not attract the eye of settlers and investors, the population and capital would go elsewhere, namely Colorado's eastern slope or the San Luis Valley.

While valley residents believed irrigation would come quickly, their optimism often overshadowed common sense and illustrated how desperately they desired good canals. In December 1882, the paper promised by spring of the following year workers would complete construction of a large ditch opening the entire valley to irrigation. From their vantage point, valley residents believed building such a ditch required only bridging gullies, "which are but few," and that once completed would open thousands of acres to farming.[17] With just a little ingenuity and hard work, residents believed, they could transform the valley into a garden. While a logical ambition for valley farmers, constructing the ditch was far more difficult from an engineering standpoint. In the end, it would require considerable technical expertise, vast financial resources, and a great deal of time.

That reality ushered in the second stage of irrigation development. In December 1882, while discussing the progress of the Pioneer and Pacific Slope Ditches, the *Grand Junction News* assured the valley that "a new ditch, to be made by local effort, [was] under contemplation; it [was] to head farther up the river, and [would] embrace a much larger scope of land, and extend further down the valley."[18] The Greeley paper opined that Grand Junction had two small ditches, "but needs a much larger one."[19] In November 1882, the Denver *Republican* predicted that promoters would build a large canal on the north side of the valley by spring, as the "prosperity of the valley" depended on it.[20]

In January 1883, Matt Arch, a young entrepreneur, bought the fledgling Grand Valley Ditch. The ditch, begun in 1881, was one of the early failed cooperative efforts. A western slope resident, Arch previously owned a ranch on the Tomichi Creek area near Gunnison, so he had some knowledge and experience with Colorado agriculture. He renamed it the Grand River Ditch and sold 20,000 shares of stock at $10 per share. He planned to divert water from the Grand River about fourteen miles above the town and build a ditch thirty-five feet wide and five feet deep that would thread through the north side of the valley.[21] Arch understood the importance of establishing irrigation in the valley: "This ditch will place a very large amount of land under irrigation, and make glad the hearts of many preemptors who have taken up land in the upper part of the valley. This will set at rest the many doubts about getting water on our land, and give new confidence to the people."[22] Undoubtedly, this meant land speculators sitting on dry land would see their land increase in value, both for farming and for resale. Arch also knew that this ditch would benefit the entire valley. True to the town's optimism, Arch promised that the ditch would pass the townsite in time for the next crop season.[23]

Arch wanted more from the ditch than just irrigation water. He designed several dramatic "drops," hoping to make the canal both aesthetically pleasing and multipurpose. The drops would feed water wheels to harness the canal's water for powering mills and other small industries, while also allowing locals to transport heavy timbers and trade goods.[24] Arch himself intended to build a flour mill on one of the drops. He also designed the drops to create some pooling, which the *Grand Junction News* thought would make excellent fishing holes, should "the company allow boats on the ditch."[25] Arch also hoped the canal would lure young lovers to walk under the Colorado moon.[26] This appealed to town promoters, who liked the idea that the irrigation canal could offer an aesthetic attraction to the valley.

Valley residents watched with interest as Arch began constructing this irrigation artery. Despite slow progress and financial difficulties, Arch raised enough money to continue ditch construction. In January 1883, the *Grand Junction News* reported that Arch had twenty-five teams of horses and nearly fifty men building the ditch, with enough funding subscribed to extend the canal twelve miles below the town.[27] However, the task at hand proved more difficult than expected. The size and length of the canal created unique problems. Occasionally workers had to blast through shale ridges with dynamite. The floor of the Grand Valley contains numerous small ravines and gullies common in an arid region. These topographic obstacles forced Arch to hire carpenters to build flumes over these ravines while he used horse-drawn excavators to dig the main canal. He also had to resurvey a portion of the canal when workers realized the first survey ran the ditch uphill.[28]

Arch was a capable promoter and understood the importance of maintaining community support. In February 1883, he toured the ditch with the local reporters. Although the mood of the report was upbeat, it was obvious that Arch faced opposition within the community. Few disagreed with the goal of irrigating the valley, but many chose to invest their money in town lots rather than putting it into a ditch venture widely considered impossible. Recognizing this, Arch invited the newspaper to observe his progress, and the *News* reported the ditch was progressing nicely and the community should rally behind it.[29] Price noted that those who wanted their town lots to increase in value should support the Grand River Ditch because only a strong agricultural base would help Grand Junction grow. He also warned that Arch's project was the valley's best chance for a professional and competent canal system. Need he remind residents of the poorly constructed and polluted ditches already built?

Valley support came more through verbal encouragement than financial assistance. Locals desired a ditch, but understood it to be a shaky investment. Struggling to maintain funding, Arch turned to another agricultural promoter for assistance in developing the ditch. William Pabor, a noted promoter, town builder, and agricultural journalist had recently relocated to the Grand Junction area in search of land for fruit growing. Pabor immediately saw the benefits of the Grand River Ditch and invested. As the author of two books on Colorado agriculture, Pabor understood well the need for irrigation.[30] He was also an excellent choice to assist Arch in completing the canal. Not only was he interested in the valley's agricultural development, he was a seasoned journalist and promoter. In fact, Pabor was more successful at those jobs than at farming: Pabor's lone attempt to farm in Greeley, Colorado, had ended in bankruptcy.[31] In addition to his experience as an irrigation promoter, Pabor also had connections to Denver capital. He first visited the Grand Valley as an agent for the Colorado Loan and Trust Company.[32] Not surprisingly, Pabor became one of the Grand River Ditch's most prominent and vocal promoters.

Another important local promoter was *Grand Junction News* editor Edwin Price. He did not work for the ditch company directly, but he served as its public defender. As newspaper editor and area landowner, he assured prospective settlers that the Grand Valley was the best place to settle. The ditch's completion would increase both land values and agricultural production. But beyond his role as booster and promoter, Price obviously felt the community was at a critical juncture in its young history. How many more failed efforts could the area absorb before settlement and investment simply moved on to neighboring towns or states? In the 24 March 1883 issue, he wrote: "In the face of all this what remains for the people of Grand Junction? Here are the facts as they are today: 1st. The Grand River Ditch or nothing. 2nd. Two-thirds of the entire work completed and all owned by men who propose to stay in this valley. 3rd. The necessity of completing an enterprise which will endure as long as there is any productiveness in this soil or sunshine over head."[33]

Price exaggerated the ditch's progress, but only slightly. By the time he wrote, workers had dug most of twenty-two miles of the ditch. What remained was to construct flumes across the ravines and to frame in the drops. The ditch would not carry water without these fixtures, but the ranchers could see a possibility of water flowing to their land. Price's assertion that the ditch was "all owned by men who propose to stay in this valley" reflected more of an appeal to locals to support the ditch than an accurate reflection of ownership. It was one last rallying effort to get the valley behind the ditch before the canal company was forced to look elsewhere for funding. Irrigation, Price felt, was the key to the community's success, and he hoped the local landowners would be the main beneficiaries. Price argued that "to the people of this valley it [the ditch] will be the purse of Fortunatus in which will always lie a shining coin."[34]

Not all landowners agreed with Price. Arch and his ditch faced further difficulties when impatient landowners below town met to develop a competing ditch company.[35] In December 1882, J. E. Walls and other farmers incorporated the Independent Ranchmen's Ditch Company, but they did not begin construction until the spring.[36] Worried the Grand River Ditch would not reach their land soon enough for the season's crops, they began construction. They planned that the ditch would meet the Grand River just west of town, right below the confluence of the Grand and Gunnison Rivers; run parallel to the river bank; and provide irrigation for land right next to the Grand River.

Faced with the competing ditch and mounting discontent, in March 1883, Arch and his backers scrambled to maintain confidence in their project. The *Grand Junction News* rallied to Arch's defense, assuring the valley it needed to support the project. It also attacked the rival ditch builders by claiming the valley would be better served if everyone pitched in on the Grand River Ditch instead of building smaller, cheaper, and less extensive ditches. These quick fixes would not serve the community's long-term interests, opined the *News,* and would simply divert precious support, labor, and resources away from the all-important Grand River Ditch.[37]

Price then encouraged local farmers to plant their spring crops as "Matt Arch's ditch will go through *sure*."[38] At the end of the month, however, the paper had to admit the ditch company was now selling bonds to raise additional capital. Details were sketchy, but the *News* reported "a certain party has taken the $30,000 dollars of bonds now issued, and has contracted to take the remainder of the $75,000 as fast as they are issued." The valley soon would learn that outside investors were now funding their ditch, but Price, Arch, and Pabor believed its successful completion was of paramount importance. This arrangement, editor Price assured his readers, "insures money enough to push matters to completion."[39]

While Price, Pabor, and other Arch supporters had self-interest at heart when they promoted the Grand River Ditch Company over the Independent Ranchmen's, many of their arguments were valid. If the valley needed agriculture to attract settlers, then irrigation was necessary to achieve that goal. The Independent, much like the earlier Pioneer Ditch, was a small and very limited project that would help irrigate a small amount of land, but would not open up the valley for agriculture. If the valley really wanted to develop arable lands on a large scale, then opting for small, locally owned projects would not accomplish that goal.

Arch's financial backing, however, proved inadequate and he soon looked outside the Grand Valley for investors. In March 1883, he led a tour of Missouri businessmen interested in investing in the ditch.[40] According to eyewitnesses, the tour seemed cursed from the beginning. Roads to the ditch were impassable, and William Pabor, in an attempt to convince the businessmen to invest, entertained them with prophetic visions of the Grand Valley.[41] His unrealistic optimism soured the visitors and they left without investing in the ditch.

On 16 May, however, the community watched in approval as Arch opened the headgate on the Grand River and turned water into the ditch for the first time. The town's leading citizens had a small festival with plentiful food and spirits. "Governor" George Crawford, the town patriarch, spoke to the assembled "jolly" crowd and then they all watched as Arch directed the workers to divert the water into the ditch.[42] The *Grand Junction News* described the ditch builder as so "quiet" and "undemonstrative" that a stranger would have been unable to pick him out of the crowd. To those in attendance, however, Arch's accomplishment seemed a giant step for the valley. As Gunnison's reporter observed the "roaring waters" began their journey through the Grand Valley, he pictured "a former desert, but soon to bloom equal, if not superior to, that of Salt Lake City."[43] Price remembered as they drove back to town and passed over the ditch, they paused, watched the advancing water, and "left it creeping down the valley bringing with it the prosperity that shall know no end."[44]

The celebration masked the fact that Arch's ditch only carried water four miles, far fewer than had been planned. But the physical reality of water in the Grand River Ditch spoke volumes to skeptical settlers anxious for a reliable irrigation system. Despite this new credibility, the ditch lacked sufficient funding to complete construction, much less sustain its maintenance. In July 1883, the *News* noted that the company would have new management, but added hopefully that this change should not delay its completion.[45] The paper did not disclose details of the new management, but it was clear that the company was in financial troubles. By August, Arch could no longer pay his workers or feed the work stock.[46]

By the end of August, the depth of Arch's financial problems became evident. Arch's trip the previous March had been to discuss the ditch's financial woes with Pabor's old company, Colorado Loan and Trust, whose president and founder, Theodore C. Henry, was deeply interested in irrigation development. Locals, and even Price, believed that Henry's involvement occurred late in August after Arch's financial problems became too much to overcome with local assistance. More likely, Arch had maintained contact with Colorado Loan and Trust since he brought Pabor into the company. During the March meeting, Henry bought $30,000 worth of bonds in the company, with the promise of buying more when they became available.[47] Arch obviously hoped this amount would finance completion of the canal and allow him to repay his debtors and retain control of the company. Instead, late summer found Arch still attempting to build the trickier portions of the canal without incoming revenue. Faced with mounting debt and with completion nowhere in sight, Arch sold his interest in the company to Henry for $200,000 and left the Grand Valley.[48] His experience with the Grand River Ditch did not sour his interest in irrigation ditches; soon after, the *Grand Junction News* reported that Arch was the main contractor on a "big ditch" near Montrose.[49]

External Capital, New Management

William Pabor took the reins of the Grand River Ditch Company and assured valley residents his loyalties lay in Grand Junction, not Denver. He began writing a weekly newspaper column that promoted the agricultural potential of the area. In September 1883, Pabor announced he would allow local workers to complete the ditch.[50] This was no small matter, crowed the

News, as the canal required extensive carpentry work alone. The necessary fluming, according to Pabor, would require "338,000 feet of lumber and 300 piles to complete it."[51] The *News* announced the company had solicited bids to furnish the lumber and the canal engineer had already made a trip to Marshall Pass to find the red cedar necessary for the piles. Pabor evidently had friends in the Denver & Rio Grande Railroad as well, because that company had sent a pile driver for the ditch company's use.[52] Farmers would win both ways: first by making a good wage working on the ditch and then again when their farmland improved in value and productivity.

With Pabor's management and Henry's financing, construction on the Grand River Ditch continued. Yet its earthen canals were prone to seepage and leaking. In 1884, the managers watched cautiously as water crept through the newer part of the ditch. The *Grand Junction News* noted its progress was as tentative and guarded as an infant's: "It [the water] bubbles up in places sixty feet away from the ditch and seems to penetrate the banks that are in place more readily than those that are made. The shale there is very open and filled in with alkali which dissolves readily. Mr. Harper [one of the ditch engineers] thinks it will settle down and stop leaking in a few days." They noted further that in some places the ditch had simply sunken in, reminding observers of a miniature chasm.[53] To remedy this, the crew "puddled" those seeping spots by soaking the ground until the soil settled and no longer leaked. In addition, the canal had difficulty handling flood waters and washed out in spots. Overflow channels had to be cut out, all of which added to the canal's construction costs.

T. C. Henry left construction to Pabor, but managed the company's finances from Denver. Although valley residents viewed Henry as a wealthy investor and thought the ditch's financial problems in the past, he was more a promoter and financier. That is, he had very little money of his own. When he assumed ownership of the Grand River Ditch, he inherited all of the Arch's financial difficulties. He still needed to finish the construction phase to have any hope of returning a profit. The company's debts were mostly in bonds, which had no value unless the ditch became profitable. To achieve that end, Henry attempted to raise the price of water to the farmer. This was a logical step for the investors, but among valley farmers it was a highly unpopular move. Along with their financial difficulties, Henry and Pabor also had to face competition from another ditch company. In the fall of 1883, angry at the higher costs and intrusive nature of the Grand River Ditch, several ranchers met to form the Pioneer Ditch Extension Company. Incorporated on 15 December 1883, "with stock set at $100,000 divided into 10,000 non-assessable shares at $10 each," the company aimed to extend the old Pioneer Ditch for irrigating and milling purposes.

These ranchers objected to Henry's attempt to sell the water rights, and they believed that soon the ditch company would own both the land and water on the north side of the Grand River, giving it too much leverage over the farmers. Those who owned land under the ditch needed irrigation water to make their land profitable. However, the investment necessary to buy the water right was more than the cost of the land itself. The critics contended the Grand River Ditch Company would sell the right for eighty acres of land for $800 dollars, or would "take a mortgage on your 160 acres for it. The result will be that in a short time they will own both the water and your farm."[54]

The *Grand Junction News* cautiously reported this conflict. As an instrument devoted to promoting the valley, the paper hesitated to alienate any potential investors, but it also needed to assuage local fears. The *News* conceded that critics of the now-outsider-financed ditch presented some critical issues, but noted the criticism was also symptomatic of "that spirit of opposition to capital that always stands up before any large enterprise." Editor Price was clearly troubled by the allegations, noting if they were true, the company "ought to be run out."[55] Yet he just as clearly believed that the outside capitalists and the Grand River Ditch Company they now owned were markers of progress. Not wanting potential settlers in Kansas or Denver reading about impending class warfare in the Grand Valley, Price continued to point out the importance of the ditch and the benefits of irrigation. "There is not an acre of level land under the present ditches that will not be worth $50 in five years and some acres will be worth $500," he wrote. When linked to a superior market and climate, the ditch would insure prospective and current farmers sizable profits if they would invest in irrigation. Price assured his readers, too, that water rental rates in California were as high as $5 per acre, which only proved that the Grand River Ditch Company's price of $2 per acre was more than fair.[56] For him, the Grand Valley was still the best opportunity in the West.

When the rebels formed their own company, they organized it quite differently from the larger Grand River Ditch. Ranchers, led by J. A. Hall, signed an agreement requiring them to perform a share of the labor needed to finish the Pioneer Extension Ditch. That share of labor was directly connected to how much land each owned under the ditch and how much water they would require. This was the formula they had hoped the Grand River Ditch Company would have adopted: a system that was inexpensive, locally controlled, and benefited only those who already owned land along the ditch, a system that did not provide easy money for speculators or enrich urban capitalists.

The Pioneer Extension Ditch was completed in the spring of 1884 and provided irrigation water for limited acreage near the Grand River.[57] Because the population of the area was still very small, most of the farm land lay under those small ditches. This explains part of the opposition to the larger Grand River Ditch; many owners of small farms questioned the need for the larger project. The small cooperative ditches provided adequate water to their lands without much expense or irritation. Farmers with lands near the river had little need to pay expensive water rates when they could easily tap the river through their small canals. Theirs was an understandable dilemma. Why should they finance irrigation for potential settlers when the existing system already served their interests?

When spring floods wiped out the headgates of the Pioneer Extension and Mesa County Ditches, however, these landowners realized the difficulty of relying on small cooperative enterprises.[58] The high water also washed out the Pacific Slope Ditch, sent water into the streets of Grand Junction, and flooded several ranches.[59] The floods emphasized the need for a professionally constructed and durable source of irrigation water and put the pressure on T. C. Henry and William Pabor to complete the Grand River Ditch.

While the other ditches were out of commission, Henry and Pabor pushed construction and completed the Grand River Ditch by early summer 1884. Canal construction had taken longer than expected, but was completed in nearly two years. The canal ran for twenty-four miles from its headgate on the Grand River to the Big Salt Wash just west of Fruita. Despite successfully completing the canal, Henry was unable to turn a profit. Moreover, although the completed ditch earned him popularity within the Grand Valley, his critics in the Travelers Insurance Company as well as other former business associates challenged his managerial abilities. Faced with this opposition and in obvious financial difficulty, on 1 November 1884 Henry sold his shares in the ditch to the Travelers Insurance Company. His replacement was Gustavus F. Davis, who represented the insurance company in the area; he, in turn, hired Julius White to manage the company, leaving him with the unenviable task of appeasing angry valley residents. Part of White's strategy for stabilizing Grand Valley's water delivery system was to unify its several small ditches into one comprehensive irrigation project. In this, he found an ally in the valley's most prominent citizen, George Crawford, who shared his belief that consolidation under single management would mitigate resentments between irrigators over water prices and availability and focus the valley's efforts on agricultural growth. They also felt the best way to address the financial needs of the ditch companies and the comparative poverty of the water users was to sell the water rights to users rather than charge them an annual rental fee. The cost would be linked to the amount of land to be irrigated, one share to equal one acre's worth of water. After paying a one-time fee, the water users would then only have to pay an occasional assessment to cover ditch upkeep and improvement.[60]

This idea proved a workable solution to irrigating the valley. There were too many small irrigation projects in the valley to sustain. If one ditch offered water at lower prices, it would cause animosity among the other farmers. The fact that the smaller ditches were constructed more cheaply and could offer less expensive water was irrelevant. Combining the canals under one Grand Valley system would, in the minds of the mediators, promote a common interest among the farmers. The plan's attempt to grant perpetual water rights also appealed to those who wanted to give the water users a more stable connection to their land. All users would share in the cost of maintaining the canals, but would also have a vested interest in protecting the water source.

However successful, consolidation did not mean that valley water conflicts disappeared, they simply went into remission. Local papers, remaining true to form, hesitated to mention any conflict that would make the valley appear less attractive. During the spring of 1885, however, the *Grand Junction News* reported that a Denver judge had been called to adjudicate a difficulty between the Travelers Insurance Company and the Mesa County

Ditch Company.[61] Evidently, the insurance company had attempted to fill in the Mesa County Ditch where it crossed Travelers's property lines. The water users threatened violence if this continued and the ditch company attempted to stop Travelers's agents from altering their water flow in any way. Local farmers viewed this as a clear example of an external company intruding on the rights of local, small landowners.[62] The judge ruled the insurance company was within its legal rights and ordered the Mesa County Ditch Company to stop interfering. Travelers's agents declined to use the judicial order, realizing that, although their actions might have been legally sanctioned, they lacked public support. The *News* noted wearily, "There is too much water here, too much work to do to spend time in ugly fights."

By 1888, the consolidation of the ditch systems was complete and uncontested, but the economic operations of the Grand River Ditch system remained troubled. One manager attempted to deepen the canal to add to the water supply, but this failed to resolve the company's financial problems. The summer of 1888 added another test of the company's strength: It was long, dry, and hot and tested all claims about the unlimited availability of water in the Grand Valley. Confronted with a shortage, the ditch manager attempted to increase the amount of water diverted from the Grand River. The extra costs involved only worsened the company's difficulties, making it even more unlikely that its investors would see any profit. An advertisement in the local newspaper in December 1888 noted there would be "no opportunity to work out assessments, or payments on contracts on the Grand River Canal system. All persons in arrears on assessments or contracts, must pay the same in cash immediately."[63]

The Grand River Ditch Company fought an uphill battle to make the operation profitable. The agricultural base of the area depended on a stable source of irrigation. The valley residents, however, had built canals to open potential farmland, rather than to supply existing farmers. As a result, the valley had more irrigation than farmers, a trend common throughout the state.[64] As the local newspaper constantly discussed, the valley needed additional farmers. During an early crisis of community support, the *News* reported that those farmers below town were welcome to plant on lots upstream if the Grand River Ditch did not reach their lands, a clear implication that there were many vacant farms under the ditch. The 1890 census reported that only 7.54 percent of Mesa County's population owned farms.

Of the 1,920,000 acres in the county, only 13,798 were under irrigation. Whereas irrigators represented only 7.28 percent of the population, they constituted 96 percent of farm owners.[65] Irrigation dominated valley farming, but the population had not caught up with the new ditch. Although this may have represented planning for the future, it made paying for existing projects impossible. Even if all farmers under the ditch paid their share, the incoming revenue would fall far short of the canal's fixed costs.

As editor Price noted in July 1888, the canal had never made money.[66] Henry and Pabor had continued to put money into the project, often between $10,000 and $25,000 annually, with only $3,500 incoming revenue. On top of that, the county commissioners still demanded the company pay property taxes. In an interview with Price, the ditch manager admitted the only reason the bondholders continued to invest money for ditch improvements was to save a portion of their investment. Many of the outside investors, including the Travelers Insurance Company, also invested in land under the ditch. If the ditch failed to produce water, those land values stayed low, depriving the company of profit from either the land or the ditch company.

The Drive for Local Control

Finally, in 1888 a district court named a receiver to liquidate the Grand River Ditch Company. Travelers Insurance Company purchased the company outright at a public sale. It changed the name to the Grand Valley Canal Company, a subsidiary of Travelers Insurance. The next year, the insurance company sold most of the water rights to the Hartford Loan and Trust, also of Connecticut. In the space of two years, the Grand Valley had gone from having some limited authority over the Grand River Ditch to having no input whatsoever, a situation that precipitated yet another conflict regarding the company's operations.

T. C. Henry, although no longer involved in the Grand Valley, still claimed some right to the ditch, so he challenged its public sale in court. The judge ruled that neither the Colorado Loan and Trust nor the Connecticut firm had legal ownership of the ditch company. He ordered the ditch and its water rights returned to the ownership prior to Travelers's involvement and that the company then be sold at public auction. A local businessman, John P. Brockway, purchased the ditch for $10,000, then immediately sold it to the local water users and their corporation, the Grand Valley Irrigation Company, for $40,000, 360 acres of land, and two Grand Junction city lots.

The newly formed Grand Valley Irrigation Company included all the combined ditches: the Independent Ranchmen's, Mesa County, Pioneer Extension, and Grand River. The company divided its shares among the water users based on their contract under the previous system. Each water user's contract was assessable for maintenance and operations fees, with those fees adjustable by a company vote. Because the water users essentially owned the company, it became a nonprofit entity charged with delivering water to its stockholders.[67]

Grand Valley's difficult experience with irrigation schemes followed the statewide pattern articulated during the 1894 National Irrigation Congress. Held in Denver, the congress discussed the development of irrigation in Colorado and reported that, in general, the initial efforts at irrigation had been cooperative ventures funded by local farmers who pooled their resources and labor to irrigate those lands closest to a river.[68] The next step came when they realized that those lands farther from the river were also valuable and should be irrigated. To do so required the formation of corporations; the investment of capital; and, to recover costs and make a profit, the charging for water delivery. It "is undoubtedly true that the highest interest of farmers is in the ownership of their own canals," one analyst noted, "but it is also a fact that the great canals, which have required millions [of dollars] of capital to construct, would never have been built if the sale of water rights had not been permitted."[69]

Outside capital also was responsible for the completion of the Grand Valley project, but this did not entail—as it did so many other aspects of the developing western economy—that that capital, and its external owners, ended with control of the water and the irrigated land. Although local farmers and community boosters had been incapable or unable to invest in the project's expansion, they had convinced outside investors to underwrite the canals' construction. To cover construction and maintenance costs, these agents sought to rent water to individual users, but this proved a difficult way to make a profit, for nineteenth-century farming in the Grand Valley did not generate the kind of revenue stream from which investors could divert a profit. External forces such as the Travelers Insurance Company could only hope to generate income through the speculative purchase of land under the ditch. But because so many locals also speculated in this prime land, companies like the Travelers could not buy enough to cover their investment in the ditch company.

In contrast, the newly formed Grand Valley Irrigation Company, which succeeded Travelers, conceded profit and simply tried to bring in enough revenue to maintain the canal. This new form of management made the local farmers happy because it allowed for the maintenance of a canal far beyond their combined financial abilities, while keeping water affordable. Town boosters gained a canal they could use to lure prospective settlers, while local farmers retained control over their land. Land speculators benefited because their lands were now irrigable at a reasonable rate. Only those who had invested in the canals' critical construction phases had lost money; they never realized the kind of return they envisioned from the Grand Valley Ditch Company.

The complexities of western irrigation projects and the realization that private ventures proved so difficult to finance, especially evident during the depression of the 1890s, led the federal and state governments to experiment with alternative financing methods. California's 1887 Wright Act, the 1894 Carey Act, and the Reclamation Act of 1902 attempted to assist arid-land reclamation. Yet despite these interventions and the resulting sense that they reflected the colonization of the West and the subordination of local concerns to national (or international) economies, small communities nonetheless played active roles in determining their future. As in the Grand Valley, sometimes local farmers benefited at the expense of outside investors.

Notes

1. Walter Prescott Webb, *The Great Plains* (New York: Grosset and Dunlap, 1931); William G. Robbins, *Colony and Empire: The Capitalist Transformation of the American West* (Lawrence: University Press of Kansas, 1994).

2. For the difficulties of private irrigation companies in the nineteenth-century American West, see Donald J. Pisani, *From the Family Farm to Agribusiness: The Irrigation Crusade in California and the West, 1850–1931* (Berkeley: University of California Press, 1984), 78–101; Donald J. Pisani, *To Reclaim a Divided West: Water, Law, and Public Pol-*

icy, 1848–1902 (Albuquerque: University of New Mexico Press, 1992), 104–8; Robert G. Dunbar, *Forging New Rights in Western Waters* (Lincoln: University of Nebraska Press, 1983), 24–28.

3. *Grand Junction News,* 28 October 1882, 4.

4. Ibid., 1.

5. *Grand Junction News,* 16 December 1882, 1.

6. *Grand Junction News,* 28 October 1882, 4.

7. *Grand Junction News,* 11 November 1882, 1.

8. *Grand Junction News,* 9 December 1882, 1.

9. Don Davidson, "The Grand River Ditch: A Short History of Pioneering Irrigation in Colorado's Grand Valley," *Journal of the Western Slope* 1 (Fall 1986): 1–30.

10. Charles W. Haskell, ed., *A History and Business Directory of Mesa County, Colorado* (Grand Junction, Colo.: Mesa County Democrat, 1886), 4–6.

11. Davidson, "Grand River Ditch," 5.

12. Georgina Norman, "The White Settlement of the Ute Reservation, 1880–1885," (master's thesis, University of Colorado, 1955), 74. In 1883, the first Colorado State Engineers reports recommended that additional water districts be created, including one to encompass the newly settled Grand Valley.

13. Nancy Blain Underhill, "Trekking to the Grand Valley in 1882," *Colorado Magazine* 8 (September 1931): 181; Norman, "White Settlement," 74. Because most landowners practiced a mixed agriculture, the terms "rancher" and "farmer" are often interchangeable. Those who raised livestock also grew a variety of crops, and farmers supplemented their crops with a small herd of cattle, sheep, or hogs.

14. Mary Rait, "Development of Grand Junction and the Colorado River Valley to Palisade from 1881 to 1931. Part 1," *Journal of the Western Slope* 3 (Summer 1988): 18. This was typical of early community ditch building. See Robert G. Dunbar, *Forging New Rights in Western Waters* (Lincoln: University of Nebraska Press, 1983), 20.

15. *Grand Junction News,* 3 March 1883, 2.

16. *Grand Junction News,* 25 November 1882, 2.

17. *Grand Junction News,* 9 December 1882, 1.

18. *Grand Junction News,* 2 December 1882, 1.

19. *Grand Junction News,* 16 December 1882, 1.

20. Norman, "White Settlement," 75.

21. Rait, "Development of Grand Junction," 18.

22. *Grand Junction News,* 6 January 1883, 2.

23. Ibid.

24. Haskell, *A History,* 71; *Grand Junction News,* 24 March 1883, 1.

25. *Grand Junction News,* 26 April 1884, 3.

26. Ibid.

27. *Grand Junction News,* 20 January 1883, 2.

28. Davidson, "Grand River Ditch," 4.

29. *Grand Junction News,* 24 February 1883, 2.

30. William Pabor, *Colorado as an Agricultural State: Its Farms, Fields, and Garden Lands* (New York: Orange Judd Company, 1883); William Pabor, *Fruit Culture in Colorado: A Manual of Information* (Denver, Colo.: Dove and Temple, 1883).

31. David Boyd, *A History: Greeley and the Union Colony of Colorado* (Greeley, Colo.: The Greeley Tribune Press, 1890), 390.

32. Haskell, *A History*, 73; Kathleen Underwood, *Town Building on the Colorado Frontier* (Albuquerque: University of New Mexico Press, 1987), 12–13; Steven Mehls, *The Valley of Opportunity: A History of West-Central Colorado,* Cultural Resources Series no. 12 (Washington, D.C.: Bureau of Land Management, 1988) 144; Morton Nolen Bergner, "The Development of Fruita and the Lower Valley of the Colorado River from 1884 to 1937" (master's thesis, University of Colorado, 1937), 14.

33. *Grand Junction News,* 24 March 1883, 1.

34. Ibid.

35. *Grand Junction News,* 10 March 1883, 3.

36. Davidson, "Grand River Ditch," 8.

37. *Grand Junction News,* 24 March 1883, 1.

38. *Grand Junction News,* 17 March 1883, 2.

39. *Grand Junction News,* 31 March 1883, 2.

40. *Grand Junction News,* 24 March 1883, 1.

41. Davidson, "Grand River Ditch," 12.

42. *Grand Junction News,* 19 May 1883, 3.

43. Norman, "White Settlement," 77.

44. *Grand Junction News,* 19 May 1883, 3.

45. *Grand Junction News,* 21 July 1883, 2.

46. Davidson, "Grand River Ditch," 14.

47. *Grand Junction News,* 31 March 1883, 2.

48. Davidson, "Grand River Ditch," 14.

49. *Grand Junction News,* 22 September 1883, 2.

50. *Grand Junction News,* 15 September 1883, 3.

51. Ibid.

52. *Grand Junction News,* 29 September 1883, 3.

53. *Grand Junction News,* 26 April 1884, 3.

54. *Grand Junction News,* 1 December 1883, 2.

55. Ibid.

56. *Grand Junction News,* 15 December 1883, 4.

57. Davidson, "Grand River Ditch," 16.

58. On 31 January 1884, the Pioneer Ditch Company changed its name to the Mesa County Ditch Company to avoid any confusion with the Pioneer Extension Ditch Company. The two shared a headgate, but remained separate corporations.

59. *Grand Junction News,* 7 June 1884, 3.

60. Davidson, "Grand River Ditch," 20.

61. *Grand Junction News,* 18 April 1885, 2.

62. Norman, "White Settlement," 78.

63. *Grand Junction News,* 22 December 1888, 2.

64. *Third Biennial Report of the State Engineer to the Governor of Colorado, for the Years 1885–1886* (Denver, 1887), 217.

65. *Eleventh Census of the United States: 1890. Report on Agriculture by Irrigation in the Western Part of the United States* (Washington, D.C.: Government Printing Office, 1894), 100.

66. *Grand Junction News,* 21 July 1888, 2.

67. Davidson, "Grand River Ditch," 24.

68. National Irrigation Congress, *Colorado as an Agricultural State: The Progress of Irrigation* (Denver, 1894), 10.

69. Ibid.

10

A Rio Grande "Brew"

Agriculture, Industry, and Water Quality in the Lower Rio Grande Valley

John P. Tiefenbacher

Here I provide a historical framework for understanding the evolution of water quality in the Lower Rio Grande Valley of Texas. The vernacular "valley" is the delta outwash plain of the Rio Grande/Rio Bravo del Norte in the southernmost tip of Texas. The relevant region includes parts of four counties to the south and east of the International Falcon Reservoir. The valley includes the more heavily populated parts of Texas's Starr, Hidalgo, Cameron, and Willacy Counties (fig. 10.1).

Ironically, the majority of the land in the valley is not actually drained by the Rio Grande. Its drainage area below the Anzalduas Dam is only about four square miles. Rather, the Arroyo Colorado and the South Laguna Madre watershed receive the runoff from the valley (fig. 10.2). The Arroyo Colorado is a naturally dry river channel that has been dredged in places and modified during the last few decades to carry water draining from farmland irrigation and municipal waste treatment plants. The arroyo roughly parallels the Rio Grande from its beginning in western Hidalgo County to its confluence with the Lower Laguna Madre and drains more than 2,780 square miles. The Arroyo Colorado is commonly considered to be two distinct segments defined by the reach of tidal flow. These sections are the estuarine arroyo within the tidal zone extending from the bay on the east to Harlingen on the west. The other section is upstream from the tidal reach west of Harlingen. Return flows from irrigation and municipal waste effluent constitute the bulk of the flow in the upper reach.

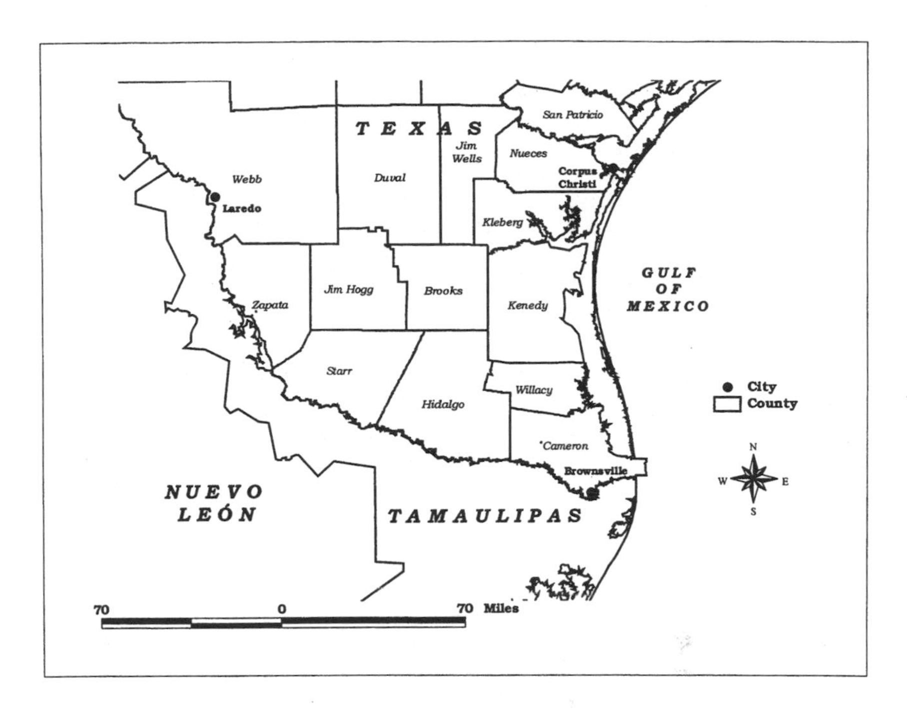

FIGURE 10.1 The counties of South Texas and the relation of the Lower Rio Grande Valley to major cities and northern Mexico. *Source:* Environmental Systems Research Institute, Inc.

Rio Grande water quality is vital to both the environmental and economic health of the valley because the river is the primary source of drinking water and irrigation. Since the environmental organization American Rivers labeled the Rio Grande "the country's most endangered river," national attention has been focused on its water quality.

Today the river supports a population of at least 750,000 permanent residents, thousands of temporary residents, and thousands more visitors. More than 34 million gallons of fresh water are supplied for domestic use. However, more than 953 million gallons per day are withdrawn from this lowest stretch of the Rio Grande and about 2 million more are used daily from surface water sources in the South Laguna Madre watershed. Nearly 900 million gallons per day are used to irrigate more than 730,000 acres of

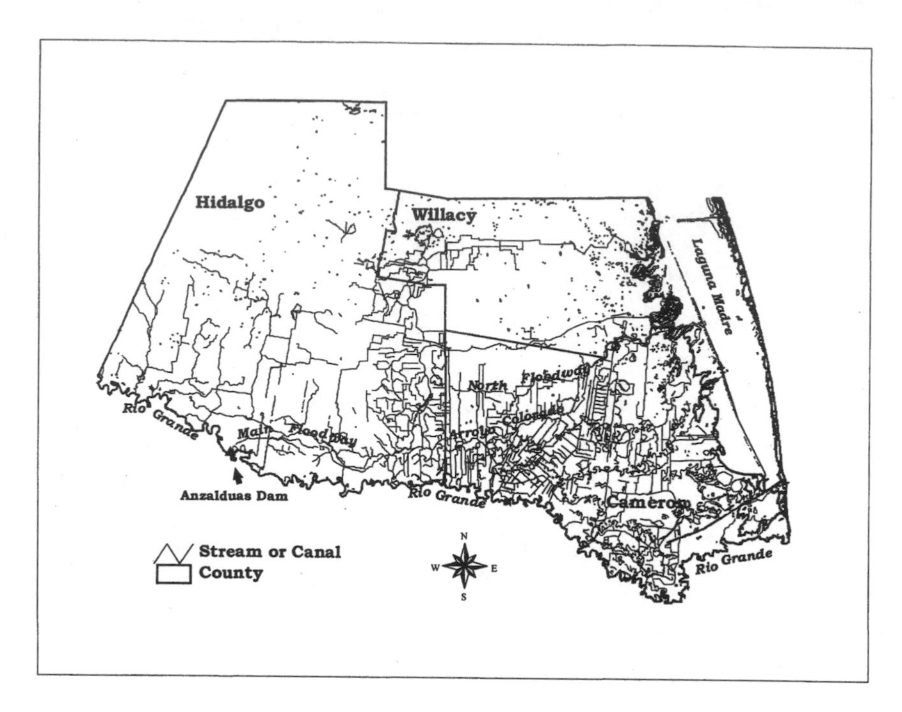

FIGURE 10.2 Surface hydrology and the major drainage features of the Lower Rio Grande Valley, Texas. *Source:* Kika de la Garza Subtropical Agricultural Research Center, U.S. Department of Agriculture.

agricultural land in Hidalgo, Cameron, and Willacy Counties. More than half of the irrigated water is consumed through conveyance loss, evaporation, transpiration, and food production.[1]

Unfortunately, the spatial relationship of agriculture to residential settlement could not be worse than it is in the Lower Rio Grande Valley today. Most irrigation is on land upstream from the bulk of population. The most fertile farmland is squeezed between the western uplands that begin west of Mission in Hidalgo County and the salt marsh coastal plain. Downstream users receive water that has recently been characterized by the Texas Natural Resource Conservation Commission as of "possible concern" with respect to fecal coliform levels, low dissolved oxygen, high levels of nutrients (nitrogen and phosphorus), heavy metals, and organic materials.[2]

So how did it come to be this way? What were the major turning points

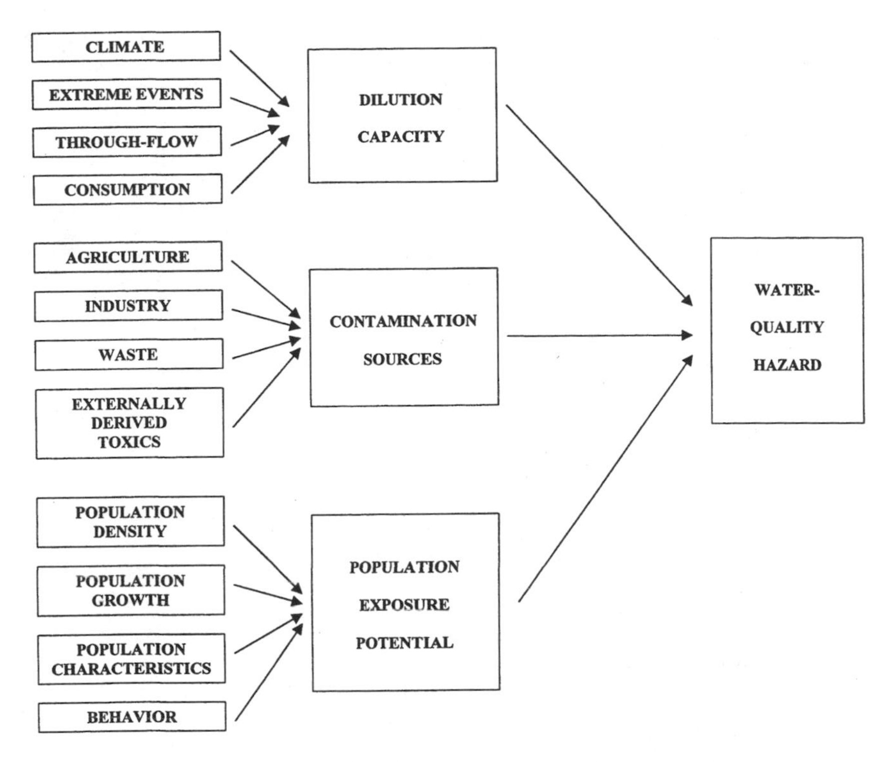

FIGURE 10.3 A simple model of regional water-quality hazard factors.

in the historical development of the Lower Rio Grande Valley? How did water quality change during this evolution? This chapter will describe the historical environmental geography of the valley as it has been defined by periods of change, beginning with Euro-American settlement, development, innovation, migration, and shifts in the now-dominant American social paradigm.

Spatial Determinants of Regional Water Quality

Human health risk from water contamination is a product of the capacity of a river to absorb and dilute pathogenic contaminants, the types and quantity of those contaminants, and the simultaneous potential for human exposure (fig. 10.3). Local climate, temporary meteorological extremes, upstream contributions to local water supply, and water demand and use interact to determine a river's threshold for "safe" disposal of contaminants. The scale of industrial development, agricultural practices, the quality of waste management infrastructure, and the frequency and severity of hazardous material spills effect the potential for contamination of surface water and groundwater. Lastly, population growth and density, demographics, and behavior influence the vulnerability of residents and the potential for exposure.

By overlaying these patterns in the region in a development or land-use chronology, one can derive a historical geography of the hazards of place. The "hazards-of-place" concept derives from hazards research in geography and denotes the spatial patterns and magnitude of risk to people of either numerous hazards at one time or a single hazard over time. The approach is used to understand how the factors involved interact spatially to determine risk, or the probability that humans will suffer loss.[3]

A Historical Framework for Water Contamination in the Lower Rio Grande Valley

The settlement history of the Lower Rio Grande Valley can be divided into seven periods. Each is characterized by one or more dominant trends and begins or concludes with particular benchmark events. The dominant trends are determined by the interaction of activities that organized the landscape of the time, the materials that affected local water quality, and/or the modus operandi of the economic activities themselves (table 10.1). I have named these periods "The Rio Bravo Wilderness" (until 1895), "Wetting the Seeds for Development" (1895–1904), "Pioneering the Lower Rio Grande Valley" (1905–1926), "Modernity and Oil" (1927–1939), "The Americanization of the Lower Rio Grande Valley" (1940–1963), "Facing South" (1964–1987), and "Hard Realities" (1988 to present). Each period is defined by distinct shifts in economic development activities, population growth, and cultural changes that influence water quantity and quality.

Table 10.1. Chronological periods in the development of the Lower Rio Grande Valley of Texas.

PERIOD	POPULATION	DOMINANT TRENDS	AGRICULTURE	CROP PROTECTION & FERTILIZATION	WATER
The Rio Bravo Wilderness (pre-1895)	fewer than 34,400	presettlement, little economic activity, cattle ranching, oriented south	none	none needed	low use, Rio Grande
Wetting the Seeds for Development (1895–1904)	approximately 40,000	irrigation projects	sugar cane and cotton	physical, biological, traditional approaches	minor amounts needed, increasing diversion to Arroyo Colorado
Pioneering the Valley (1905–1926)	growth to 85,400 in 1920	railroad built, towns established, land development	winter vegetables, cotton, citrus	organic and inorganic materials, dusts and powders	water for domestic use acquired through same canals as irrigation water
Modernity and Oil (1927–1939)	growth to 215,900 in 1940	modernization, natural gas imported, economic boom, oil discovery, agricultural produce, canneries	citrus, cotton, vegetables	early innovations toward technological, synthetic, and extensive applications	water utilities emerge to provide untreated surface water to residential and commercial areas
Into the Fold: The Americanization of the Valley (1940–1963)	growth to 369,200 in 1960	industrial development, water-supply monitoring	highest number of farms in history; acreage reaches maximum	increasingly complex chemicals, mechanized application, fertilization increases significantly	water treatment developed for cities; rural users used untreated water
Facing South (1964–1987)	growth to more than 650,000 in the late 1980s	industrialization of northern Mexico, development of *colonias*, water-quality monitoring	citrus, cotton, sorghum, winter vegetables, new products like aloe vera	higher toxicity of chemicals, more intensive applications	increasingly efficient municipal treatment, but increased contamination from agriculture and industry
Hard Realities (1988 to present)	more than 700,000 by 1990	concerns about toxic chemicals in the water from industrial development, growth, and agriculture	maintenance of trends; attempt to eradicate boll weevil	large-scale efforts at control of pests, emergence of "organic" agriculture	intensive water improvement efforts; Water Development Board to provide safe drinking water to all

The Rio Bravo Wilderness

Until 1895, the Lower Rio Grande Valley was essentially a wilderness. A few nascent settlements had been established along the river in the form of missions and villas to serve religious outposts ministering to the indigenous Tamaulipecan Indians and as Mexican lines of defense against marauding Apaches and Comanches from the north. So that although Frederick Jackson Turner's frontier may have closed with the 1890 census, *la frontera's* did not. This is exemplified by the publication of a pamphlet for the Southern Pacific Railroad during the 1920s that proclaimed this valley "the richest frontier in the world today."[4] This era was characterized by minimal settlement and small, sparsely distributed towns; in 1900, the valley's population was only 34,400. Economic activity in the Lower Rio Grande Valley was nearly limited to cattle ranching. The southern reaches of the King and Kenedy ranches extended into what was then northern Cameron County; in 1911, Willacy County was established from parts of Cameron and Hidalgo Counties. A few small settlements existed along the river, but little economic activity other than the military outpost at Brownsville (population 7,000) and minor amounts of trade with the more developed Mexican towns of Reynosa and Matamoros, Tamaulipas, occurred. Before 1895, the residents were oriented toward northern Mexico, and the region might have been more aptly named the "El Valle del Río Bravo Bajo," or the Lower Rio Bravo Valley.[5]

The coastal plain was salt marsh and the vegetation had reverted to mesquite and cactus where cattle had earlier grazed the native grasslands. The uplands were too dry and vegetated to provide for economically gainful agriculture. Although fertile, the land required extensive labor to clear and prepare it, and isolation and the lack of infrastructure to deliver produce to any market further dissuaded agricultural development. Corpus Christi was the nearest "large" Texas settlement (population about four thousand) outside the Lower Rio Grande Valley; however, mail delivery by stage was the only external northward connection.

Wetting the Seeds for Development, 1895–1904

In 1895, John Closner, a valley immigrant stagecoach driver, constructed the first irrigation network branching off of the Rio Grande to water his 45,000-acre San Juan Plantation in Hidalgo County. Closner grew sugar cane and helped to establish a nascent sugar industry in the Lower Rio Grande Valley, but it lasted barely a decade. For example, the Ohio and Texas Sugar Plantation was established in 1898, but the quality of locally grown sugar was significantly below that of the competing regions (Cuba and Hawaii), and the local demand for molasses, the alternative to refined sugar, was not great enough to justify the industry's existence. In addition to sugar cane, corn and pinto beans were grown for local consumption and small amounts of cotton—some six hundred bales in 1900—were exported. By 1904, attempts to grow rice, alfalfa, and citrus were under way.[6]

The next few years brought other irrigation projects to aid in the agricultural expansion. The Santa Maria Canal Company (1897), La Blanca Ranch system (1900), the Brownsville Irrigation System (1901), and the Hidalgo Canal Company (1902) provided a foundation for agricultural development, but the region's isolation and lack of infrastructure hampered the growth of an export economy. Projects such as the Llano Grande system (1904), the San Benito Land and Irrigation Company (1904), the American Rio Grande Land and Irrigation Company (1905), and the Tresquilas Plantation (1905) were tied to the development of and boom in real estate.[7]

Salinization of soils caused by evaporation of irrigation water further hampered the development of plantation agriculture during the early years of development. This problem would be solved during the next few decades as more effective drainage systems were designed and orchards and farm fields were reconstructed.

Pioneering the Lower Rio Grande Valley, 1905–1926

The valley's renaissance arrived aboard the St. Louis, Brownsville, and Mexico Railroad (the SLB&M). Valley railroad projects had been planned for several decades before one came to fruition. The SLB&M was the brainchild of Uriah Lott, a financier from Corpus Christi who recognized

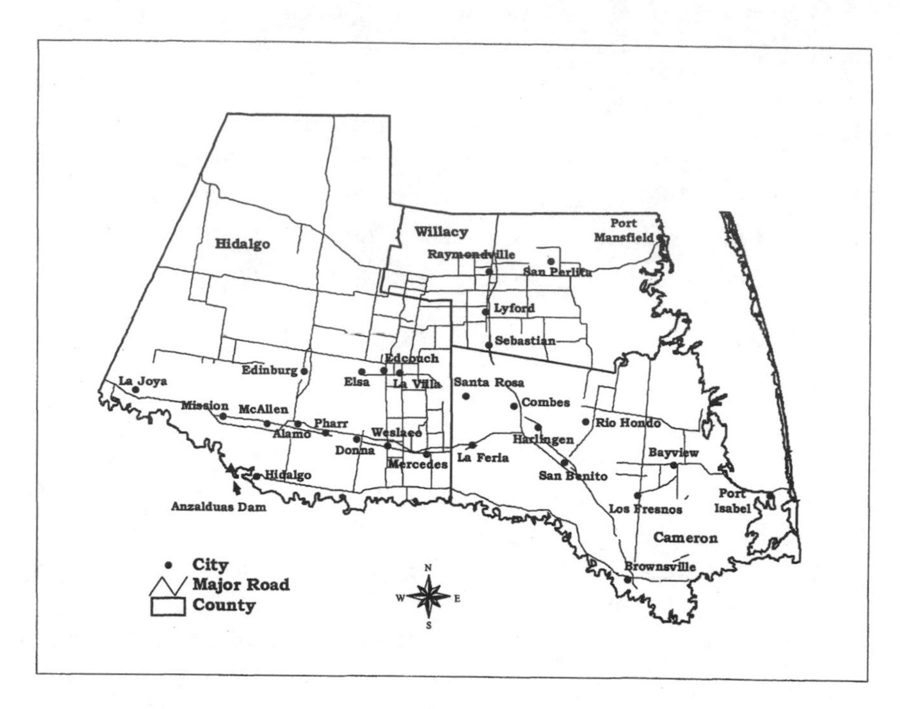

FIGURE 10.4 The locations of the major municipalities of the Lower Rio Grande Valley, Texas. *Source:* Environmental Systems Research Institute, Inc.

the agricultural potential of the region. The company received its charter in January 1903, and construction began in July at the location of what would eventually become Robstown, west of Corpus Christi. There were no settlements for 160 miles to the south. In fact, every town in the Lower Rio Grande Valley today, with the exception of Brownsville and Hidalgo, dates its birth to the arrival of the railroad.

The rails reached fledgling Raymondville (in Willacy County) in April 1904 and by May they had passed through the new towns of Harlingen and San Benito (fig. 10.4). The connection of Brownsville to the rest of the United States was made on 7 June 1904, thus marking the closing of this frontier. The ensuing two decades were a time of conquest and subjugation of the wild nature of the valley.

While the rails were being extended to Brownsville, a branch of the railroad was being built westward from Harlingen. Known as the Hidalgo and

San Miguel Extension (later called the Sam Fordyce branch), its progress and eventual completion in 1905 gave birth to the "string of pearls" that paralleled the river from Harlingen to La Feria, Mercedes, Weslaco, Donna, Alamo, San Juan, Pharr, McAllen, and Mission. The railroad established these towns, which then served as growth poles for the valley's agricultural and industrial development.[8]

The railroad also promulgated new land and water companies. Besides the aforementioned San Benito Land and Irrigation Company and the American Rio Grande Land and Irrigation Company, development was pushed by the Harlingen Land and Water Company, which was founded in 1907, beginning the proliferation of irrigation development companies. For instance, the following year the Santa Maria Canal Company became the Mutual Canal Company. The Rio Grande Valley Reservoir and Irrigation Company was established. The Llano Grande Plantation was irrigated. The Lomita Co-operative Irrigation Company of Mission formed (it later became the United Irrigation Company), and the Indiana Co-operative Canal Company (east of Brownsville) began irrigation. The Louisiana and Rio Grande Canal Company began irrigating the Alamo tract in 1911.[9]

From 1915 to 1920, many developers began to experience financial problems due to the costs associated with improvement and extension of the canal projects, irrigation systems, and pump stations. Mergers and buyouts reduced competition. To ensure adequate water supplies, farmers organized irrigation districts and bought the systems from their developers. For example, in 1920, the Hidalgo Water Improvement District no. 2 took over the Louisiana and Rio Grande Canal Company. By that year, there were only four privately owned irrigation systems left. Also, by 1924, most electricity suppliers, ice companies, and other public services had been bought up and unified into public utility companies.[10]

Agriculture had established a foothold in the valley. In 1907, 761 railroad cars full of vegetables were shipped from the valley. By 1915, the cars numbered more than 1,000. By 1925, the shipments were more than 16,000 carlots. In 1920, nearly 1.5 million bushels of corn were produced, more than 13,000 bales of cotton were shipped, and more than 600 acres of land were planted in citrus. The number of farms had tripled from 1,372 in 1900 to more than 3,800 in 1920. Agriculture following World War I was dominated by what one observer has called a "crop triumvirate" of winter vegetables, cotton, and citrus.[11]

The Lower Rio Grande Valley's population had more than doubled to 85,400 in 1920, with the bulk of the growth occurring since 1910. The spurt in population required increased supplies of water for personal consumption, but systems to provide water to residents lagged behind those developed for irrigation. For example, in 1907, San Benito acquired its water in tank cars. The following year, its water was delivered via irrigation canal. By 1909, the town had installed pumps to provide water through a separate system. In 1914 and 1915, both McAllen and Mission pumped water via windmill from the river and stored it in water towers. Not until 1916 did Rio Grande City, in Starr County, build a water supply system. By 1920, however, most systems delivered untreated river water to towns for personal consumption.[12]

Because water was commonly consumed untreated from the river, it was fortunate that the major potential source of water pollutants, at this time, was agriculture. Fertilizers were certainly not yet needed in the valley—it would be years before the land started to demonstrate the stress of nutrient consumption by agriculture. Pesticide use was the other potential agricultural pollutant, yet pest control was undergoing a shift to scientific approaches; it was still in its infancy and farmers used naturally occurring ingredients to treat infestations and blights. The pesticides commonly used during this period were naturally occurring organic compounds and inorganic metals, for example, fumigants like naphthalene and carbon bisulfide and fungicides and insecticides including lead arsenate, calcium arsenate, and nicotine. The methods of applying these materials were primitive. Hand application of powders and dusts would involve use of whisk brooms and hand-pumped dust ejectors. Mechanized application by tractor-pulled fans blowing powders and dusts across fields certainly contributed to greater distribution and drift of the poisons. But not until the 1920s did pesticide application "take off." The first aerial application of pesticides occurred in Ohio in 1921, beginning the modernization of pest control and the intensification of health risks and the potential for water contamination problems. The hazardousness of pesticides and their use greatly increased during the new era.

From 1920 to 1926, improvements in transportation within the valley began. The first hard-surfaced roads were built in 1921 and in 1926 the Southern Pacific Railroad arrived in the Lower Rio Grande Valley. The railroad provided more links to Corpus Christi, Falfurrias, and San Antonio and assisted in the growth of other extractive industries like oil, which had been discovered in Starr County in the mid-1920s.[13]

Modernity and Oil, 1927–1939

The rising demand for more modern conveniences was signaled by the arrival in McAllen of a natural gas pipeline in 1927. An economic boom was in full swing in the valley. The population had more than doubled from 1920 to 1930 (from about 86,000 to more than 176,000). Irrigation had increased to more than 257,000 acres in 1930, five times the acreage of two decades earlier. By 1930, citrus acreage was more than 22,000. More than 23,000 rail cars full of vegetables were being exported annually, and by 1939 there were more than forty canneries operating in the Lower Rio Grande Valley. Secondary sector activity was growing during this period as energy supplies derived from local oil were being developed. The number of power plants in the valley reached a high of ninety in the 1930s.[14]

A third mode of import and export was developed during the Depression. Deep water ports and channels were dredged and opened to San Benito from Port Isabel and to the Port of Brownsville in 1936. Intercoastal shipping greatly increased with easier passage to New Orleans, the Mississippi River, and points north.

Perhaps the greatest threat to residents' health via water quality involved developments in crop protection that were occurring nationwide. Two important innovations took place during the 1930s that eventually led to the industrialization of agriculture. The first was the discovery in 1930 of the insecticidal qualities of the compound dichlorodiphenyltrichloroethane (DDT). A synthetic organochlorine first formulated in the 1890s, DDT's adoption as a pesticide signaled the beginning of the engineering of nonnatural substances to combat crop depredation. DDT offered one particular advantage over the traditional and "natural" substances, an advantage that was also its eventual undoing: It persisted in the environment. The second revolution involved a shift away from dusts and powders to deliver the active ingredient to crops and their surfaces. Liquid formulations made applications more efficient, more effective, and potentially more harmful. These three revolutions raised the risk of surface- and groundwater contamination by increasing the spatial distribution of airborne pesticides.[15]

Americanization of the Lower Rio Grande Valley, 1940–1963

South Texas rode the wave of industrial development begun by America's entry into World War II and the Lower Rio Grande Valley was brought "into the fold" of the American way of life. Heavy industry arrived with the building of the Stanolind Oil and Gas Company and the U.S. Chemical Company plants in 1951. In 1952, three gasoline plants were built in the region: Carthage Hydrocol near Port Brownsville; Continental Oil Company in the Rincon Field, and Sun Oil Company, both in Starr County. Oil refineries were located at Port Isabel, La Blanca, Rio Grande City, Brownsville, and McAllen. The Chlorophyll Chemical Corporation, the Paris Gum Corporation of America, and Pan-American World Airways Overhaul Base were established. In 1953, the Taylor Refineries and the Pan-American Oil Company were founded.[16]

However, the growth was beginning to present water-quality problems. For example, canneries, spurred on by the growing fruit juice market, were making more than $9 million a year. The disposal of citrus peel and fruit liquor waste into public sewers was beginning to cause significant problems. Eventually a new industry emerged that converted the waste into cattle feed.

Valley population during this industrialization phase swelled from 215,900 in 1940 to 320,500 in 1950 to 369,200 in 1960. The cities of Brownsville, Harlingen, Edinburg, McAllen, and Mission were the foci for residential growth, and it was this demographic boom that made the 1940s a high point of water supply management in the valley. The United States had secured 58.6 percent of the Rio Grande water south of Fort Quitman, Texas, via a water allocation treaty it signed with Mexico in 1944. This treaty required the International Boundary and Water Commission to monitor discharge to determine the size of the allotments—its records provide abundant historical flow data. As luck would have it, the treaty was just in time to secure water to the valley during the most severe drought to hit South Texas. Beginning in 1949, the drought intensified until it hit its nadir in 1953. Below-average rainfall lasted for five more years.[17]

Between 1900 and 1939, the Rio Grande had overflowed 23 times and hurricanes had hit the valley in 1910, 1913, and 1933. Because of two consecutive decades of above-average rainfall and occasional flooding, the International Boundary and Water Commission responded with flood-control projects. By 1950 it had planned and completed 75 percent of these, including the building of Falcon Dam, 145 miles of floodways, and 285 miles of levees along the river channels and floodways. The September 1933 hurricane caused $12 million in damages and forty deaths. Not until Beulah in 1967 would the valley experience greater economic devastation by hurricane; the third largest hurricane in Texas history, Beulah caused more than $150 million in damages and fifteen deaths.[18]

In just five years, the acreage of farmland irrigated in the four-county region grew from 260,000 in 1941 to 536,000 in 1946. Irrigated farmland acreage in the Lower Rio Grande Valley was 20 percent of all the irrigated farmland in the state of Texas. Corn, cotton, citrus, and the "world's largest vegetable patch"—10,000 acres of parsley, broccoli, and carrots—were some of the legacy of this farming explosion. Irrigated acreage held steady for the next two decades, but the number of farms began to decline, as it did nationally, from more than 11,000 in 1950 to merely 5,500 in 1964. As farm size grew, the demand for agricultural chemicals grew.[19]

Fertilizer use during the 1950s increased from 278,000 acres in 1954 to more than 440,000 in 1964, while the number of farms held steady. Pesticide use also gained momentum during agricultural industrialization. During this time more new synthetic pesticide products had entered the commercial market than during any previous period. Little regard was paid to the long-term impacts of the contamination generated by such extensive use. In fact, the Department of Agriculture did not report pesticide use data until 1969. In 1962, Rachel Carson's *Silent Spring* presented evidence of environmental and human health risks generated by pesticide use, but the immediate responses to her warnings were neither changes in practice nor heightened concern. This period is best characterized as the beginning of agriculture's "pesticide treadmill."[20]

Facing South, 1964–1987

In 1963, the U.S.–Mexico *Bracero* Program of internationally contracted agricultural labor was formally ended. To hold the rapidly increasing population in Mexican border towns on that side of the line, the two countries began a program of industrialization in the border region that laid the groundwork for the establishment of American industrial assembly plants in Mexico. These *maquiladoras* were intended to provide Mexicans with employment in Mexico to undercut the attractiveness of illegal migration for agricultural or other work in the United States. These opportunities, some argue, combined with changes in Mexican economic and agricultural policies, backfired and intensified rural-to-urban migration in the border region and cross-border traffic as well. Although steady during the 1960s, the valley's population exploded, doubling to more than 580,000 residents during the 1970s. By the 1990s, the population of the four valley counties would top 700,000, with many additional uncounted semipermanent residents, tourists, and illegal immigrants.[21]

The population growth of the 1970s and 1980s pressured land-use changes. Suburbanization and exurbanization of the Lower Rio Grande Valley caused farmland conversion within municipal limits, and demand for housing sparked the beginnings of problematic extramunicipal residential developments known as *colonias.* Housing occupancy rates exceeding 98 percent created a housing crisis that was not met by conventional means. Owners of farmland beyond municipal limits occasionally subdivided their property into very small lots and sold these to the growing, young Mexican-American families (fig. 10.5). Low down payments, high interest rates, and long-term mortgage agreements provided land to first-time middle- and low-income homebuyers with promises of services like water supply, electricity, sewerage, and pavement. These services were rarely provided. Unfortunately, the opportunities afforded to those who otherwise could not realize the American dream led them into a tremendously hazardous landscape—particularly with respect to areas vulnerable to inundation, extremely poor water quality, and in close proximity to industrial agricultural activities.[22]

Allen, which hit in August 1980, was the Lower Rio Grande Valley's most recent hurricane. The storm produced McAllen's all-time record rain-

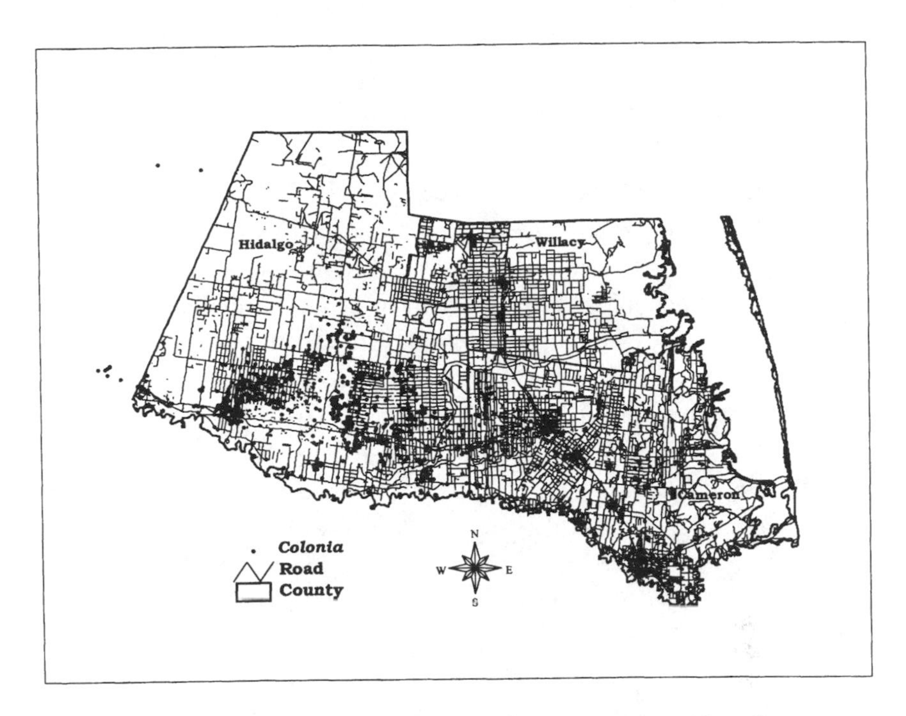

FIGURE 10.5 The locations of *colonias* in the Lower Rio Grande Valley, Texas. *Source:* Texas Water Development Board.

fall and caused $600 million in damages. Major flooding also occurred in September 1984, generating an additional $30 million in damages. Combined with freezes that decimated citrus in the valley, these disasters led to the reduction of farmland acreage by 25 percent. The valley has not gained that acreage back and seems destined for greater industrial and transportation development generated by the North American Free Trade Agreement (NAFTA) of 1994 and less agriculture.

Although painful, this shift might not be such a bad thing. Pesticide expenditures had topped out at a maximum of more than $36 million in 1984, having swelled from only $8 million in 1969. Likewise, fertilizer costs had reached a high of $21 million in 1987. These costs have been declining as well.

Until recently, solid measurements of water quality for the Lower Rio Grande Valley were very difficult to find. Recent high-profile studies conducted by state and federal agencies have piqued public attention.[23] Federal agencies began monitoring the quality of water in the Rio Grande and its tributaries in the mid-1950s. Their investigations focused on the concentrations of heavy metals, bicarbonates, nutrients, and dissolved solids. In 1966, monitors on the Arroyo Colorado and the North Floodway were added, but only discharge and sedimentation data were gathered. In 1968, testing for pesticides was begun in two locations: below the Anzalduas Dam at Mission and in the Arroyo Colorado to the south of Mercedes. Although no pesticides were detected in the Rio Grande, samples from the Arroyo Colorado during those first years indicated detectable levels of Endrin, Lindane, 2,4-D, and 2,4,5-T in 1968 and dichlorodiphenyldichloroethane (DDD), dichlorodiphenyldichloroethylene (DDE), DDT, Dieldrin, Endrin, Lindane, and 2,4,5-T in 1969. Other chemicals have also been detected at this site in subsequent years (table 10.2), thus corroborating data gathered by the U.S. Fish and Wildlife Service that documented high levels of DDT, DDE, and toxaphene in Rio Grande fish from 1967 to 1979.[24]

In 1972, DDT was banned in the United States. The reason for this was its persistence in the environment leading to biomagnification that ultimately threatened wildlife. A number of other compounds would eventually be withdrawn either for similar reasons or because evidence clearly showed links of exposure to cancer. The pesticide industry's response to these developments was to work toward the development of more acutely toxic materials that were not as persistent. Thus, the industry has collapsed in space the toxicity that was previously spread over time. The more toxic materials of the 1970s and 1980s were required to do their work in significantly less time, which, unfortunately, has led to the development of "superpests" whose survival reflects natural selection based on resistance to the pesticides applied. The industry response was to step on the treadmill and escalate the war: intensified toxicities, increased frequency of treatment, and greater and greater quantities.

Hard Realities, 1988–Present

What effect might this most recent shift toward industrialization and international trans-shipment have? This period is shaping up to be the era of

*Table 10.2. Pesticides detected in U.S. Geological Survey water-quality testing of the Arroyo Colorado at El Fuste Siphon, South of Mercedes, Texas, from 1967 through 1980. Testing was discontinued at the site in 1981. X, pesticide detected at 0.01 micrograms per liter; #, water not tested for chemical during period; *, chemical testing information not reported.*

WATER YEAR / CHEMICAL	67	68	69	70	71	72	73	74	75	76	77	78	79	80
Aldrin							*	*	*					
Chlordane	#	#					*	*	*	X				
DDD	X	X	X		X		*	*	*			X		
DDE	X	X	X	X	X	X	*	*	*	X	X	X	X	X
DDT	X	X					*	*	*	X				
Diazanon	#	#			X	X	*	*	*	X	X	X	X	X
Dieldrin	X	X	X	X	X	X	*	*	*	X	X	X	X	X
Endosulfan	#	#	#	#	#	#	#	#	#	#	X			
Endrin	X	X				X	*	*	*	X	X	X	X	
Ethion	#	#	#	#	#	#	#	#	#					
Heptachlor							*	*	*	X				
Heptachlorepoxide							*	*	*					
Lindane	X	X	X			X	*	*	*					X
Malathion	#	#					*	*	*					X
Methyl parathion	#	#	X		X	X	*	*	*	X			X	X
Methyl trithion	#	#	#	#	#	#	#	#	#					
Mirex	#	#	#	#	#	#	#	#	#	#				
Parathion	#	#	X		X	X	*	*	*	X	X	X	X	X
PCBs	#	#	#	#	#	#	#	#	#		X			
Polychlorinated naphthalenes	#	#	#	#	#	#	#	#	#					
Silvex						X	*	*	*					
Toxaphene	#	#	#	#	#	#	#	#	#					
Trithion	#	#	#	#	#	#	#	#	#					
2,4-D	X					X	*	*	*		X	X		X
2,4,5-T	X	X	X		X		*	*	*			X		X

SOURCES: U.S. Geological Survey, 1968–1981.

growth management in the Lower Rio Grande Valley, as the mistakes of the past have become the problems of the present. Since 1987, the Environmental Protection Agency has required the reporting of toxic substances emitted, transferred, or disposed of from all industrial sources. The Toxic Release Inventory provides us with an indication of the environmental burden of nonagricultural industries. In 1995, for instance, the valley was the home of only thirty-two potential emitters of toxic materials: in Brownsville (twelve), McAllen (eight), and Harlingen (six), with the others in Edinburg, Pharr, Olmito, San Benito, Alamo, and Rio Hondo. Only ten companies, however, actually emitted materials into the environment (eight into the atmosphere and four transferred wastes off-site for processing). The compounds released into the atmosphere were ammonia, styrene, trichloroethylene, chromium, and copper. Most releases were generated by factories that either made ice or froze fresh fish or fruit.[25] Greater threat may ultimately derive from materials transported through the Lower Rio Grande Valley to Mexican assembly plants and back to the United States. The nature and extent of transboundary movement of hazardous materials is not well documented. However, improvements are being made under the auspices of NAFTA, the North American Development Bank, the Border Environmental Cooperation Commission, and the Border XXI Program.

Interest in the problems generated by toxic contamination flared in the 1990s. Unusually high frequencies of birth anomalies, specifically neural tube defects, were detected in Cameron County beginning in 1991 and prompted investigations into their causes. Since the case of the "Mallory babies," sensitivity to birth defects had been heightened in the valley. This tragedy involved claims by workers at the R. R. Mallory capacitor plant in Matamoros that miscarriages, stillbirths, and extensive physical and mental defects were the result of mismanagement and exposure to toxic materials during the mid-1970s. In the wake of the 1991 cluster of birth anomalies, it was determined that more than 130 cases of anencephaly (babies born without a cranial vault and brain) occurred in the county between 1980 and 1993.[26]

The Lower Rio Grande Valley's population had grown to more than 700,000 by 1990, and there seemed to be no end to the trends of the previous decade. Despite positive signs of improvements in the region's economy as a result of the "NAFTA-sization" of the borderlands, the valley's population remains among the poorest in the nation. Poverty makes for a vulnerable population. Health and environmental quality are highly dependent on economic wherewithal. As the popularity of examining cases of "environmental inequity" demonstrates, class and ethnicity are determinants of the hazardousness of one's space, as the residents of colonias can attest.[27]

The agricultural landscape in which the colonias are located in the valley underwent one last, grand, lethal assault during 1996 growing season. The agricultural industry in Texas made a once-and-for-all attempt to rid the state of the cotton farmer's scourge, the boll weevil, by staging an all-out eradication campaign of pesticide application. The dousing of cotton fields was undertaken in the all-or-nothing involuntary program. However, the results of the program were less than spectacular. The numbers of boll weevils did not decrease, although cotton yields in Texas fell; across the border in the unsprayed cotton fields of northern Mexico, the yields soared. The long-term consequences of this dousing on the region's environmental health is unclear, as water quality studies for 1996 and 1997 are not yet available.

The History of Water Quality in the Lower Rio Grande Valley

This sweeping review of the sequential development of the Lower Rio Grande Valley is intended to illustrate that chronological periods can be useful to reflect the trends of a region's water quality. The periods identified here may be representative of similar noncontemporaneous periods of other regions of the country. Whether along the U.S.–Mexico border (for example, the Imperial Valley) or in other settings (the Olympic Peninsula of Washington State or the San Luis River Valley of Colorado), such regional development chronologies could be useful.

The Lower Rio Grande Valley has undergone major changes in its first century of development. The seven periods characterized here not only demonstrate that the agricultural and industrial development of the region has hinged significantly on the availability of water, but that concern about the quality of water trailed, by several decades, the contamination of the developed sources. The physical and economic characteristics that make the region attractive to agricultural and industrial interests have also made the region attractive to migrants from other regions. These factors greatly enhance the potential for harmful side effects from the economic development by associating an increasing population with unregulated pollution.

The trends represented in the development of the valley confirm the commonsensical notion that initial settlement of a region presented low potential for water-borne negative impacts, but as population has grown and water supply systems are developed, the risk of exposure to disease increases exponentially. When agricultural land uses and industrial practices intensify on both sides of the border, they have combined to present a much more formidable challenge to the health of the resident population. Only as the hazards are discovered can measures be taken to mitigate the risks. Adjustments, over time, will reduce the total risk and produce a landscape that harbors only the most insidious and complex health threats. Recent efforts by the agencies charged with protection of health and resources have moved to solve the most apparent health threats to vulnerable populations living in and moving through the valley. Future improvements in water quality will be dependent on the cost and judged fruitfulness of attempts to reduce risk further. That dependence is also the greatest justification for spatial environmental histories because they offer opportunities to connect economic processes and activities with environmental change and to connect those changes with their impact on natural and human landscapes.

Notes

1. U.S. Geological Survey, *Water Resources Data—Texas: Water Years 1967 to 1980* (Washington, D.C.: Government Printing Office, 1968–1981).

2. *1994 Regional Assessment of Water Quality* (Austin: Texas Natural Resource Conservation Commission, 1996).

3. Susan L. Cutter and William D. Solecki, "The National Pattern of Airborne Toxic Releases," *Professional Geographer* 41, no. 2 (1989): 149–61; John P. Tiefenbacher, "Pesticide Drift and the Hazards of Place in San Joaquin County, California" (Ph.D. diss., Rutgers University, 1992); John P. Tiefenbacher, "Mapping the Pesticide Driftscape: Theoretical Patterns of the Drift Hazard," *Geographical and Environmental Modeling* 2, no. 1 (1998): 75–93.

4. Daniel D. Arreola and James R. Curtis, *The Mexican Border Cities: Landscape Autonomy and Place Personality* (Tucson: University of Arizona Press, 1993); Milo Kearney and Anthony Knopp, *Border Cuates: A History of the U.S.–Mexican Twin Cities* (Austin, Tex.: Eakin Press, 1995); Julia Cameron Montgomery, *A Little Journey through the Lower Valley of the Rio Grande: The Magic Valley of Texas* (Houston, Tex.: Rein, 1928).

5. J. Lee Stambaugh and Lillian J. Stambaugh, *The Lower Rio Grande Valley of Texas* (San Antonio, Tex.: The Naylor Company, 1954).

6. Ibid.; C. Daniel Dillman, "Transformation of the Lower Rio Grande of Texas and Tamaulipas," *Ecumene* 11, no. 2 (1970): 3–11.

7. Ibid.

8. Ibid.

9. Stambaugh and Stambaugh, *The Lower Rio Grande Valley of Texas.*

10. Ibid.

11. Ibid.; Dillman, "Transformation of the Lower Rio Grande."

12. Stambaugh and Stambaugh, *The Lower Rio Grande Valley of Texas.*

13. Ibid.

14. Ibid.

15. B. F. Stanton, "Changes in Farm Size and Structure in American Agriculture in the Twentieth Century," in *Size, Structure, and the Changing Face of American Agriculture*, ed. Arne Hallem (Boulder, Colo.: Westview Press, 1993), 42–70; Tiefenbacher, "Mapping the Pesticide Driftscape."

16. Stambaugh and Stambaugh, *The Lower Rio Grande Valley of Texas.*

17. George W. Bomar, *Texas Weather* (Austin: University of Texas Press, 1995).

18. Stambaugh and Stambaugh, *The Lower Rio Grande Valley of Texas;* Bomar, *Texas Weather.*

19. Stambaugh and Stambaugh, *The Lower Rio Grande Valley of Texas.*

20. U.S. Bureau of the Census, *Censuses of Agriculture* (Washington, D.C.: U.S. Government Printing Office, 1954–1992); Robert Van den Bosch, *The Pesticide Conspiracy* (Berkeley: University of California Press, 1978); Thomas R. Dunlap, *DDT: Scientists, Citizens, and Public Policy* (Princeton, N.J.: Princeton University Press, 1981).

21. Dianne C. Betts and Daniel J. Slottje, *Crisis on the Rio Grande: Poverty, Unemployment, and Economic Development on the Texas-Mexico Border* (Boulder, Colo.: Westview Press, 1994); John P. Tiefenbacher, "Environmental Hazards of the Lower Rio Grande Valley," in *A Geographic Glimpse of Central Texas and the Borderlands*, ed. James F. Petersen and Julie A. Tuason (Indiana, Pa.: National Council for Geographic Education, 1995), 85–94.

22. Robert Lee Maril, *Poorest of Americans: The Mexican-Americans of the Lower Rio Grande Valley of Texas* (Notre Dame, Ind.: University of Notre Dame Press, 1989); Betts and Slottje, *Crisis on the Rio Grande;* Rebecca Ingram, "Pesticide Exposure Vulnerability Assessment for Southwest Hidalgo County, Texas" (master's thesis, Department of Geography and Planning, Southwest Texas State University, 1997); Tiefenbacher, "Environmental Hazards of the Lower Rio Grande Valley."

23. Texas Natural Resource Conservation Commission, *1994 Regional Assessment of Water Quality in the Rio Grande Basin Including the Pecos River, the Devils River, the Arroyo Colorado and the Lower Laguna Madre* (Austin: Texas Natural Resource Conservation Commission, 1994); International Boundary and Water Commission, *Binational Study Regarding the Presence of Toxic Substances in the Rio Grande/Rio Bravo and Its Tributaries along the Boundary Portion between the United States and Mexico: Final Report* (El Paso, Tex.: International Boundary and Water Commission, 1994); Mary E. Kelly and Salvador Contreras, *The 1994 Rio Grande Toxics Study: An Evaluation and User's Guide* (Austin: Texas Center for Policy Studies, 1995); Jack R. Davis, Leroy J. Kleinsasser,

and Roxie Cantu, *Toxic Contaminants Survey of the Lower Rio Grande, Lower Arroyo Colorado, and Associated Coastal Waters* (Austin: Texas Natural Resources Conservation Commission, 1995); Texas Natural Resource Conservation Commission, *1996 Regional Assessment of Water Quality in the Rio Grande Basin including the Pecos River, the Devils River, the Arroyo Colorado and the Lower Laguna Madre* (Austin: Texas Natural Resource Conservation Commission, 1996).

24. International Boundary and Water Commission, *Binational Study;* U.S. Geological Survey, *Water Resources Data.*

25. Environmental Protection Agency, *Surf Your Watershed: Watershed Environmental Profile* (1998), available via http://www.epa.gov/surf/HUCS/hucinfo/

26. Texas Department of Health, "A GIS Approach to Assessing the Role of Environmental Exposures in Neural Tube Development" (draft of research proposal to the U.S. Environmental Protection Agency, 1996); Tiefenbacher, "Environmental Hazards of the Lower Rio Grande Valley."

27. Maril, *Poorest of Americans;* Betts and Slottje, *Crisis on the Rio Grande.*

11

Specialization and Diversification in the Agricultural System of Southwestern Kansas, 1887–1980

Thomas C. Schafer

To fly over southwestern Kansas on a sunny summer day can be an exhilarating experience. You rise up from a small airfield through a few light cumulus clouds and, after a few minutes circling to gain altitude, you gaze down at the countryside, with nothing between you and the face of the Earth but a few thousand feet of air and sunshine. Certainly, there is usually a little turbulence to remind you that you have not really "slipped the surly bonds of Earth," but that, for the enthusiast, is only a reminder of the buoyant forces that make flight possible. All in all, it is a pleasant way to spend a summer afternoon.

The face of the countryside is pleasing, too. From such a vantage point, one can see the pattern of river and stream, road and fence spread out below like a great choropleth map. But the most prominent sight will certainly be the great crop circles—not the supposed alien artifacts seen by moonlight in English fields, but the great green circles created by the center-pivot irrigation systems. Spread out below you will be a profusion of circle after circle of plant growth, as green as a well-watered golf course—irrigated wheat, corn, soybeans, sorghum, and even sunflowers. It is a beautiful sight to a person who loves growing things—so beautiful, and yet so familiar, that one may forget how strange this sight would have seemed to a farmer in the region just fifty years ago. Circular fields! Who ever heard of such a thing, especially in a nation that, up until a few years ago, seemed to be dedicated to the proposition that all fields should be square? Obviously something strange is going on in this unusual man-made environment of giant lawn sprinklers, circular fields, and underground rain.

There has been an overall trend toward regional specialization in American agriculture in the middle third of the twentieth century. This is in accordance with the main line of agricultural geographic theory, which postulates that increasing mechanization and modernization in agriculture leads to greater regional specialization and that a given geographic region should, over time, see an increase in overall uniformity of crops grown at the regional level, as measured in percentage of acres planted.[1] But as Morton Winsberg and other scholars have noted, this overall trend toward specialization can obscure smaller regional and subregional trends toward greater crop diversification.[2] In particular, the role played by local factors such as climate, settlement history, and character of the people involved in agriculture has often been ignored—perhaps because it is difficult to quantify such things. However, the human and environmental factors such as climate, perceptions of the environment, and economic factors were crucial in the shaping of cropping patterns in southwestern Kansas over the period 1887–1980.

First, the significance of crop specialization as a concept will be discussed, followed by a description of the human and natural environment of the region in 1887, when accurate statistics first became available for population and crop type. Then the dramatic changes in cropping patterns and practices that occurred in southwestern Kansas between 1887 and 1929, 1930 and 1949, and 1950 and 1980 will be described. The trends seen during these periods are far more complex and variable than predicted by current agricultural geographic theory and are in many cases quite at variance with that theory.

The Significance of Crop Specialization

The economic impacts of agriculture are obvious. What is less often realized is that changes in the basic agricultural practices can produce dramatic alterations in the entire economy of a region. An excellent example of this is the transition from corn and soft wheat to hard winter wheat in the Flint Hills–Bluestem Prairie portion of Kansas. As James Malin has pointed out in his classic study of agricultural transition, *Winter Wheat in the Golden Belt of Kansas,* this alteration brought with it dramatic changes in flour milling, machinery sales, retail development, and even the world economy.[3] These economic consequences tended to spread out in a ripple effect, leaving no part of the economy untouched:

> Purchasing labor-saving devices required large cash flows. Substituting a grain binder for a sythe increased the land required for future production. A farmer could not use a binder economically on the same amount of acres as were harvested with a scythe. That additional land could be acquired either by renting or buying nearby acreage, or by taking it from some other line of production. Often both were done, and both resulted in production specialization. A basic economic change thus occurred in the structure of Kansas agriculture as the advantages of specialization increased.[4]

Less obvious, but equally important, are the social changes that specialization often brings in its wake. Commentators as ideologically diverse as James Malin and Donald Worster have agreed that the incorporation of farmers into the market economy can have dramatic social effects.[5] One of the most notable of these effects is the reduction in self-sufficiency that specialization can bring. As agricultural historian John Sjo has noted, this certainly has had an effect on Kansas in the past: "As farmers moved into the marketplace, the degree of self-sufficiency was reduced. Farmers had cash to buy farm-produced products that other farmers had marketed. Growth of non-farm processing firms reduced on-farm processing and released family labor to produce more of the product that the farm could market best."[6]

The ecological impacts of crop specialization are more controversial and have been made more so by the sudden rise to importance of nonlinear dynamics, or chaos theory. It is argued by some researchers that complex systems are better able to withstand drastic change, whereas others argue that, in fact, complex interrelated systems are more liable to chaotic breakdown.[7] However, regardless of how this debate is finally resolved, replacing a complex system with a more simple monocultural has at least some ecological consequences, especially with regard to susceptibility to disease and genetic diversity.

Twelve counties located in the southwesternmost corner of Kansas, Stanton, Grant, Haskell, Gray, Ford, Kiowa, Morton, Stevens, Seward, Meade, Clark, and Comanche Counties, are appropriate to assess for a number of reasons. These counties have all undergone dramatic change in cropping practices in the last 110 years.[8] All were close to the heart of the dust bowl during the "dirty thirties" and so would tend to reveal the effects of this phenomenon on crop diversification. Also, all have since made dramatic agricultural recoveries with the help of irrigation.[9] Precipitation in this region is generally close to what is considered marginal for most nonirrigated agriculture, that is, around twenty inches of rainfall per year.[10] The region is far enough away from any potential large urban market for its agricultural products that transportation cost, if important, would show a detectable effect.[11] And, perhaps most importantly, the region is underlain by the largest source of fresh groundwater in North America—the Ogallala aquifer.[12]

The counties chosen for study lie primarily within the land resource area known as the Central High Table Land, although the more southerly portions of Morton, Stevens, and Seward Counties lie partially within the Southern High Plains. The average elevation is typical of the High Plains, averaging between two thousand and four thousand feet, with the lower elevations generally lying in the east and sloping gradually upward toward the Rocky Mountains. Precipitation varies from fifteen to twenty-two inches per year, with averages decreasing by about one inch for every seventeen miles of westward travel.[13] The precipitation totals demonstrate a strong seasonal component, with about 75 percent of precipitation falling between April and September. Growing season ranges from 180 to 190 days on average, but unseasonable killing frosts are common, especially in the higher elevations.[14] The underlying geology is composed of gently dipping sedimentary rock, limestones, shales, and sandstones. The most important geologic formation for agriculturalists in this region is, of course, the great Ogallala aquifer.[15] In many ways, the region is a typical High Plains environment.

The vegetative cover of this area prior to Euro-American settlement was primarily what Sen. John J. Ingalls referred to in 1872 as the "Universal beneficence of grass."[16] In this portion of Kansas, short grasses such as buffalo and blue grama were the dominant form of vegetation, with some taller grasses and trees in the few watercourses. It should be noted that while this vegetative cover is often referred to as "natural," a growing body of evidence indicates that it was an environment actively managed and shaped by

American Indian practices.[17] While the extent of this management is still controversial, one should remember that human interaction with the environment in this region did not begin with the coming of the white man.

The Human Environment, 1887

The history of Euro-American settlement in this region prior to 1887 had been one of a gradual extension into the region during wet years and a pulling back in the face of recurrent drought and grasshopper infestations, which in the "hopper year" of 1874 were compared by settlers in the region to a biblical plague.[18] However, the 1880s had generally been good years for southwestern Kansas, in terms of agricultural conditions. Rain had fallen consistently since 1884. These wet years gave rise to theories in certain quarters that cultivation was actually leading to an increase in total rainfall; the fact that previous attempts at cultivation had not led to any consistent increase in no way dampened the optimist's rosy hopes.[19] A great influx of settlers had flowed into the region during the 1880s, and by 1887 populations had reached a peak that would not be matched for several decades after this. Cropping patterns in 1887 reflected the belief that a more humid climate was here to stay. Such was not the case.

It would be a mistake, however, to conclude that the farmers of southwestern Kansas were ignorant or that they did not do what they could to maximize their chances of success in their new environment. A typical example of these tough-minded pioneers was Elam Bartholomew, who seems to have been atypical only insofar as he kept a diary of remarkable completeness.[20] Bartholomew attempted to maximize his profit by marketing his crops throughout the year, waiting for the best price with as much forethought and business acumen as any commodities speculator in Chicago. This strategy was sometimes pre-empted by the need to pay off debts, but overall it seems to have been a success. Bartholomew also sought relief from the fluctuations of the market by crop diversification, growing "rice corn" as well as wheat, and selling eggs and garden produce. This strategy did not produce invariable economic success, and Bartholomew had to borrow money when times were hard—and they were often hard, it seems. Still, the ingenuity of Elam Bartholomew illustrates that High Plains agriculturalists, while perhaps occasionally overoptimistic in their assessment of the nature of the environment they sought to master, certainly did not march blindly to economic destruction.

Elam Bartholomew was not alone in his attempt to grow humid-area crops on the dry plains of southwestern Kansas. The agriculture pattern found in this area was one that might have been expected for a far more humid environment. A remarkable 99,694 acres were in corn—a crop less suited for nonirrigated agriculture in western Kansas can hardly be imagined, due to its high moisture requirements and rapid transpiration rate. Other common crops were hard winter wheat and sorghum, which, given the alternatives, were not bad choices. Of the 255,301 cultivated acres in the twelve-county area, only 10,395 acres were planted in hard winter wheat; 26,627 in sorghum; and only 21,471 acres in oats, barley, and rye, crops suitable for feeding domestic animals. Other crops and fenced pasture occupied the remaining acreage. In sum, almost 50 percent of the agricultural area of these semi-arid counties were occupied by humid-area crops.

In terms of specialization (defined here as the percentage of land in any given crop, with higher percentages of land in a single crop equaling greater specialization), the numbers are also instructive. About 4 percent of all cropland was in wheat; 39 percent was in corn; 10 percent was in sorghum; and 8.4 percent was in oats, barley, and rye. Corn and sorghum occupied about half of all cropland.

In figure 11.1, the major crops for the area are listed by county. In figures 11.2–11.4, wheat, sorghum, and corn will be consistently listed, but other crops will be noted as necessary to display the changing agricultural nature of the region. For figure 11.1, oats, barley, and rye make up the fourth category.[21]

It seems clear that this cropping pattern essentially represented an attempt to impose a humid agricultural pattern upon a semi-arid area. This becomes even more apparent when one examines the things that were said by the people of the region at this time. The editor of the *Dodge City Times* predicted that "a few years more will see every quarter section in Ford County occupied and blooming like a garden," and this budding optimism was seconded by other sources.[22] A report by the State Board of Agriculture, in what might seem to be a blatant assertion of boosterism, asserted that "the country is improving, the rainfall is increasing, and the time is coming when crops in western Kansas will be nearly as certain as in the

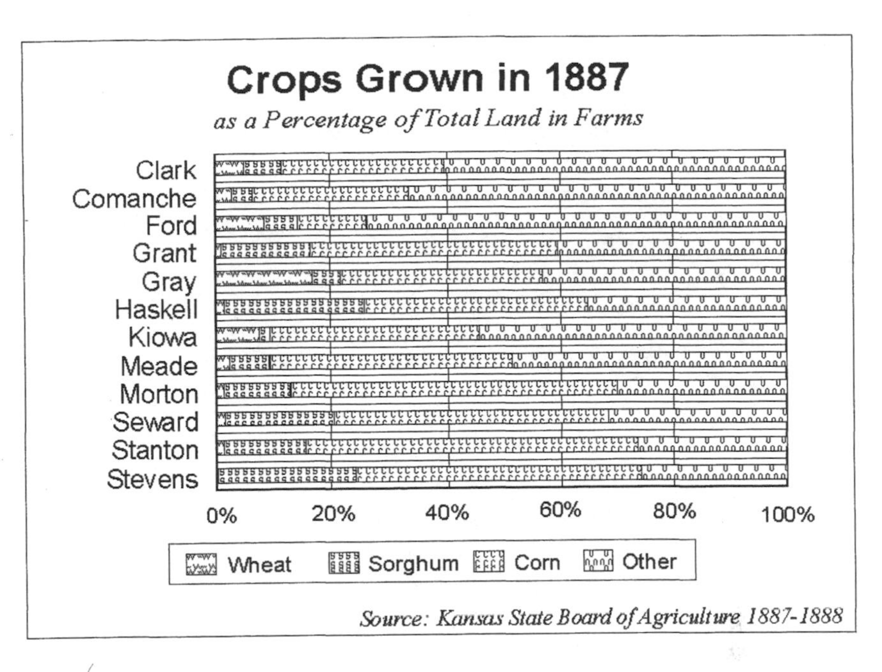

FIGURE 11.1 Crops grown in southwestern Kansas in 1887.

eastern part of the state . . . with intelligent treatment, this vast domain may be converted into a veritable paradise, a world in itself, possessing the highest possible environments [sic] ever realized by man."[23] The attempt to farm western Kansas as if it were Illinois had, it seems, the sanction of the civic boosters and state authorities.

Perhaps the most obvious indication of this environmental bullheadedness was the attempt to grow corn. However, during the nineteenth century, corn was far more to American farmers than just another crop—it was a symbol of civilization and of American patriotism. At least one commentator has stated that a veritable "corn cult" existed in the United States during the nineteenth century.[24] The attachment of American farmers to this crop was profound. Aside from ideological considerations, many of the farmers who settled in this region had originally come from corn-growing regions, and they tended to plant those crops with which they had the most familiarity.[25] James Malin noted the stubbornness of farmers in central Kansas in persisting in corn cultivation long after conditions had shown

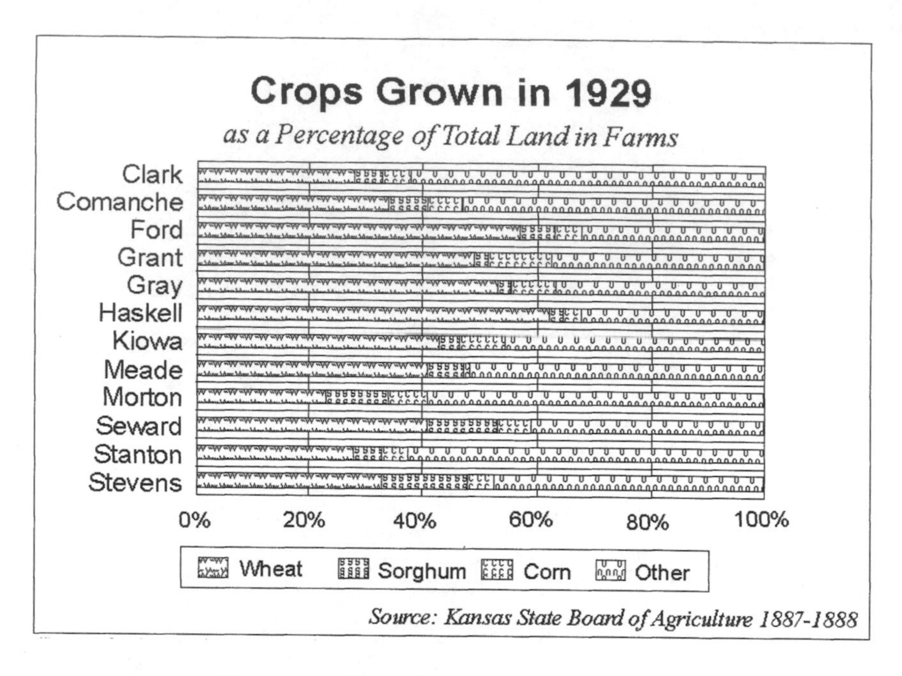

FIGURE 11.2 Crops grown in southwestern Kansas in 1929.

themselves to be unsuitable; conditions in western Kansas were even more unsuitable. It was the presence of corn, more than any other crop, that indicated the human belief that the climate would become more humid with cultivation. In a phrase quoted numerous times by Paul Travis: "Rain would follow the plow."[26]

Adaptation and Illusion

The decade immediately following 1887 was not a good one for the farmers of western Kansas. In the wake of drought, only about 5 percent of the corn crop was harvested in 1890, thus demonstrating the total unsuitability of this crop for western Kansas conditions. Although some continued to parrot the line that rainfall would surely permanently increase due to agriculture, such optimists were primarily newspaper editors and other boosters, not farmers.[27] Most farmers in western Kansas were quick to adopt new dry-land cultivation techniques and to change crops in the face of a climate that showed two faces with regularity. Whereas corn had failed in the terri-

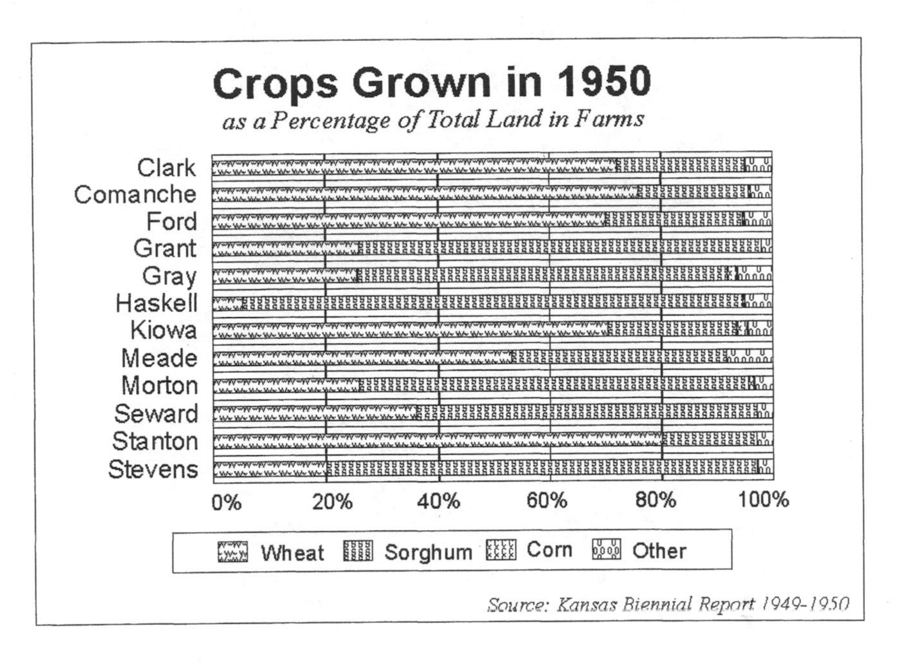

FIGURE 11.3 Crops grown in southwestern Kansas in 1950.

ble drought year of 1890, 90 percent of seeded acreage in winter wheat was harvested, with an average yield of fourteen bushels per acre.[28] This lesson was not lost on the enterprising farmers of western Kansas, and acreage planted in corn dwindled as that of wheat increased. The lessons of drought were hammered home once again by yet another drought lasting from 1893 until 1898, with even the winter wheat crop taking a beating from heat and dryness. By 1900, some of the twelve counties had been virtually depopulated. The attempt to produce a diversified, humid-area agriculture in a semi-arid region was, for the time being, over.

The lessons to be learned from this sad record of failure were apparent to some as early as 1889. Professor Edward M. Shelton of Kansas State College traveled through most portions of the region during the fall of 1889. After due consideration, Shelton issued a report stressing the following four points: (1) climatic conditions in western Kansas were dramatically different from conditions in eastern Kansas; (2) these conditions remained constant, and although the extreme natural variability of the climate might

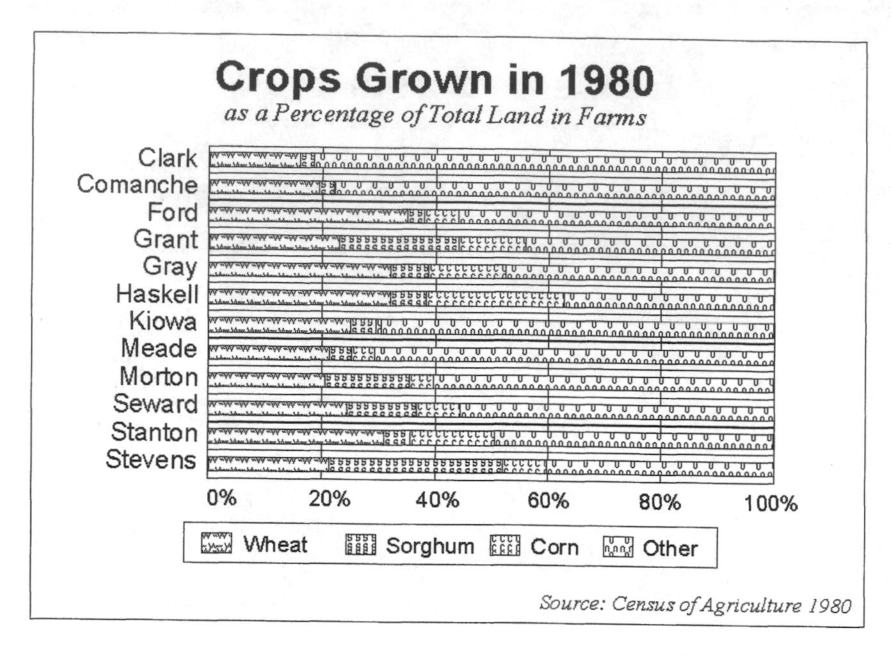

FIGURE 11.4 Crops grown in southwestern Kansas in 1980.

produce a short-term impression of change, the overall climate remained semi-arid; (3) a general system of irrigation was not possible in western Kansas due to the lack of dependable sources of water; and (4) agricultural success in the region would not come until crops and systems of farming were adapted to the environment.[29] Although some debated his conclusions, the overall development of agriculture for the next several decades would be roughly in the direction Shelton laid out.

In the first decade of the 1900s, a distinct trend toward winter wheat cultivation arose, along with a consequent decrease in acreage devoted to other crops. By 1908, kharkov and imported turkey wheat seed were available for planting; being hardier than previously planted varieties, they were much in demand.[30] Not until 1917, under the wartime demand for bread, did wheat become the dominant crop of the state. Sorghum also found increasing favor at this time, because it resisted conditions that could wither even wheat and it thrived in saline soil. Farmers also displayed an increased interest in dry-land farming. Overall, the trend in agriculture was toward a greater accommodation of the environment, rather than its modification.[31]

The major exception to this trend in the twelve counties was a growing interest in irrigation schemes involving the Arkansas River and Ogallala aquifer water. Irrigation in western Kansas has had a long history; in fact, the Garden City Irrigation Company was financing the digging of irrigation canals as early as 1880. The best technology of the 1890s could not harness the river, but interest increased as technology offered more possibilities for water management. When the years 1908–1914 proved to be drier than usual, interest increased even more.[32] The major roadblock to use of the Arkansas River was thought to be the amount of water siphoned off by large irrigation projects in Colorado, so Kansas sued to prevent diversion, an effort that was ultimately unsuccessful.[33] However, far more important than any lawsuit was the cropping possibilities that were opened up by the use of Arkansas River water. The combination of water and soil found in the Arkansas Valley led to the introduction of a crop that was both new and totally unsuited to a nonirrigated environment, sugar beets.

Farmers first began growing sugar beets in western Kansas in 1900, and their acreage increased rather rapidly.[34] In 1906, a processing plant was constructed at Garden City, and by the late 1920s, nine thousand acres were devoted to sugar beets.[35] This crop certainly did not compete with either wheat or corn in terms of acres planted, but its presence was significant nonetheless; as a crop that could thrive only in a human-engineered environment, it represented a trend directly counter to the adaptation to prevailing climatic conditions of wheat and sorghum cultivators.

The years between 1914 and 1929 brought both prosperity and trouble to farmers in southwestern Kansas. The coming of war in 1914 led to a tremendous increase in wheat acreage, along with profits for wheat growers. However, when the war ended the bottom dropped out of the market, and sorghum replaced wheat on a good deal of land, due to its greater drought tolerance and resistance to wind damage.[36] Mechanization also made notable inroads during this period. In 1914, there were almost 1.3 million horses and mules on Kansas farms, but no tractors or combines. The first combines were introduced in 1918. By 1929, there were 53,000 tractors and 21,000 combines in the state; the number of horses and mules had declined by one-half.[37] In terms of technological progress, the period 1914–1929 was in many ways the best era western Kansas had yet seen or, at least, the period of most rapid growth.

Despite increasing yields, benign weather conditions, and seeming prosperity, all was not well in southwestern Kansas agriculture. Spurred on by patriotism and phenomenal wartime prices, farmers all over the United States had increased production to record levels. This had entailed bringing into production marginal land that would have been better left unplanted. This massive overproduction ensured a tremendous drop in prices once world production returned to normal.[38] Prices declined by one-half between 1919 and 1921, and this decline continued throughout the 1920s. This increased the push for greater production to counteract the lower prices, which led to increases in surpluses and even lower prices—a vicious circle that could only result in depressed prices and fewer farmers on the land. In many ways, the farm sector of Kansas entered the Great Depression a full decade earlier than did the rest of the United States.[39]

The Human Environment, 1929

Kansas is the "wheat state," if the Kansas Department of Agriculture is to be believed. It is often not appreciated, however, that up until 1907, corn acreage was greater than that of wheat and in terms of total production corn was the predominant crop in the state up until World War I.[40] This was apparent even in the study area, where conditions did not favor "King Corn." However, with the coming of war in 1914 hard red winter wheat, introduced in the 1880s by Mennonite immigrants, began to assume greater importance, especially in southwestern and central Kansas.[41] In 1929, southwestern Kansas still displayed the effects of this wartime boom. Wheat as a percentage of total farmland would never be higher. Grain sorghum also displayed rapid growth as a crop in the years immediately preceding 1929 and supplanted corn on even more acreage. Crops used as fodder for horses and mules, such as oats and barley, were in steep decline by this time due to growing mechanization.[42] In 1929, wheat and sorghum were the major crops in the area, with corn a declining but still important third. The disc plow started to appear on the more affluent farms, as did other technological innovations that made greater production possible for individual farmers.[43]

Like many other farmers in southwestern Kansas, Lawrence Svobida, who wrote of his experiences in his classic 1940 work *Farming the Dust Bowl,* sought to make up for declining wheat prices and occasional crop failure by plowing more land and even by hiring others to plow for him: "My tractor roared day and night and I was turning eighty acres every twenty-four hours, only stopping for servicing every six hours."[44] Svobida's hired man drove a twelve-hour shift all day and Svobida drove all night, breaking the land for yet more wheat. Disasters due to hail and blizzards did not discourage him; rather, they only spurred him on to greater efforts, more furious exertions: "Often I could hardly keep my eyes open, and when I would doze off for a few seconds, only an instinctive but vicelike grip on the steering wheel kept me from falling and being ground to pieces under the plow." Svobida's almost heroic exertions in the face of what must have seemed to be a unremittingly hostile environment may stand as representative of the southwestern Kansas farmer of those years.

Lawrence Svobida's primary crop was wheat, although he raised some livestock and feed grain. In this, he was also much like his fellow southwestern Kansas farmers, as the 1929 county-by-county record of crop distribution, expressed as a percentage of total land in farms, reveals (see fig. 11.2).[45]

Wheat had become the dominant crop in southwestern Kansas. Corn was the third largest crop, but it should be noted that it was, surprisingly, the second crop in several of the individual counties. The sorghum totals had actually fallen since 1887, losing ground to wheat. Fallow acreage was higher for the study area than for the state as a whole. This was not surprising, given the semi-arid nature and repopulation that had occurred in the region. However, the amount of fallow land was lower at this point for both the state and the area than at any other time in the study period, perhaps reflecting pressures to plant all available acreage, as Lawrence Svobida had done. Southwestern Kansas displayed a fairly high degree of specialization as of 1929, which indicated that a good deal of adaptation to the climate and soil characteristics had taken place.

Bust, Dust, War, and Recovery

The agricultural history of this region between the years 1930 and 1949 reflect the impacts of three overwhelming events: the dust bowl, the Great Depression, and World War II. During this period the unsuitability of dryland corn for this area became apparent; the corn crops of 1934 and 1936 were virtual failures.[46] Topsoil loss was dramatic. The bill on the marginal and submarginal land plowed up in the 1910s and 1920s came due, and the "black blizzards" began to roll. Acres in the "land idle or fallow" category also reached very high levels, as a percentage of total farmland, and in some counties this category actually took up more land than wheat or sorghum. The once indomitable Lawrence Svobida left the High Plains forever, his efforts unavailing against the twin scourges of drought and economic depression.

Prosperity only gradually returned to the twelve counties and the state. The key events were the coming of rain and World War II. As great a catastrophe as the war was for the world as a whole, it did at least have one salutary effect: It finally pulled Kansas agriculture from its two-decade doldrums. With farmers in southwestern Kansas being encouraged to plant "fencerow to fencerow," they responded as patriotically as had their predecessors in World War I.[47] The experience was not precisely the same, however. New technology was introduced, and the amount of horsepower at the disposal of the individual farmer increased dramatically.[48] The Office of Price Adjustment and other wartime agencies controlled prices, which led to much wry comment among Kansas farmers, who allegedly wanted price controls on what they bought, but not on what they sold. The U.S. government guaranteed high supports for farmers for up to two years after the end of the conflict, however, and this helped to make the price controls more palatable.[49] Despite the grim nature of the conflict and the chafing of individualistic farmers under federal controls, the period 1940–1945 seems to have been one of recovery.

The two most significant developments in cropping practices between 1930 and 1949 were the early beginnings of the practice of summer fallow and, above all, the rise of grain sorghum as an alternative to wheat. Sorghum production increased 400 percent, and the large amounts of fallow land decreased to more normal levels. The practice of alternating wheat-sorghum-fallow began about this time.[50] Grain exports increased rapidly from about 5 percent of all Kansas grain in 1940 to 35 percent by 1950.[51] Corn production continued to fall throughout the state, however. Farmers seem to have given up the attempt to cultivate it as a dry-land crop.

The Human Environment, 1950

The study region looked very different in 1950 than it had in 1929, in terms of crops grown. The trend toward adaptation to the climate evident in the years since the dust bowl of the 1930s probably reached its apogee at this time. Corn had dwindled to a minute percentage of farmland, whereas sorghum and wheat reigned supreme. Needless to say, this also represented the high noon of specialization in the agriculture of southwestern Kansas, because only those crops suitable for the climate were grown there and there were few crops that could tolerate such dry conditions.

As mentioned above, one of the most significant developments in agriculture for the period 1930–1949 was the advent of the summer fallow method of cropping in semi-arid environments. Variants of this technique, which involved leaving land fallow to increase buildup of subsoil moisture and often involved the use of crop rotation methods, had been used by some farmers as early as the 1920s.[52] By 1947, summer fallowing had become widespread enough to be recognized by the Department of Agriculture as a separate type of fallowing, and was hereafter recognized in a separate category in the agriculture census. Summer fallowing was found to be effective in raising yields and, to a lesser extent, in reducing variability in yields on semi-arid lands. Aiding the practice was the growing use of the one-way disc plow, which did not bury as much straw and significantly improved water storage efficiency.[53]

However, other changes were occurring at this time that would reverse the past 50 years. The most important of these changes was a growing interest in the subject of irrigation, spurred by the development of highly efficient centrifugal pumps in the 1940s. The discovery of natural gas in the area, coupled with the improved pumping technology, lightweight aluminum pipe, and new chemical fertilizers, enabled irrigation to assume new importance. In 1950, this revolution in agricultural practices was only in its earliest stages.[54] This becomes evident in the 1950 county-by-county record of crop distribution (see fig. 11.3). Overwhelmingly, the land was planted in wheat and sorghum, that is, in primarily dry-land crops.[55]

The study region had experienced a dramatic increase in specialization between 1930 and 1949. The percentage of land in wheat increased slightly, but the amount of acres devoted to sorghum rose dramatically, as farmers switched to this more drought-resistant crop, and the wheat-sorghum rotation pattern became ascendant. Corn reached its lowest recorded point in the study region. As to specialization, over 90 percent of all cultivated land was sown in just two crops—the crops that were best adapted to the environment—that is, adaptation to the region's climate had also led to greater crop specialization.

This period also witnessed the beginning of the end of the sugar beet boomlet. According to environmental historian James Sherow, "Before 1950 the industry was in serious trouble, and by 1980 all the factories in the [Arkansas River] valley had closed their doors."[56]

The Center-Pivot Era, 1950–1980

The period 1950–1965 marked a watershed in agricultural practices throughout Kansas. On the whole, these changes stemmed from the massive strides made in science and technology during the war years, on the farm as well as in the agricultural experiment stations. Under the spur of necessity, Kansas farmers had invented a wide array of new machinery, which implement companies would bring to market.[57] Simultaneously, chemical fertilizers, pesticides, and herbicides were introduced, increasing yields dramatically, as did new strains of hybrid wheat and sorghum, while the amount of horsepower available to the average farmer increased again.[58] U.S. grain exports also leaped in the early 1950s, a result of the activist foreign policies of the Eisenhower and Truman administrations and of a reviving world economy. These factors coincided with record wheat production in Kansas, especially during the "bin-buster" year of 1952.[59] Land in farms increased for both the study region and the state, as marginal land was brought back into production. A drought in the mid-1950s resulted in the "little dust bowl," which led to some distress. However, unlike the 1930s, the influence of climate was mitigated by technology and the use of chemicals on the farm. Leo Hoover, in his history of Kansas agriculture, notes "the amount [of fertilizers] used was very small until World War II. The increase in use after the war was so rapid that the amount applied in 1950 was almost four and a half times the amount applied in 1945."[60] However, the application of fertilizers was significantly higher in the eastern portion of the state because the lack of rainfall in the western third of the state limited fertilizer's role in western Kansas crop production.

The drought of the 1950s served as proof to many farmers in the area that a replay of the dust bowl was inevitable if technology was not employed. As John Opie remarks in his *Ogallala: Water for a Dry Land:* "Momentum [for irrigation] grew during the 'filthy Fifties,' and in the 1960s a man-made deluge of 'walking water' [the mobile center pivots] began to be pumped from underground."[61] But the period of greatest growth in irrigation in the study area was to lie in the future. In the 1950s, irrigation in southwestern Kansas held out a great deal of promise—but that promise still exceeded the reality.

The years between 1965 and 1980 saw an intensification of chemical use, mechanization, hybridization, and irrigation. George E. Ham notes that mechanization had increased so dramatically that a farmer in the early 1970s used almost ten times as much horsepower as had his counterpart at the beginning of the century. Ham also noted that the average Kansas farmer used fifteen times as much fertilizer as did a farmer in 1929, and input of farm labor had decreased to only 30 percent of what it was fifty years earlier. Yields had also skyrocketed.[62] In the early 1970s, new strains of hybrid wheat were introduced into Kansas, but, the adoption of these new strains was held back to some degree by questions about hybrid vigor. This was significant because it undoubtedly made selection of crops other than wheat all the more appealing for farmers in the study area, and many of these crops needed irrigation. The key to this transition was the development of the center-pivot irrigation system.

Described as "a sprinkler irrigation lateral that is mounted on wheeled structures [towers], anchored at one end [pivot point] and which automatically rotates in a circle when irrigating" and working on the same principles as a giant lawn sprinkler watering an average of 130 acres per pivot, this system was found to be optimal for use on the flat, somewhat sandy soils of southwestern Kansas.[63] This irrigation technology, when coupled with the ready availability of natural gas with which to power turbine pumps, led to ever-increasing demands on the chief source of subterranean water for the study region, the great Ogallala aquifer. Between 1959 and 1974, according to the U.S. Census of Agriculture, the amount of irrigated land in Kansas almost tripled, and many of the counties in southwestern Kansas were the leaders in this development. This led to the reappearance of crops such as corn, which had essentially been abandoned as unsuitable in the 1930s, along with the introduction of humid-area crops such as soybeans.

Clarence J. Gigot was perhaps the best example of this "center-pivot revolution" in action. Purchasing a single center pivot as an experiment in the early 1960s, Gigot used it to water a single quarter-section of corn. It its first year, the section yielded 120 bushels per acre. By 1965, Gigot was using eleven center pivots to water 1,760 acres of land. Between 1972 and 1974, the Gigot family drilled more than one hundred wells, just in time for the great irrigation boom of the 1970s. By 1980, the value of the Gigot "empire" was estimated to be around $13 million. Irrigation had come to be big business on the High Plains of Kansas.

This trend toward ever-greater crop diversification and groundwater use abated somewhat in the late 1970s as rising energy prices imposed limits on the amount of water that could be "lifted" economically from the Ogallala aquifer. Even the mighty Gigot "empire" was forced to reduce the number of pivots in operation and to more closely monitor those that remained.[64] These economic factors made water conservation of prime concern in the region for the first time. By 1980, new technology was being introduced to limit water use, along with some minor changes in crop selection. But these concerns only served to emphasize the magnitude of the changes the area had undergone in the last thirty years.

The Human Environment, 1980

Since 1950, contrary to what had been predicted by agricultural geographic theory, considerable diversification had taken place, at least on the regional level, although this may not have been the case at the level of individual farmers. Corn had made quite a remarkable resurgence, and irrigated soybeans had entered the cropping system (see fig. 11.4). In 1980, in fact, the area had begun to look a good deal as it had in the wet years of the 1880s, but this time the years were wet because of underground rain—the Ogallala aquifer.[65]

These statistics are an indication of a massive revolution in agricultural practices fueled by the turbine pump, the center-pivot irrigation system, the growing use of pesticides and herbicides, and the planting of hybrid crops. Wheat was only slightly more prevalent as a crop than it had been in 1887; sorghum continued its long decline from its apex in 1950; and corn rebounded significantly, actually surpassing 1929 levels. Soybeans became statistically significant in both the state and the study region for the first time.

It should be noted, however, that greater diversification in the region did not necessarily correlate to diversification on individual farms. It is entirely possible that some individual farmers became more specialized, but that different farmers specialized in different crops, thus leading to overall regional diversification. There is some anecdotal evidence that larger operators, such as the Gigots, have diversified in response to market conditions and the greater possibilities opened up by center-pivot irrigation.[66]

Conclusion

I do not claim that this study is a comprehensive look at the theory and practice of crop diversification. It may be that the trends that apparently occurred in this region at this time are actually making a deeper, more significant trend, which more research will in time uncover. However, several conclusions may be drawn.

There has been no notable trend toward greater specialization in agriculture. In fact, the trend seems to have been in the opposite direction, at least at the scale examined here. As measured by crop acreage, the twelve counties were at their most specialized in the middle of the period, with diversity increasing as time wore on. The dust bowl era of the 1930s made this trend less apparent, because it seemingly led to greater diversification (or perhaps simply less cropping), but the general direction is clear. Agriculture, as measured in crop percentages, has not become more specialized.

The pattern in 1980 shows a resurgence in some of the crops common to the region in 1887, especially corn. This make sense, when one considers that the net impact of technology has been to decrease the importance of the climate; in the case of irrigation technology, to decrease the importance of natural rainfall (at least for a time). This enables farmers to plant crops that fit the area's natural conditions and those desired by the market, within the economic constraints. This has the net impact of increasing crop diversity, and of mimicking, to some extent, the wet years of the 1880s.

Theoretical models of the type often used by geographers are not adequate to explain this crop diversification on the regional level. This may well be a matter of scale. At the national or global level, the models may be effective, but as scale is reduced, other factors, such as technological innovation and cultural factors, may override transportation costs as the determining factor in crop selection. The twelve counties in southwestern Kansas certainly represent such a case. No theoretical model of crop specialization will be successful unless it takes such special cases into account.

Notes

1. O. E. Baker, "The Increasing Importance of the Physical Conditions in Determining the Utilization of Land for Agriculture and Forest Production in the United States," *Annals of the Association of American Geographers* (1921): 11.

2. Morton D. Winsberg, "Concentration and Specialization in United States Agriculture," *Economic Geography* 56 (1980): 79.

3. James Malin, *Winter Wheat in the Golden Belt of Kansas* (Lawrence: University of Kansas Press, 1944), 3.

4. John Sjo, "The Family Farm Becomes a Business Enterprise: 1860 to 1980," in *The Rise of the Wheat State: A History of Kansas Agriculture, 1861–1986*, ed. George E. Ham and Robin Higham (Manhattan, Kans.: Sunflower University Press, 1987), 115–23.

5. Donald Worster, *Dust Bowl* (New York: Oxford University Press, 1977), 238.

6. Sjo, "The Family Farm," 120.

7. James Gleick, *Chaos: The Making of a New Science* (New York: Penguin Books, 1987), 175; Alston Chase, *In a Dark Wood* (New York: Houghton Mifflin, 1995), 277.

8. David E. Kromm and Steven E. White, *Conserving Water in the High Plains* (Manhattan: Kansas State University, 1990), 3.

9. Leo Hoover, *Kansas Agriculture after 100 Years* (Manhattan: Kansas Agricultural Experimental Station, 1957), 1; E. G. Heyne, "The Development of Wheat in Kansas," in Ham and Higham, *The Rise of the Wheat State*, 41–57.

10. Orville W. Bidwell and William E. Roth, "The Land and the Soil," in Ham and Higham, *The Rise of the Wheat State*, 1–19.

11. Peter E. Lloyd and Peter Dicken, *Location in Space: A Theoretical Approach to Economic Geography* (New York: Harper and Row, 1977), 183.

12. Bidwell and Roth, "The Land and the Soil," 16.

13. Ibid., 17.

14. L. Dean Bark, *When to Expect Late-Spring and Early-Fall Freezes in Kansas* (Manhattan: Kansas State University, 1979), 2.

15. Bidwell and Roth, "The Land and the Soil," 16.

16. Clarence Leo Petrowski, "Kansas Agriculture before 1900" (Ph.D. diss., University of Oklahoma, 1968), 18.

17. William Denevan, "The Pristine Myth: The Landscape of the Americas in 1492," *Annals of the Association of American Geographers* 82/83 (October 1992): 369–85.

18. Craig Miner, *West of Wichita* (Lawrence: University Press of Kansas, 1986), 52.

19. Lynnell Rubright, "Development of Farming Systems in Western Kansas, 1885–1915" (Ph.D. diss., University of Wisconsin–Madison, 1977), 37.

20. Elam Bartholomew, Diaries, 1 January 1871–21 December 1896, MS 1069–70, Kansas Historical Society, Topeka.

21. Kansas State Board of Agriculture, *Biennial Report, 1887–1888* (Topeka, Kans.: State Printing Office, 1889).

22. Paul Dee Travis, "Charlatans, Sharpers, and Climatology: The Symbolism and Mythology of Late Nineteenth Century Expansionism in Kansas" (Ph.D. diss., University of Oklahoma, 1975), 41.

23. Ibid., 39.

24. William J. Rorabaugh, *The Alcoholic Republic: An American Tradition* (New York: Oxford University Press, 1979), 113.

25. Travis, "Charlatans," 42.

26. Paul Travis, "Changing Climate in Kansas: A Late 19th-Century Myth," *Kansas History* 1 (1978): 48–58.

27. Rubright, "Development of Farming Systems in Western Kansas," passim; Travis, "Charlatans," 44.

28. Petrowski, "Kansas Agriculture before 1900," 76.

29. Ibid., 211–23.

30. Fred L. Parrish, "Kansas Agriculture before 1900," in *Kansas: The First Century*, ed. John D. Bright (New York: Lewis Historical Publishing Co.), 401–27.

31. Ibid., 115–16.

32. James Sherow, *Watering the Valley: Development along the High Plains Arkansas River, 1870–1950* (Lawrence: University Press of Kansas, 1990), 28; Parrish, "Kansas Agriculture before 1900," 115.

33. James K. Koelliker, "Water," in Ham and Higham, *The Rise of the Wheat State*, 93–103.

34. Kansas State Chamber of Commerce, *Kansas Year Book, 1937–1938* (Topeka, Kans.: Capper Printing Co., 1939), 210.

35. Ibid.

36. Parrish, "Kansas Agriculture before 1900," 125.

37. *Kansas Year Book, 1937–1938*, 211.

38. George E. Ham, "An Agronomist's View," in Ham and Higham, *The Rise of the Wheat State*, 15–19.

39. Paul Johnson, *Modern Times: The World from the Twenties to the Eighties* (New York: Harper and Row, 1983).

40. Hoover, *Kansas Agriculture after 100 Years*, 4.

41. Malin, *Winter Wheat*, 7.

42. Ham, "An Agronomist's View," 17.

43. Gustave E. Fairbanks, "The Mechanization of the Kansas Farm," in Ham and Higham, *The Rise of the Wheat State*, 103–14.

44. Lawrence Svobida, *Farming the Dust Bowl: A First-Hand Account from Kansas* (Lawrence: University of Kansas Press, 1986).

45. Kansas State Board of Agriculture, *Biennial Report, 1929–1930* (Topeka, Kans.: State Printing Office, 1931).

46. Frank G. Bieberly, "Other Crops in the Wheat State," in Ham and Higham, *The Rise of the Wheat State*, 57–73.

47. Parrish, """Kansas Agriculture before 1900," 151.

48. Ham, "An Agronomist's View," 19.

49. Ibid., 20.

50. Hoover, *Kansas Agriculture after 100 Years*, 49.

51. Harvey L. Kiser and Frank Orazem, "Marketing and Exporting of Kansas Agricultural Products," in Ham and Higham, *The Rise of the Wheat State*, 149–59.

52. Hoover, *Kansas Agriculture after 100 Years*, 52.

53. Koelliker, "Water," 93–103.

54. Ibid., 97.

55. Kansas State Board of Agriculture, *Biennial Report, 1951–1952* (Topeka, Kans.: State Printing Office, 1952).

56. Sherow, *Watering the Valley*, 170.

57. James C. Malin, *The Grassland of North America: Prolegomena to Its History* (Glouchester, Mass.: Peter Smith, 1967), 197.

58. Ham, "An Agronomist's View," 21.

59. Hoover, *Kansas Agriculture after 100 Years*, 33.

60. Ibid., 34.

61. John Opie, *Ogallala: Water for a Dry Land* (Lincoln: University of Nebraska Press, 1993), 143.

62. Bieberly, "Other Crops in the Wheat State," 63.

63. Kromm and White, *Conserving Water in the High Plains*, 11.

64. Opie, *Ogallala*, 144.

65. Kansas State Department of Agriculture, *Report, 1980* (Topeka: State Printing Office, 1980).

66. Opie, *Ogallala*, 145.

12

John Wesley Powell Was Right

Resizing the Ogallala High Plains

John Opie

Environmental historians are, or should be, map consumers—even map junkies. Admittedly, as with other historians, our field is set in time, and we must speak of historical particularity. But environmental historians also emphasize that human affairs are set in identifiable physical places and thus are site specific. Without a geographical particularity, environmental history is meaningless. However, this identification and description of location, which is often done through mapping, has its own contrariness that deserves our attention. I propose a shift from map information based on counties and sections to natural resource and ecosystem-wide strategies.

Confusion and misinformation due to mapping conventions do more to shape historical inquiry of the High Plains than is usually acknowledged. This is notably the case of the old dust bowl region of western Kansas and the Texas-Oklahoma panhandle. This region was one of the last to be settled into ranches and wheat farms between 1880 and 1920. Beginning in the 1930s, the region became the epicenter of debates over the heavy consumption of topsoil. Since the 1960s, the region's fate has been determined in large part by heavy consumption of groundwater for the production of high-yield crops of wheat, corn, and sorghum grains. The mapping of soils, water, and agricultural production has little to do with section boundaries and county lines. The need for continuous resizing is also suggested by a firestorm that broke out in the 1990s over the invasion of concentrated animal production, notably hogs. The impact of waste materials, pollution, and smells, which create "dead zones" around each hog unit, requires different mapping still, one tied neither to ecosystems nor county lines. We have little knowledge of long-term consequences in part because of the lack of appropriate mapping.

Mapping the Plains: Are We Using the Best Information?

The Limits of Survey and Section

By looking at the mapping of the state of Kansas, for example, we can see historic contradictions and misleading information. The relatively serene physiography of Kansas, compared to section geometry, portrays the divergence between natural landforms and the grid of survey and section (fig. 12.1). The classic rectangular land survey, which goes back to the Ordinance of 1785, may have had a geometric rationality so highly praised by Thomas Jefferson, but it was badly flawed in not recognizing the uneven physical features of even the most ordinary landscape. In Ohio, a farmer found himself in deep trouble if his quarter section of fertile land was badly split by a geological ridge or wetland. In the West, his situation was fatal if his quarter section did not include water. Despite these devastating disadvantages, the rectangular land survey continued to march westward. It deliberately fostered the interchangeability of the nation's regions regardless of terrain—"one size fits all." The Homestead Act of 1862, which urged settler-farmers to find their living on the classic quarter section of 160 acres, was based on humid regions east of the ninety-eighth parallel, where thirty to forty inches of precipitation was common. By the time settlers reached the semi-arid High Plains, where even a scant twenty inches of precipitation was joyously received, they needed a full section of 640 acres and still might not have ready access to surface or groundwater.

This flawed information based on section and counties is not likely to be dislodged. Most such census and agricultural data are set in stone. Add to this the absolute power of private property ownership based on section lines. Today on the central High Plains, property ownership is still shaped by sections, half sections, and quarter sections (fig. 12.2).

Demographic Scale by Counties. Mapping and measurement by counties rather than physiography has also led to controversial and sometimes flawed conclusions about the Plains; yet it is the most typical standard for

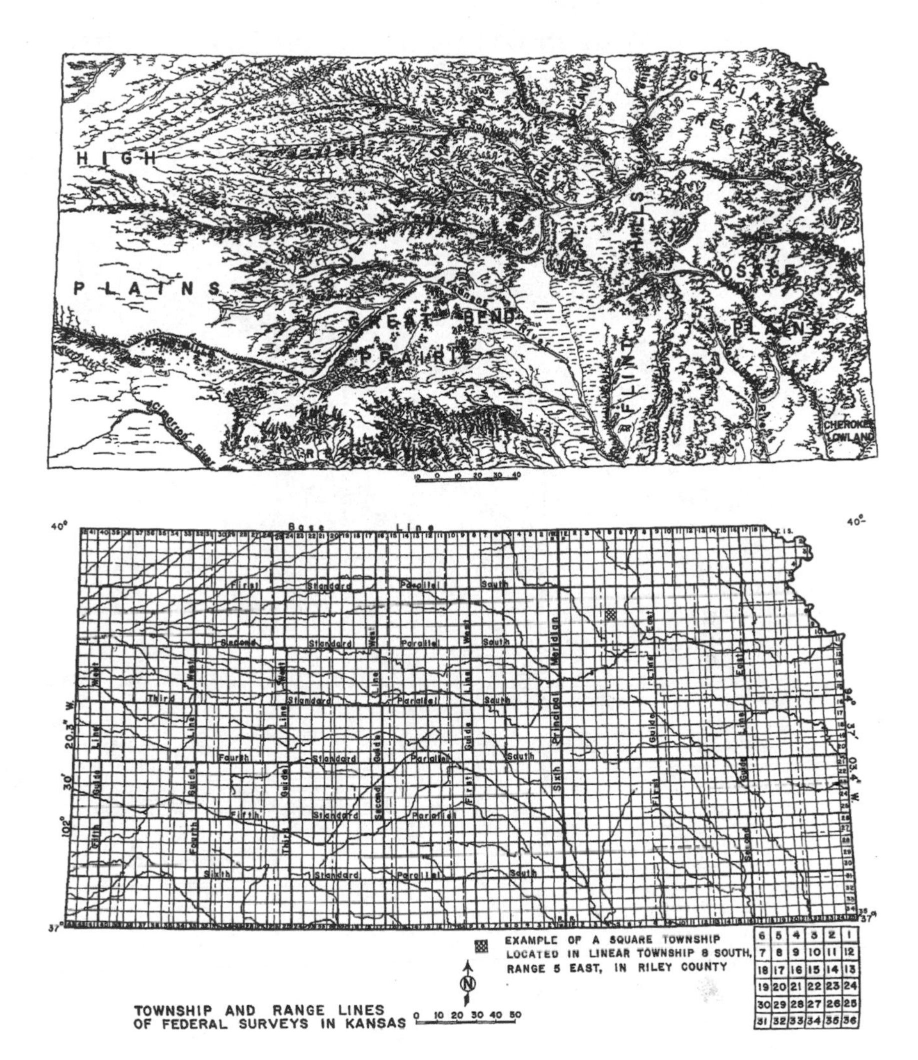

FIGURE 12.1 The natural terrain of Kansas (top) was not a land of extremes, but it included prime farmland, windy sandhills, limestone outcroppings, and widely dispersed flowing rivers. These differences, all-important to agricultural prosperity, were ignored when the public domain was surveyed and sold to farmers and speculators. The indifference of the geometric survey lines to actual physiography (bottom) was typical of the survey and sale system that began with the Land Ordinance of 1785. From Homer E. Socolofsy and Huber Self, *Historical Atlas of Kansas* (Norman: University of Oklahoma Press, 1972). Reprinted with permission.

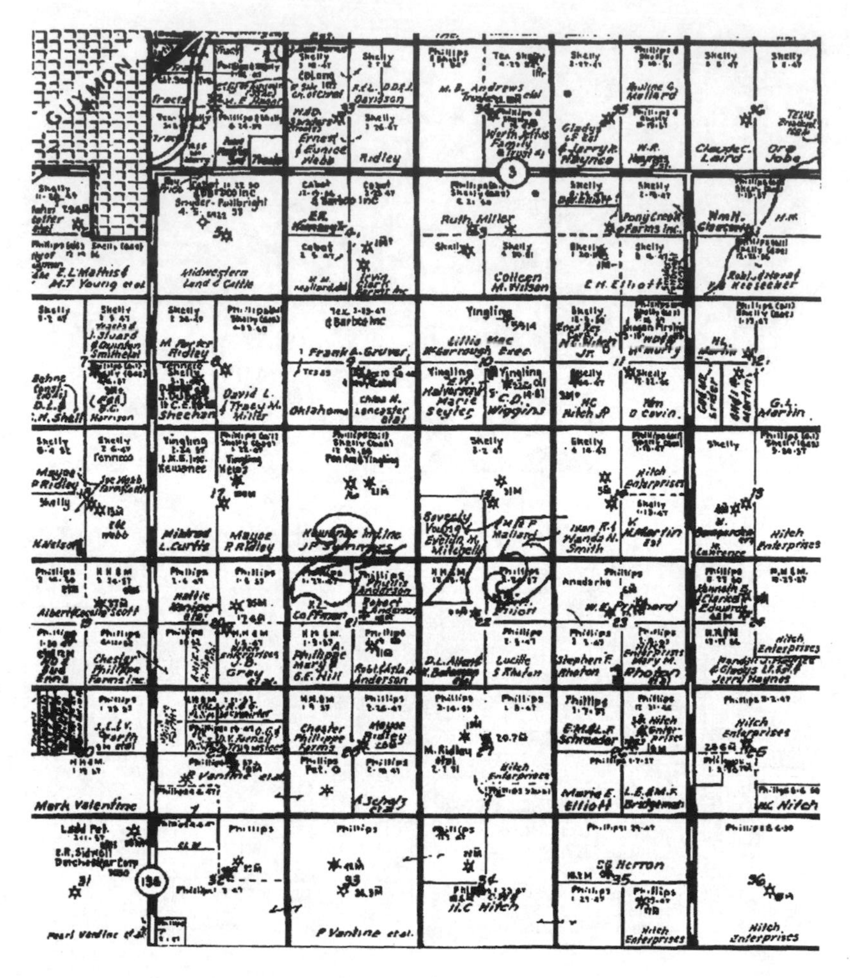

FIGURE 12.2 Detail of the 1990 hand-lettered property map near Guymon in central Texas County, Oklahoma. The square grid depicts section lines of 640 acres, some of which are divided into 160-acre quarter sections or even 40-acre quarter-quarter sections. Each owner is identified. Many have natural gas wells leased to corporate energy companies such as Skelly, Texaco, and Philips. Note that the lower right identifies several properties (not all) owned by Hitch Enterprises, one of the most influential cattle and hog ranchers in the Oklahoma panhandle. Local farmers, entrepreneurs, and inhabitants considered such property maps as essential as government soil maps (see fig. 12.9) to their lives and work. From "Texas County, Oklahoma," (Wichita: Kansas Blue Print Co., Inc., copyright 1987). Reprinted with permission.

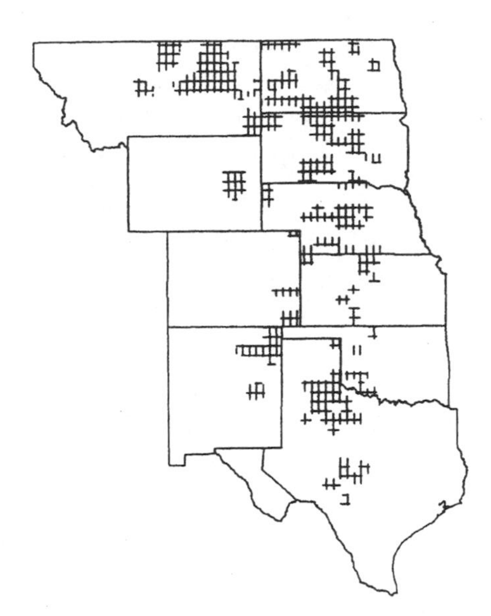

FIGURE 12.3 The cross-hatching identifies Plains counties threatened with permanent economic and social collapse according to at least three of six criteria (such as population loss, median age, poverty, and number of building permits as an indicator of economic growth) used by Frank J. Popper and Deborah E. Popper in a 1992 study. As a result, the Poppers argued for a programmed shift of the affected counties into a depopulated land reserve, the Buffalo Commons. From "Plains Counties in Land-Use Distress: Counties with at Least Three Indicators," in a report received by the author from Frank J. Popper and Deborah E. Popper, 1992. Reprinted with permission.

data gathering and interpretation, whether for population, land use, economic patterns, or other demographics. Measurement on the scale of counties or even townships is too gross for local agricultural or environmental decisions. Prosperity around Garden City, Kansas, does not necessarily depend upon other locations in Finney County, but all are lumped together in most databases. One of the most controversial uses of such data was the assertion first made in the 1980s by Frank and Deborah Popper that the Plains were collapsing into an uninhabited "Buffalo Commons" (fig. 12.3). They looked carefully at a county-by-county census report and discovered declining populations across a broad swath of the region. There was no alternative database. In a 1992 article, they devised six indicators of county-wide decline, any three of which put an entire county at risk: (1) long-term (thirty-year) population loss; (2) short-term (ten-year) population loss; (3) density per square mile (four or fewer people); (4) median age of thirty-five or higher; (5) poverty rate at 20 percent; and (6) number of building permits.

By default, such countywide criteria are widely used to shape public policy; the questions are whether alternative mapping would revise the above indicators and whether such results would offer alternative policies. The Poppers' conclusions were immediately castigated as wrong and irresponsible; they came near to being stoned whenever they visited the Plains. But if the Poppers were wrong, on what grounds could their conclusions be corrected? This essay seeks to explore options to measurement of the Plains by means other than land survey or county delineation.

Regional Steps in Alternative Mapping: The Example of Groundwater

Countywide data do not distinguish between different county operations based on the availability of water, such as dry-land farming, irrigated acres, or grazing land, but aggregates them into the dominant operation. The immense but declining groundwater of the Ogallala aquifer has become a major issue on the Plains. Fully one-third of the aquifer was consumed in the last forty years to support tens of thousands of irrigation applications. The supply of this groundwater, its rate of consumption, pump sites, and irrigated fields cannot be aggregated along county lines. Groundwater location and its declining availability do not stop at the edge of a section or county (fig. 12.4).

The mapping agendas of different groundwater management districts can diverge remarkably. For example, Kansas Groundwater Management District no. 1, in the far west-central part of the state, includes large-scale weather modification (cloud seeding) that often impacts a region larger than the district itself. Its neighbor to the south, district no. 3, while supporting weather modification, also records well-spacing directly onto U.S. Geological Survey topographic maps. The survey maps themselves depict topography as well as section or county lines.

Two hundred miles to the south, in Texas, High Plains Underground Water Conservation District no. 1 is addicted to maps. Its multicounty district shows patterns of center pivots that would be lost if reported only by county (fig. 12.5). The outline of the Ogallala aquifer is the most critical limit in the district's maps, not section or county lines. The district also provides similar maps of groundwater levels, annual consumption, and local soil moisture levels on a seasonal basis (fig. 12.6). Headquartered in Lubbock and in operation since 1952, Texas Groundwater Management District no. 1 has the most comprehensive mapping operation. A. Wayne Wyatt, manager of the district since it opened its doors, developed a detailed plan for county-by-county mapping of depth to the aquifer and its saturated thickness over his entire district.[1] The district also provides critical information for decision-making by its irrigators by identifying seasonal moisture levels and ground moisture needs for corn, wheat, grain sorghum, and wheat (fig. 12.7).

When such groundwater management districts began to appear in Texas, Kansas, and Colorado from the 1950s to the 1970s, they were a complex and contradictory combination of counties, townships, and Ogallala boundaries. This complexity, which is highly informative for understanding regional irrigation, led Stephen White and David Kromm, geographers at Kansas State University, to create overlays that included state and county lines, groundwater management districts, and Ogallala aquifer boundaries (fig. 12.8). Kromm and White also include the all-important water-level changes from 1980 to 1990. However, natural systems like watersheds and soil types are not included.

Powell Revisited for Sustainability Mapping

The history of alternative mapping on the Plains is not new. We need to begin with John Wesley Powell, who published his official *Report on the Lands of the Arid Region of the United States* in 1878. He urged Americans to reconsider their existing pattern of frontier settlement. Powell was wrong when he concluded that only about 3 percent of the American West was available for successful farming (he used Utah as his model).[2] In his U.S. Geological Survey report of 1889–1890, Powell admitted that earlier he had been far too pessimistic about the future of agricultural settlement.[3]

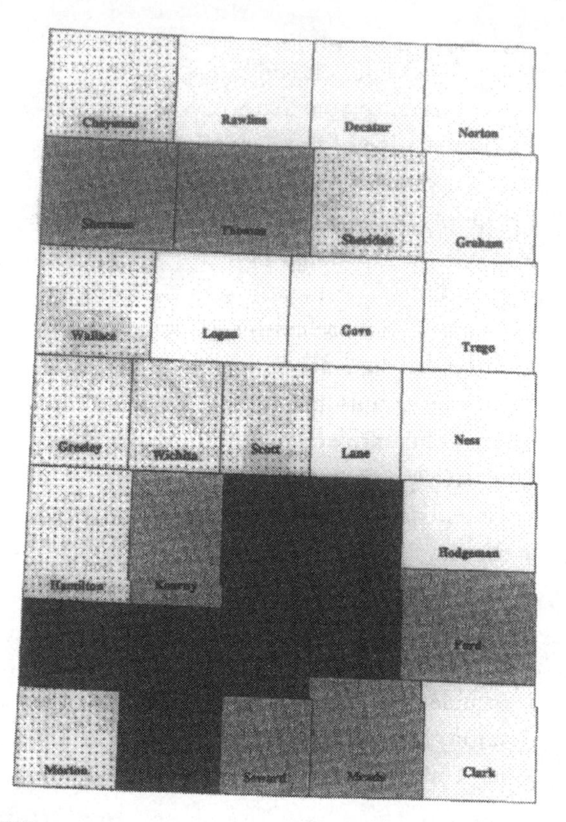

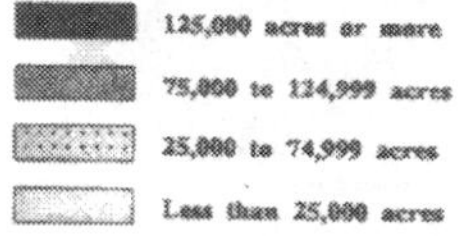

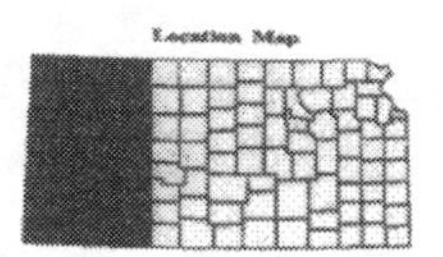

FIGURE 12.4 The likelihood of distorted irrigation data (left) when it is averaged out countywide. Compare this to water pumping rates along natural lines (right). Scott County, for example, is listed with only moderate irrigation (left), but a large part of the county is heavily irrigated (right). In another case, Wallace County appears as a fair region for irrigation (left), but a different look suggests wide swatches of unirrigated farmland (right). From Homer E. Socolofsy and Huber Self, *Historical Atlas of Kansas* (Norman: University of Oklahoma Press, 1972). Reprinted with permission.

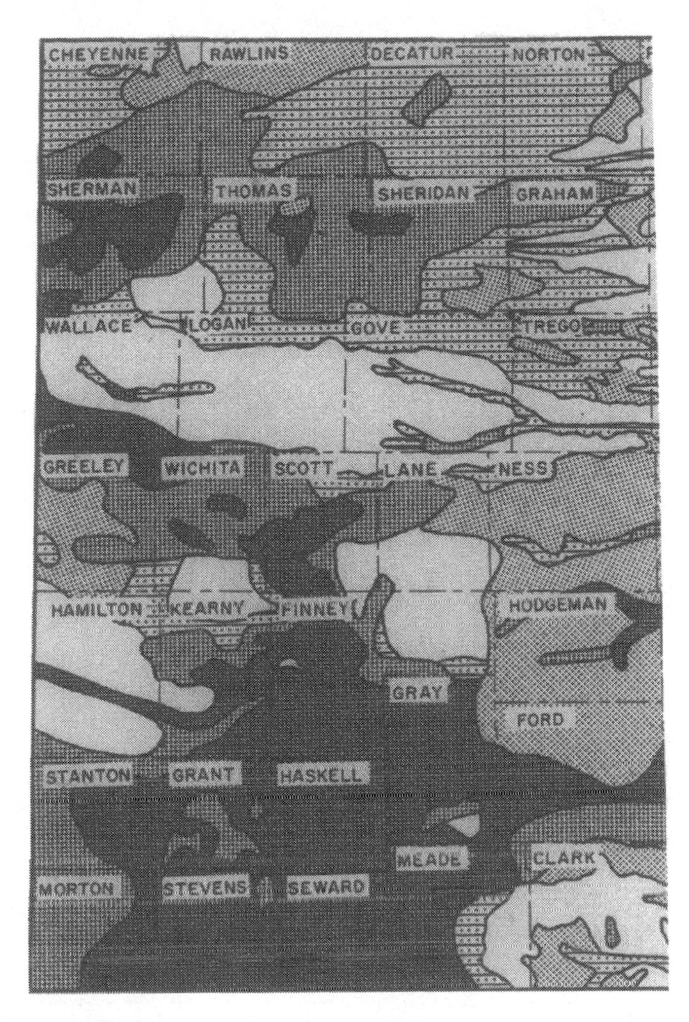

Over 1000

500 to 1000

100 to 500

10 to 100

0 to 10

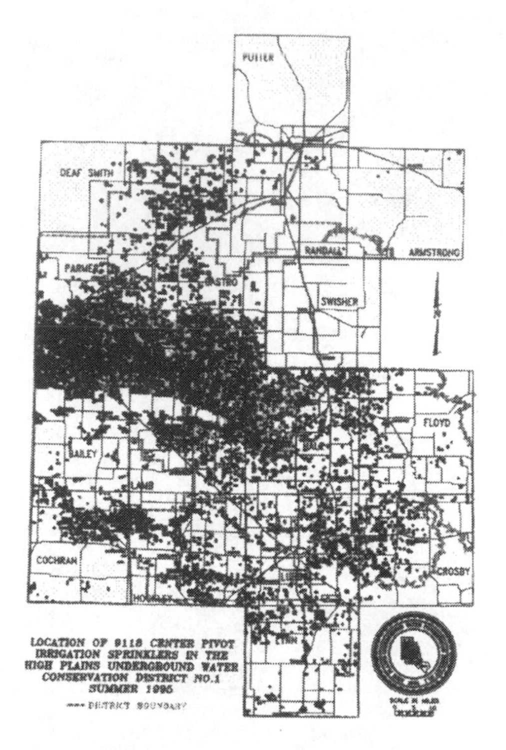

FIGURE 12.5 The center-pivot irrigation circles that dominate irrigation from the Ogallala aquifer in the High Plains of the Texas panhandle. The patterns reflect availability of groundwater depending on its depth, amount available, and the equipment used. This map would inform potential irrigation farmers and groundwater district managers as to likely overcrowding and overconsumption, as well as locations unlikely to produce irrigation. From "Location of 9118 Center Pivot Irrigation Sprinklers in the High Plains Underground Water Conservation District No. 1" (Lubbock, Tex.: High Plains Underground Water Conservation District no. 1, 1995). Reprinted with permission.

Nevertheless, Powell was surely on target when he said that available water rules the development of the West, beginning at the ninety-eighth meridian that cuts Kansas and Nebraska roughly in east-west halves.

But how does one map water and its potential? In 1878, Powell got the ball rolling when he recommended that "the entire arid region be orga-

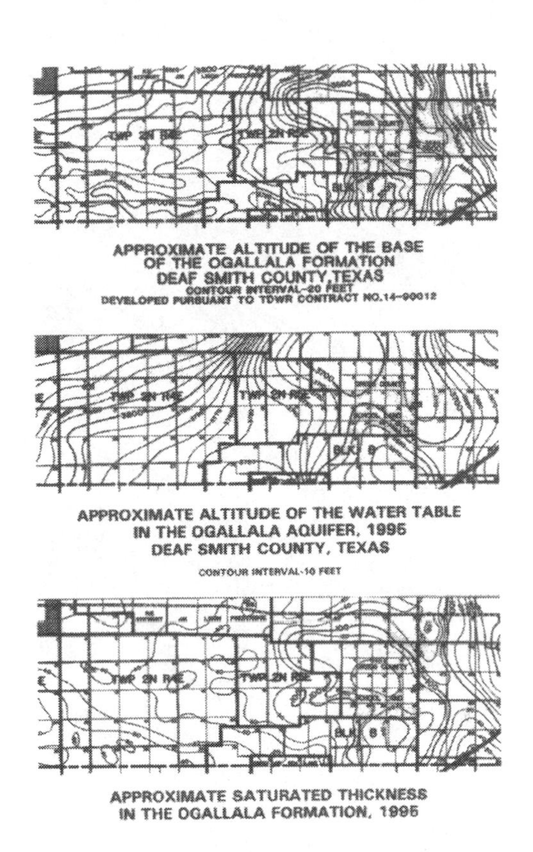

FIGURE 12.6 Sample maps (partial with labels) published for irrigation farmers by the High Plains Underground Water Conservation District no. 1, a state agency headquartered in Lubbock, Texas. Irrigators can locate their land and identify accessibility and amount to plan current and future pumping. Details from Don McReynolds, *Hydrological Atlas for Deaf Smith County, Texas* (Lubbock, Tex.: High Plains Underground Water Conservation District no. 1, 1996). Reprinted with permission.

nized into natural hydrographic districts" instead of states, counties, townships, or other political units.[4] Powell strongly urged the reorganization, not because he was an early environmentalist, but because it was the foundation for the survival of farm families moving into the dry-land West. Each district was to be largely self-governing, both to protect water re-

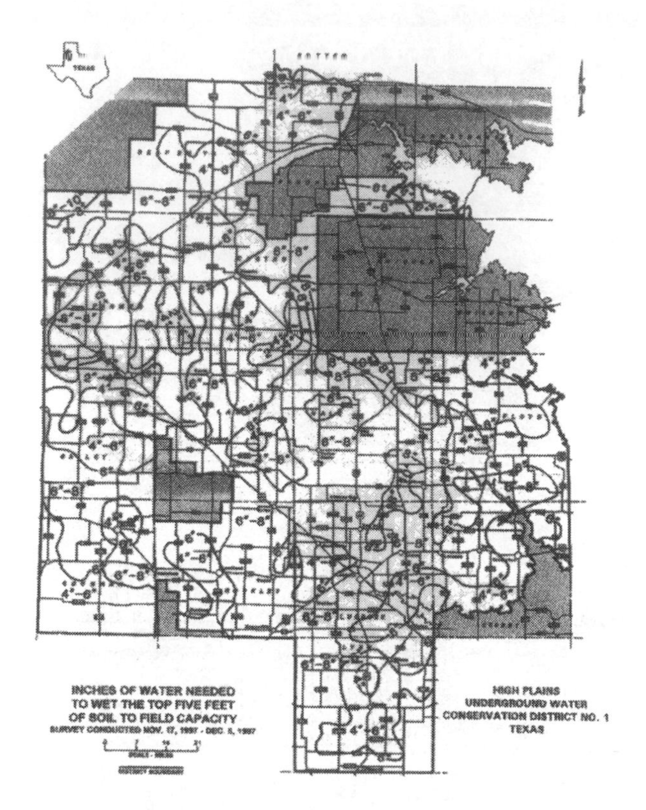

FIGURE 12.7 High Plains Underground Water Conservation District no. 1 regularly provides its irrigators with critical information, based on rainfall and soil types, that determines saturation up to five feet. Irrigators locate their land on the map and then they can pump and irrigate to make up the difference for successful crops. From High Plains Underground Water Conservation District no. 1, "Inches of Water Needed to Wet the Top Five Feet of Soil to Field Capacity. Survey Conducted Nov. 17, 1997–Dec. 5, 1997," *The Cross Section* 44, no. 2 (February 1998): 2. Reprinted with permission.

sources and to establish democracy. In his 1889–1890 report, Powell also laid out "drainage districts" that were to him the most important physical units across the arid region. These districts disregarded state and territorial borders.[5]

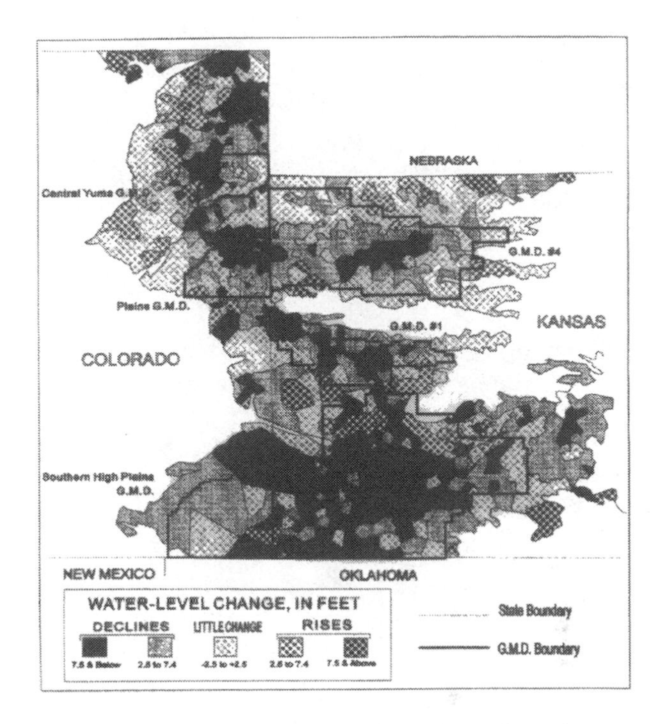

FIGURE 12.8 Geographers David Kromm and Stephen White of Kansas State University ignored county and section lines entirely when they reported on the critical water-level changes in the Ogallala aquifer between 1980 and 1990. The map also locates six groundwater management districts (no label on Kansas District no. 3 in southwestern Kansas). This information would aid irrigation farmers and groundwater management officials, especially as to areas of rapid decline where immediate responses were required. "Changes in Ground Water Level, 1980 to 1990," in Stephen E. White and David E. Kromm, "Who Should Manage the High Plains Aquifer? The Irrigators' Perspective," paper no. 94105, *Water Resources Bulletin* 31, no. 4, (August 1995): 719. Reprinted with permission from the authors.

Alternative Mapping According to Natural Physiography and Resources

For decades the U.S. Department of Agriculture (USDA) provided soil maps that cut across sections and counties to provide a large variety of agricultural soil information that can be directly applied by local farmers (fig.

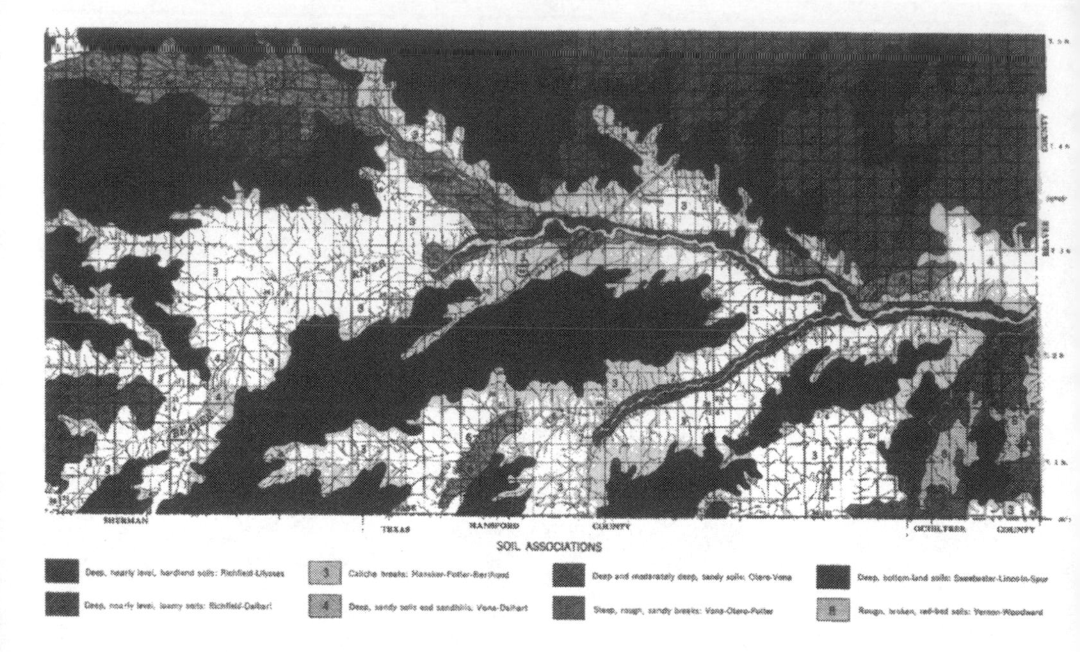

FIGURE 12.9 Even more than property maps (see fig. 12.2), agricultural land purchases depended on soil maps produced by the U.S. Department of Agriculture. Although the potential for irrigation helped set land prices, the nature and quality of the soil remained the primary criterion for land value. One of the most productive soil types on the central plains was Richfield sandy loam (no. 2 in the legend, medium-dark hatching), and it commanded premium prices on the local land market. Detail from *Soil Types in Texas County, Oklahoma* (Washington D.C.: U.S. Department of Agriculture, circa 1960).

12.9). The department categorized soils according to their physical and chemical properties into classes and subclasses. The USDA identified which soils would support or limit different types of plantings, including land capability and yields per acre of crops and pasture. The department recommended the actions farmers might take to properly use, manage, and engineer the soils mapped on their property and defined prime farmland and less-than-prime farmland. Such mapping allowed farmers to identify their soils, potential crops, and extrapolate growing seasons (temperature and precipitation, freeze dates in spring and fall, water management) and land management strategies. Thus, the information provided by soil mapping could help to produce appropriate, sustainable farming.

The USDA's 1960 soil mapping of Texas County in Oklahoma's panhandle, for example, labels a "mature soil" as Richfield sandy loam, after its "type area" near Richfield in Morton County, Kansas.[6] The soil type covers about 10 percent of Texas County, is listed among the most productive soils in the county, and commands premium prices on the local land market. A deep, dark, clayey soil, Richfield sandy loam is about six inches thick and was less affected by the windstorms of the dust bowl because of its medium texture. Coarser-textured, sandier soils blew away faster and to greater depths. But almost 80 years of off-and-on cultivation burned up a good deal of the fertility of the virgin sod. Richfield sandy loam is Soil Conservation Service (SCS) class III soil, which is superior for a region with little rain and significant wind erosion. (Class I and class II are more productive, but because of climate conditions do not exist in Texas County.) According to national SCS standards, class III includes "soils that have severe limitations that reduce the choice of plants, or that require special conservation practices, or both."

Local farmers are also reminded that water is the key to successful Plains agriculture. Their use and management of the soil hinges on the realities of High Plains climate: low rainfall, strong winds, high temperatures in the summer, and low humidity. If the rains come late in spring, they are right for planting sorghum, but too late to help wheat mature. Rains late in summer are good for planting wheat, but too late to save a sorghum crop. In the summer, hot winds and low humidity bring high rates of evaporation. Without irrigation, Richfield sandy loam can grow dry-land wheat, and, when the rains are good, sorghum. Texas County farmers on Richfield are told in their soil survey manuals that:

> These soils need to be protected by a growing crop or a heavy stubble to help control wind erosion. . . . Use a cropping system that fits the moisture conditions. Fallow ¼ to ⅓ of the field. . . . Wheat is likely to fail if it is sown in soil that is moist to a depth of less than 24 inches. Delay tilling fields that have been left fallow, until the danger of soil blowing has passed in spring . . . strip-cropping will help to reduce wind erosion, and the stubble will help conserve moisture by catching snow. Avoid excessive tillage and tillage that will leave the surface soil loose and powdery.[7]

The 1960 edition of the USDA/SCS *Soil Survey of Texas County, Oklahoma* said that few farmers "have access to an ample supply of [irrigation] water and have enough time and money to irrigate all of their land."[8] It added, "Farmers cannot control the weather, but they can adjust their farming methods to protect the soil from extremes of climate, to keep moisture in the soil" to depths at planting time between two and four feet. Techniques from 1960 included crop residue (leaving crop debris on the fields), stubble mulching, delayed fallow, strip-cropping, terracing, and contour farming.

Other important resizing began with the SCS in the mid-1930s; this was probably the first attempt to rethink the old rectangular survey and county data approach. It anticipated the major land resource area (MLRA) mapping of the USDA of the late 1970s (fig. 12.10). These maps emphasized land use as a productive resource that would have made Gifford Pinchot, who pioneered efficient land use, happy. On the High Plains the distinguishing MLRA characteristics were: (1) land use (farms, ranches, dry-land farming, irrigation farming); (2) elevation and topography; (3) climate (precipitation, temperature, freeze-free period); (4) water (precipitation, rivers and streams, groundwater); (5) soils (texture, moisture/temperature regimes); and (6) potential natural vegetation (grasses, awns, trees). Using different criteria based on natural or nonhuman features, the Nature Conservancy, an environmental organization, established its own national ecosystem mapping to assist in its major strategy of acquisition of at-risk land.[9] The Nature Conservancy ecosystems are based on the recent work of Robert G. Bailey, leader of the Ecosystem Management Analysis Center of the U.S. Forest Service (fig. 12.11).[10]

Bailey's ambitious project seeks the convergence of land and water management on the same unit of land, regardless of the different interests of players. He also emphasizes the importance of scale in devising different ecosystems that are often nested within each other or have significant overlaps. Macroecosystems, which are defined by climate and landforms, would include the entire High Plains. In contrast, mesoecosystems can be measured by vegetation or plant productivity, such as short grasses or dry-land wheat. Finally, microecosystems (also called sites) can be defined by a complex of localized conditions and a highly particularized combination of climate and landforms, soils, and vegetation.[11] Although Bailey gives his at-

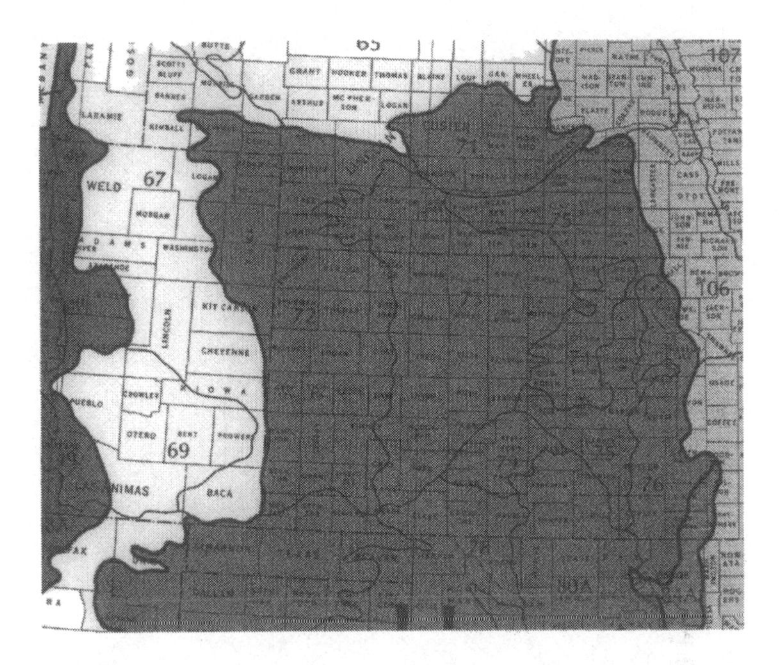

FIGURE 12.10 By the 1970s, new information about natural ecosystem boundaries, together with government policies that sought improved land management, led to multifaceted agricultural land resource studies. This 1981 handbook emphasized conservation for human consumption, with categories including land use (farms, ranches, dry-land farming, irrigation farming); elevation and topography; climate (precipitation, temperature, freeze-free period); water (precipitation, rivers and streams, groundwater); soils (texture, moisture and temperature regimes); and potential natural vegetation (grasses, awns, trees). In the central plains, area 67 is designated "Central High Plains," 69 is "Upper Arkansas Valley Rolling Plains," and 72 is "Central High Tableland." Detail from "Major Land Resource Area (MLRA) Boundaries," Agricultural Handbook no. 296 (Washington, D.C.: Soil Conservation Service, U.S. Department of Agriculture, 1981).

tention to natural ecosystems with little thought to human intervention, his categories can be applied to agriculture, whether frontier, dry-land, rural, irrigated, or industrial. He also concludes that ecosystems will invariably cross jurisdictional, ownership, and state boundaries, thus making sections and counties far less relevant.[12]

Although they both are free from right-angled counties and sections,

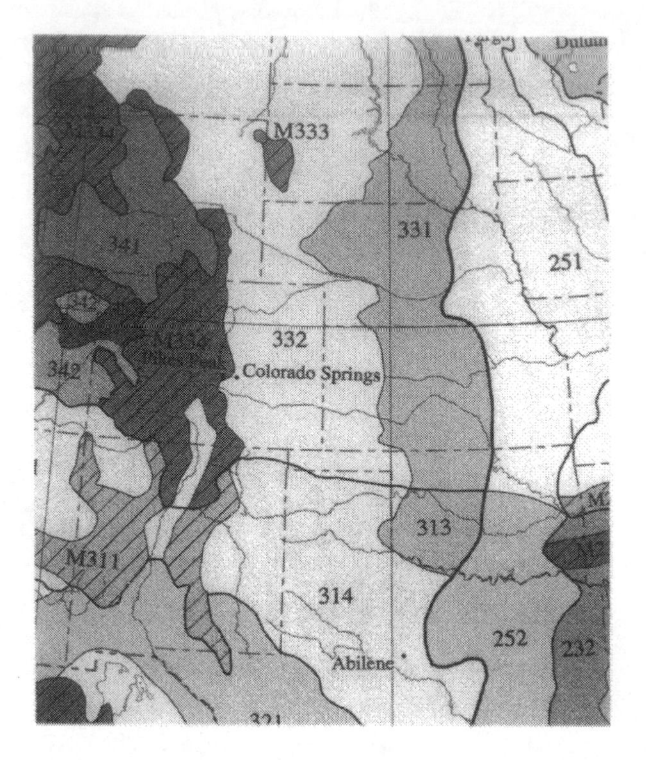

FIGURE 12.11 Unlike the MLRA criteria (see fig. 12.10), U.S. Geological Survey scientist Robert G. Bailey took a strictly environmental approach that emphasized natural soil, terrain, and climate conditions. In this light, the central plains were largely contained in region 332, "Dry Steppes." The southern plains were region 314, "Shortgrass Steppes." Bailey concluded that if one starts with such ecosystems, they invariably cross jurisdictional, ownership, and state boundaries, and make counties and sections far less relevant. Detail from Robert G. Bailey, "Ecoregions Map of North America," Miscellaneous Publication no. 1548 (Washington, D.C.: Forest Service, U.S. Department of Agriculture, revised 1997).

MLRA mapping and Bailey's ecosystem mapping often contradict each other in the units they identify. Their agendas, while commendable in their recognition of natural features, are significantly different, the first dealing with economic development and the latter with the integrity of natural systems.

Mapping Modern "Dead Zones" on the Plains

Are we going through an important shift in agriculture that might be clarified by a resizing of the Plains? A heated debate filled with great passion, intensive rhetoric, and heavy conflict erupted in the mid-1990s about industrial hog-confinement operations. As recently as 1990, a Plains county in Kansas, Oklahoma, or Texas produced only a few thousand hogs in widely dispersed locations—mostly as a sideline by wheat farmers. Such limited production barely showed up on the horizon of any map. But in less than a decade, Texas County, for example, became inhabited by hundreds of thousands of hogs bred and raised in industrial sheds, sometimes on the unirrigated corners of center-pivot quarter sections. A hundred thousand hogs produce waste comparable to 170,000 additional people, a significant number in a region where a town of 25,000 is a metropolis. The management and disposal of animal wastes, first in lagoons and finally liquid dispersal on a neighboring field, became major issues. Not the least was the real insult of strong smells that local farmers and townspeople complained destroyed their quality of life. In 1997, geographer Owen Furuseth concluded that confined animal production had created a new agricultural geography—an imploded landscape. The impacts are highly localized, but divide the agricultural landscape into habitable or uninhabitable zones.[13]

Seaboard, Inc., a multinational corporation, is a major hog producer in the central Oklahoma panhandle, with numerous operations in and near Guymon, Texas County (fig. 12.12). Buffers between hog-confinement sheds, their adjacent lagoons, and waste treatment fields might depopulate entire neighborhoods or keep human habitation from entering such "dead zones." Local residents are urging a buffer of two to three miles between a habitable structure and a hog-confinement operation because of smells and real air pollution; the current regulated distance is between one-quarter and three-quarters of a mile. Comprehensive mapping of industrial hog-confinement operations and their public impacts is yet to be done. Such mapping might identify the following two items: (1) location of permitted facilities, considering property and operations ownership and liability, acreage, animal capacity, water source, potential risk to freshwater sources, water consumption, liquid waste holding ponds, and liquid waste application acreage; and (2) setbacks from other swilling units, habitation struc-

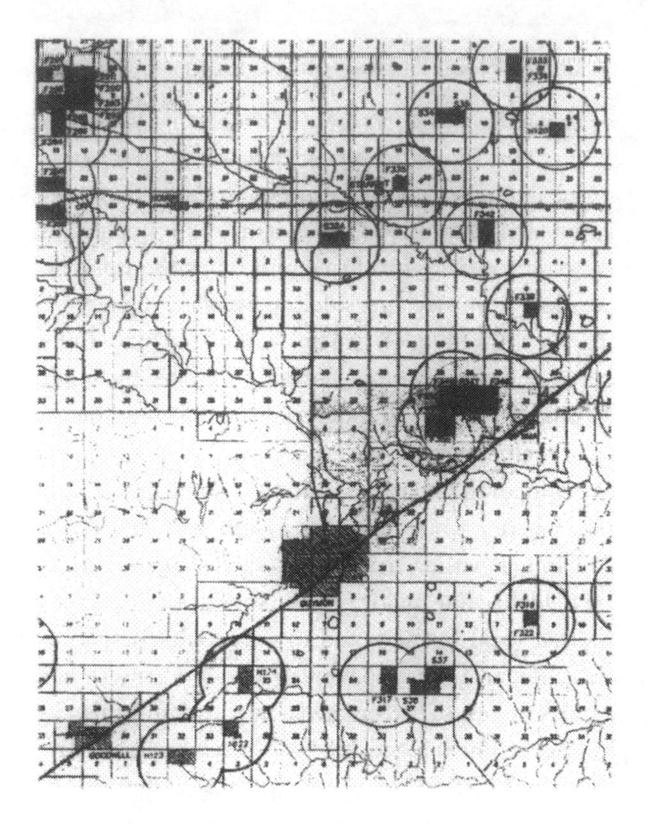

FIGURE 12.12 Author's overlaid circles of approximately 1.5-mile diameter, each centered on concentrated hog farms of Seaboard Corporation around Guymon, Oklahoma. These identify nonresidential zones around each hog operation and suggest the extent of "dead zones" that are emerging where the hog industry is concentrated. The debate in the 1990s was over the spread and state regulation of these zones. Detail of "Seaboard Hog Operations in Texas County, Oklahoma," Seaboard Map GUSP 012F, dated 28 January 1997. Reprinted with permission.

tures, public and private drinking water, public and private nonfarm activities, municipality limits, and pollution plumes (air and water).

In mid-1998, legislation that regulated hog pollution was being debated and passed in the Plains states of Texas, Oklahoma, Kansas, and Colorado. This legislation can restructure the region. A 1997 Kansas map shows distribution of hog facilities in the state (fig. 12.13). Instead of sections and

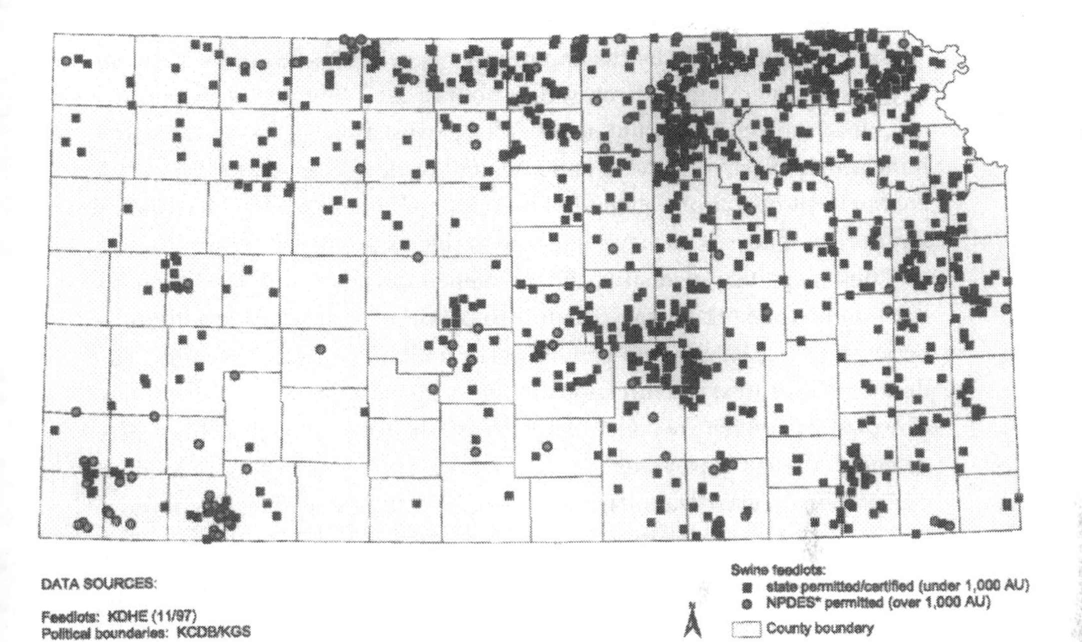

FIGURE 12.13 This map depicts the extent of industrial hog operations in Kansas in 1997 and where such operations are concentrated. There is little relationship with traditional county lines, which have historically been the basis for agricultural data. The map also suggests the extent of nonresidential "dead zones" that are being institutionalized by the hog industry and resulting state regulation. "Permitted Hog Facilities," (Wichita: Kansas Department of Health and Environment, November 1997).

counties, the legislation defines its mapping in terms of groundwater well-spacing and the distance of hog-confinement lagoons from residences and communities. For example, Kansas Substitute House Bill 2950 requires set-backs up to five thousand feet between a residence and a hog-confinement operation. Where liquid waste is applied to cropland, which is mostly the case, it could be sprayed no closer than three hundred feet from a water well (even the owner's well) and one thousand feet from a habitable structure.[14] Waste lagoons cannot be closer than three hundred feet from a water well, and wells themselves must be spaced one-quarter mile apart. The well-spacing is being fought by the hog producers, who need concentrated and large-scale water supplies twenty-four hours a day.[15] The actual geographical impacts are currently unknown. These regulations acknowledge large buffers that effectively create unpopulated, or dead zones, wherever hog-confinement operations abound. The Poppers' predicted depopulation of large areas of the Plains may come true in this surprising turn.

Thus, mapping can identify the threatened spread of water pollution. In both Kansas and Oklahoma, debate also continues on the vertical distance between a hog lagoon and groundwater underneath. Not currently addressed is the threat and direction of migrating underground "water pollution plumes," which are under regulation of the Environmental Protection Agency and the Clean Water Act. The Clean Air Act does not cover odors except when they contain harmful substances such as hydrogen chloride. However, local inhabitants would find the mapping of "odor pollution plumes" important, based on intensity, wind direction, and distance.

Conclusion

Nature's patterns have often been ignored or distorted in High Plains mapping. Economists have long spoken of the "invisible hand" of the marketplace in shaping material life. We need to better understand that the natural world is a powerful "visible hand" that only seemed invisible because we ignored it. One challenge is to estimate values of the services provided by nature's capital (soil, water, natural gas) on the Plains.[16] Nature and farming are better understood as environmental or economic units than political units, which is why a remapping (resizing) of the Plains will better enable us to properly conceive of that landscape.

Notes

1. According to A. Wayne Wyatt ("The Nuts and Bolts of the Work Effort Necessary to Construct the Maps Included in the Hydrological Atlas Series," 20 February 1998, email

to John Opie), to construct saturated thickness of the Ogallala formation maps, maps illustrating the elevation (above sea level) of the base of the formation first have to be constructed. The High Plains Underground Water Conservation District no. 1 (Texas) uses well driller's logs to obtain this data. The district has about 50,000 well driller's logs on file in its permit files. The permit shows the location of the well in yards from two, nonparallel property lines. For all wells drilled, the permits issued for each square mile of the district are filed in the same folder. Each folder is pulled and a well log is selected that appears to best represent the geology on the section of land. The depth-to-the-base of the Ogallala formation in feet is written on the work map at the location the well was drilled. One well driller's log is selected for each square mile. The next step is to determine the elevation of the land surface at the well location. This is determined from 7-1/2-minute U.S. Geological Survey land surface elevation maps. The elevation of the land surface at these well sites is written on the face of the map. The depth-to-the-base of the Ogallala aquifer as recorded from the well log is subtracted from land surface elevation, which provides the elevation of the base of the Ogallala formation. These data are then contoured to illustrate the elevation of the base of the Ogallala formation. (A contour line illustrates equal values along the line for the subject matter being contoured.)

The elevation of the base of the Ogallala formation maps only have to be made one time. Most of the district's base maps were made in 1980. Counties or portions of counties added to the water district since the date have to be made. In some areas, when additional data become available (new wells drilled), a portion of a county might be revised to improve the accuracy of the maps. The U.S. Geological Survey 7-½-minute land surface elevation maps were transferred to the district's base map in 1980. Only those for the newly annexed counties have had to be made since 1980.

The next step in the process is to construct maps illustrating the elevations of the water table. The locations of all the water level observation wells are plotted on a map and the land surface elevation is noted on the map at the well site. The depth-to-water below land surface value is subtracted from the land surface elevation, resulting in the elevation of the water table.

The density of the annual water level observation well network is about one well per nine square miles (1,196 wells in the 10,728-square-mile area of the water district). For the year's new saturated thickness maps to be made, supplemental measurements are made in a larger group of wells to improve the accuracy of the maps. The depth-to-water is measured in one well per square mile in the thinner saturated thickness areas of the district (fifty feet or less saturated thickness) and about one well per three square miles is measured in the remaining portion of the district. (The district attempts to measure the same group of supplementary wells every fifth year. Sometimes pumps are pulled or changed or the district employees can no longer get the tape into the well, which makes it impossible). The supplementary well measurements are plotted on the face of the map and corrected to sea level elevation data. The maps are contoured to illustrate the elevations of the water table using the annual water observation well data and the supplementary water level data.

The saturated thickness map is then constructed by placing the elevation of the depth-to-water level map over the elevation of the base of the Ogallala formation map and subtracting the difference in the values where the lines cross each other. These are noted with an x, with the value written at the site. Additionally, the wells for which the district has both a depth-in-feet below land surface to the base of the Ogallala formation and a depth-in-feet below land surface to the water level are subtracted and noted with an o, with the value written at the site. (These actual well sites are given a higher value by the person contouring the map.) The values are then contoured, which yields a saturated thickness map. This map plus the elevation of the water table map, the elevation of the land surface, and the base of the Ogallala formation maps are used in combination to make the hydrologic atlases published by the district. The annual depth-to-water level measurements are also used for various other programs.

2. John Wesley Powell, *Report on the Lands of the Arid Region of the United States,* (1879; reprint, Boston, Mass.: Harvard Common Press, 1983), 37–39.

3. John Wesley Powell, *Eleventh Annual Report of the Director of the United States Geological Survey to the Secretary of the Interior, 1889–1890*. Pt. II. *Irrigation*. (Washington, D.C.: Government Printing Office, 1891), 203–5.

4. John Wesley Powell, "Water for the West," in *Revised Management Program III: Rules and Regulations, and Policies and Standards* (Garden City, Kans.: Southwest Kansas Groundwater Management District no. 3, 1986), frontispiece.

5. Powell, "Arid Region of the United States, Showing Drainage Districts," *Eleventh Annual Report,* x.

6. The basic soil handbook used by farmers in Texas County and the local offices of state and federal agencies is Hadley C. Meinders, Maurice Mitchell, Edward S. Grover, and Jimmie W. Frie, *Soil Survey of Texas County, Oklahoma* (Washington, D.C.: SCS, USDA, and Oklahoma Agricultural Experiment Station, 1961, 1984). Most of the following information is based on this handbook, which offers a general survey of soil science, soil maps, and aerial photographs. A farmer is readily able to identify the soil types on his land and plant and irrigate accordingly. For Kansas, see George N. Coffey and Thomas D. Rice, "Reconnaissance Soil Survey of Western Kansas," in *Field Operations of the Bureau of Soils* (Washington, D.C.: USDA, 1910); "Soil Survey of the Garden City Area," in *Field Operations of the Bureau of Soils* (Washington, D.C.: USDA, 1904); *Soil Survey of Finney County, Kansas* (Washington, D.C.: SCS, USDA, 1965). See Meinders et al., *Soil Survey of Texas County, Oklahoma,* 57.

7. Meinders et al., *Soil Survey of Texas County, Oklahoma,* 19.

8. Ibid., 23, 27–33.

9. See the map "The United States of Nature," *Nature Conservancy* May/June 1998, 8–9, described as Bailey's ecosystem plan.

10. See Robert G. Bailey, *Ecosystem Geography* (New York: Springer-Verlag, 1998); Robert G. Bailey, *Ecoregions* (New York: Springer-Verlag, 1998).

11. Bailey, *Ecosystem Geography,* 23ff, 83ff.

12. Ibid., 14–18.

13. Owen J. Furuseth, "Restructuring of Hog Farming in North Carolina: Explosion

and Implosion," *The Professional Geographer,* 49, no. 4 (November 1997): 391–403; John Fraser Hart, "A Map of the Agricultural Implosion," *Proceedings of the Association of American Geographers* 2 (1992): 68–71.

14. Substitute for H. Bill 2950, Committee on Environment, Kansas Legislature, 1998 sess., 11 March 1998.

15. Articles in the *Daily Oklahoman,* 3 April 1997, 2 May 1997, 10 March 1998.

16. See the innovative discussion by Robert Constanza et al., "The Value of the World's Ecosystem Services and Natural Capital," 387 *Nature* (13 May 1997): 253–60; Stuart L. Pimm, "The Value of Everything," 387 *Nature* (13 May 1997).

Dam those Waters!

13

Private Initiative, Public Works

Ed Fletcher, the Santa Fe Railway, and Phoenix's Cave Creek Flood Control Dam

Donald C. Jackson

With the emphasis that historians place on federal influence over western water projects, the role of the private sector in promoting and financing such activity is frequently overshadowed. Of course, the huge projects associated with the New Deal are all closely involved with the federal government, and the patterns of political financing that fostered these projects proved enormously influential during the dam-building boom that followed World War II.[1] But despite the surge of federal activity in the mid-twentieth century, the importance of the private sector in western water use and allocation has remained strong and will continue as such for the foreseeable future.

My intent in this chapter is not to formulate a grand synthesis of private water development in the West, but to elucidate revealing factors and professional relationships underlying the construction of one innovative dam in central Arizona. Completed in 1923 about twenty miles northwest of downtown Phoenix, Cave Creek Dam exemplifies the multiple arch dam technology pioneered by the California-based engineer John S. Eastwood. Until superseded by a larger dam in the late 1970s (Cave Butte Dam), the 1,700-foot-long Cave Creek Dam provided flood protection to thousands of acres of land in eastern Maricopa County and the environs surrounding the Arizona state capital. Whereas the floods that prompted construction of the dam are not so remarkable, the forces that promoted Eastwood's Cave Creek design appear quite distinctive, especially in light of a prevailing narrative that places the federal government at the center of western water history. Financed by a consortium that included the State of Arizona; the City of Phoenix; Maricopa County; the Salt River Valley Water Users' Association; Standard Oil Company; Union Oil Company; and—most significantly—the Atchison, Topeka, and Santa Fe Railway, the Cave Creek Dam undoubtedly constituted a major "public-works" project.[2] By considering how Cave Creek Dam (which did not involve any federal agencies) came to be built, historians can gain insight into the character of western water development and into plausible, yet hardly radical, alternatives to the federally financed system of water projects that dominated the post–New Deal West.

In recounting this story of the Cave Creek Dam, the emphasis is not so much on local factors (i.e., concern over flooding) or the character of political and economic interests within greater Phoenix; these certainly constitute topics of historical interest, nonetheless they are essentially ancillary to events underlying how Eastwood's design came to the attention of the Cave Creek Flood Control Board and how it was selected for construction. Although the importance of "local history" is readily acknowledged, much of the "local" attention in this chapter shifts about four hundred miles east of Phoenix to San Diego, California, and focuses on businessman/real estate developer Ed Fletcher and his interactions with the Atchison, Topeka, and Santa Fe Railway. As part of this, the role of the Santa Fe in the development of the Southwest is relevant. Similarly, Eastwood's work on Sierra Nevada hydroelectric projects and its subsequent impact on his engineering career influenced the form of the structure built across Cave Creek.[3] As will be described, the Fletcher–Santa Fe relationship precipitated Eastwood's design for Lake Hodges Dam in San Diego County and made it possible that Arizona officials would consider a multiple arch design for Cave Creek in the fall of 1921.

Cave Creek Dam, like Lake Hodges Dam, represented a project in which safety and cost savings generated by innovative design proved to be of great local interest. And it was in the arena of cost minimization that John Eastwood came to prominence in the world of western water. By its very nature, a study focussing on Eastwood's Cave Creek design draws attention to the economic underpinnings of large-scale construction projects and raises issues extending across the range of dam designs funded by public agencies and private entities in the early twentieth century. At its core, the story of

Cave Creek Dam is deceptively simple—at least if one views modern American technology as a manifestation of culture defined and mediated by a network of business, government, and personal relationships. But the complexity of such relationships is often obscured by approaches to western water history that obsessively situate the federal government at the center of analysis. In this light, the story of Cave Creek Dam may not be so simple after all.

Santa Fe Railway

The origins of the Atchison, Topeka, and Santa Fe Railway (hereafter termed the Santa Fe Railway, or simply the Santa Fe) date to the late 1850s, but it was not until the mid-1880s that its rail lines reached San Diego.[4] Once there, the Santa Fe quickly became a major force in the southwestern economy, and the company acted to expand its reach into central Arizona. Separated from northern Arizona (and the main line of the Santa Fe) by the imposing escarpment of the Mogollon Rim, Phoenix had been settled in the late 1860s by a few pioneering farmers who diverted water from the Salt River and—using ditch lines excavated by the long-departed Hohokam Indians—irrigated crops for sale to the nearby army encampment. By the late 1880s, the irrigation economy of Phoenix and surrounding Maricopa County supported a population of about 10,000 and residents eagerly sought a transcontinental rail connection. In the late 1870s, the Southern Pacific completed a southern route reaching El Paso via Tucson, but this passed about thirty miles south of Phoenix; a connecting spur line completed in 1887 (and controlled by the Southern Pacific) provided only minimal satisfaction to Phoenix boosters. Strong local support soon developed for an extension northward to the main line of the Santa Fe that would offer an alternative to the Southern Pacific.[5]

The Santa Fe, Prescott, and Phoenix Railway completed a 150-mile connection between the main line at Ash Fork and Phoenix in 1895.[6] Descending across the Mogollon Rim, this route entered the city from the northwest and crossed the Cave Creek flood plain on its approach to downtown. While this Santa Fe–controlled spur line competed with the Southern Pacific for the Phoenix market, its steep gradient did not present a particularly desirable alternative. Between 1903 and 1910 the Santa Fe constructed a more efficient route from the Pacific Coast to Phoenix that retained the right-of-way across Cave Creek, but branched westward about halfway to Prescott, Arizona. Using a new crossing of the Colorado River, the line shortened the travel distance by more than one hundred miles and allowed daily overnight runs of the *Phoenix Express* between Los Angeles and central Arizona. With this, the Santa Fe solidified its interest in the commercial affairs of Phoenix and the Southwest as a whole.

Just as the Santa Fe Railway's economic ties to southern California and Phoenix are central to the history of Cave Creek Dam, so too is the company's interest in dam-building that dated to the 1880s. In particular, two dams warrant special notice because of their design and location: the Sweetwater Dam in southern San Diego County and the Ash Fork Dam in northern Arizona.

Built across the Sweetwater River less than ten miles southeast of downtown San Diego, the Sweetwater Dam used an innovative masonry arch design to store water for the National City settlement located at the terminus of the California Southern Railroad Company. Financed by the San Diego Land and Town Company (a subsidiary of the Santa Fe Railway), this ninety-foot-high structure was completed in 1888.[7] Although not quite as daring as the thin arch Bear Valley Dam completed by irrigation interests in 1884 near San Bernardino, the Sweetwater Dam featured a cross-section far less bulky than conventional gravity designs. Consequently, the thin profile used at Sweetwater significantly reduced the expense of quarrying, hauling, and placing masonry blocks.[8] During heavy rains in early 1895, flood waters poured over the top of the structure for more than twenty-four hours and offered a convincing demonstration—apparently not lost on Santa Fe officials—that innovative dam design did not necessarily reduce structural safety.[9]

About ten years after completion of Sweetwater Dam, the Santa Fe funded yet another innovative dam design, this time to supply water for its main-line railroad facilities at Ash Fork. Completed in 1898, the forty-two-foot-high Ash Fork Dam used a buttress design with an inclined upstream face consisting of steel plates (heretofore, buttress dams had used either wood or masonry for the upstream face).[10] Steel buttress dams were soon superseded by reinforced concrete designs, but nonetheless Ash Fork Dam proved to be a remarkably durable structure—it still holds back water—and provided Santa Fe officials and engineers with confidence in any future association with buttress dam technology.[11] Although records documenting John Eastwood's interactions with Santa Fe engineers contain no explicit references to the Sweetwater or Ash Fork Dams, these earlier successes would have only encouraged the company to embrace his multiple arch designs.

John S. Eastwood

Born in Minnesota in 1857, John Eastwood studied as an engineer at the University of Minnesota before heading West in 1880.[12] In 1883, he settled in Fresno, California, and for the next three decades worked in the San Joaquin Valley and the southern Sierra Nevada. Eastwood's interest in water storage derived from work involving irrigation, lumber flumes, and, starting in the early 1890s, the nascent technology of hydroelectric power. As chief engineer for the Fresno-based San Joaquin Electric Company, in 1895–1896 Eastwood built a hydroelectric power system that, at the time operations began in April 1896, constituted the longest commercial power system in the world.[13] However, a serious drought hit central California in the late 1890s that forced the electric company into bankruptcy before it could finance construction of a storage dam.[14] With this, Eastwood experienced firsthand the economic importance of water storage in the arid West, and it spurred him to find ways of reducing dam construction costs.

Eastwood remained involved in the development of large-scale hydroelectric power plants and during the first decade of the twentieth century worked for Henry Huntington's Pacific Light and Power Corporation in planning the Big Creek system in the Sierra Nevada.[15] As part of this project, Eastwood conceived plans for a new design that would impound a key reservoir associated with the system (now known as Huntington Lake). His intent was to devise a form that would be less expensive to build than conventional massive gravity dams (whether made of earth, rockfill, or masonry), yet equally strong. In 1906, Eastwood first developed multiple arch designs requiring remarkably small quantities of concrete that could achieve this goal.

Financial uncertainties caused by the panic of 1907 and corporate machinations of the Pacific Light and Power Corporation kept Eastwood from building any multiple arch dams at Big Creek. However, in 1908 he demonstrated the efficacy of his ideas by building a sixty-four-foot-high dam for a logging company in the Sierra Nevada about fifty miles east of Fresno.[16] Completed in 1909, the Hume Lake Dam comprised the world's first reinforced concrete multiple arch dam. The following year, Eastwood began work on the ninety-two-foot-high Big Bear Valley Dam in southern California to be used by irrigation farmers to increase crop production in the Redlands–San Bernardino region.[17] After completing the Big Bear Valley Dam in 1911, Eastwood began working on a major project for the Great Western Power Company in northern California.

As part of Great Western's Feather River hydroelectric power system, the company planned a large storage dam at Big Meadows. Originally, the company intended to erect a concrete gravity design, or at least this had been the hope of the firm's engineering consultant, John R. Freeman. Freeman had helped Boston, New York City, and Los Angeles plan their municipal water supply systems and was highly esteemed by America's business and technical elite.[18] Nonetheless, Eastwood's success at Big Bear Valley and cost savings promised by his multiple arch design won him the Big Meadows commission. Work on Eastwood's dam was well underway by the fall of 1912, but a death in the Great Western's corporate leadership provided Freeman with an opportunity to reassert his influence over the company. In early 1913, Freeman convinced the company to abandon Eastwood's design in favor of a massive earthfill structure. Significantly, the heart of Freeman's objection to Eastwood's design did not rest on technical arguments, but derived from nontechnical concerns about the appearance of the multiple arch design and "psychological" disquiet it would supposedly engender among the general public.[19]

In the wake of the Big Meadows controversy—and the associated dispute with Freeman—Eastwood found himself a professional outsider within the world of engineering and high-level finance. Rather than abandon interest in multiple arch technology, Eastwood's response was to concentrate all his energies on the goal of developing inexpensive—yet structurally sound—dam designs to further economic development in the West. He even went so far as to rhapsodize in a 1914 speech that "the California Slogan e'er should be, that t'is a crime to let our rivers reach the sea."[20]

Between 1913 and 1915, Eastwood struggled to find commissions, building a small irrigation dam north of Sacramento for less than $25,000 and a mining debris dam in Jackson, California, for a comparable amount. In early 1916, Salt Lake City officials engaged him to design their municipal dam at Mountain Dell, but Eastwood's most important break came when the ambitious San Diego County businessman Ed Fletcher embraced his dam design skills.[21] With Fletcher, Eastwood found a patron who could counter the influence of Freeman and help disseminate his ideas within the western business community.

Ed Fletcher

As early as 1911, Fletcher became associated with William G. Henshaw, owner of the Riverside Cement Company and Warner Ranch in northern San Diego County.[22] In 1914, Fletcher and Henshaw asked Eastwood to develop a preliminary dam design to impound Lake Henshaw within Warner Ranch. This design was never built, but it soon fostered construction of an Eastwood dam across the nearby San Dieguito River that brought the Santa Fe Railway into financial partnership with Fletcher and Henshaw.[23]

The San Dieguito River flows westward out of northern San Diego County into the Pacific Ocean. By 1915, Fletcher had surveyed the stream and obtained an option on a dam site about seven miles from the Pacific. This site offered the possibility of storing more than 70,000 acre-feet of water that, without pumping, could irrigate valuable land in the "frost-free belt" hugging the coast. Included in this land was the 12,000-acre San Dieguito Ranch (later named Rancho Santa Fe) owned by the Santa Fe Railway. In 1916, Henshaw, Fletcher, and the Santa Fe Railway pooled their interests into the San Dieguito Mutual Water Company; Fletcher became president, Henshaw served on the board of directors, and the Santa Fe placed three employees on the board including W. E. Hodges, the railway's vice president. In 1918, symbolic acknowledgment of the Santa Fe's financial support occurred when the structure was officially designated Lake Hodges Dam.

In early 1917, Eastwood developed detailed plans for Lake Hodges Dam.[24] Later that spring George L. Davenport, a Santa Fe structural engineer based in Los Angeles, assessed Eastwood's design to determine if it would be suitable for the company to finance. Davenport's review considered the multiple arch design strictly on its structural merits and gave no credence to any "psychological" issues akin to those John Freeman had raised at Big Meadows. Davenport's endorsement of Eastwood's design convinced Santa Fe officials of its structural integrity, and in May 1917 the mutual water company negotiated a contract for the 550-foot-long, 136-foot-high Lake Hodges Dam.[25] Construction took a little less than a year, and by the spring of 1918 water was rising in the reservoir.[26] Along with Lake Hodges, the company's system included an auxiliary "distribution" reservoir close to Rancho Santa Fe. Formed by the 52-foot-high San Dieguito Dam designed by Eastwood, a concrete conduit connected this supplementary facility to the main reservoir.[27]

Almost contemporaneously with the beginning of work on Lake Hodges Dam, a new dam safety law came into effect in California—a law that placed many projects (including Lake Hodges) under the supervision of the California state engineer.[28] With this new law, Eastwood's design came under a bureaucratic scrutiny quite distinct from the review offered by Davenport and the Santa Fe Railway. At Lake Hodges, this state supervision became manifest in the large overflow spillway that the San Dieguito Mutual Water Company was forced to excavate across the north end of the dam. By late 1918, Eastwood designs for other dams in California (most notably the Littlerock Dam in northern Los Angeles County) encountered opposition from California State Engineer Wilbur McClure and signaled an antipathy to multiple arch technology within California state government that remained in place until Eastwood's death (and long afterwards).[29] The details of this antipathy need not concern us here, but what is important to note is that neither Fletcher nor the Santa Fe backed down from supporting Eastwood in reaction to the state engineer's efforts.

In addition to the Lake Hodges and San Dieguito Dams, Eastwood designed two other dams for Ed Fletcher in 1917–1918. The largest of these was the 117-foot-high, 900-foot-long Murray Dam constructed east of San Diego for the Cuyamaca Water Company (which Fletcher controlled in partnership with James Murray).[30] The other was a small structure erected at the Fletcher family retreat (called "Eagles Nest") about sixty-five miles northeast of San Diego in the hills east of Warner Springs.[31] The Eagles Nest Dam received little publicity, but nonetheless it highlighted Eastwood and Fletcher's distinctive professional relationship. From Eastwood's perspective, Fletcher constituted an excellent client who could advocate his work within the western business community. Conversely, Fletcher found in Eastwood a technologically adept ally who offered the possibility of developing major water projects with modest capital outlay (Murray Dam cost less than $120,000 and Lake Hodges Dam cost less than $330,000—including a special $68,376 cement freighting expense billed to the Santa Fe Railway and $38,138 for spillway excavation demanded by the state engineer).[32] After 1918, the two men continued to collaborate on San Diego County projects (which, for political reasons, remained unbuilt before Eastwood's death in 1924) and on other projects—including two associated with the Santa Fe Railway outside of California.

Phoenix and the Cave Creek Dam

Beginning in 1895, the Santa Fe Railway contributed to Phoenix's ongoing growth as a commercial center. The prominence of the city was reflected in its selection as territorial capital in 1890 (it became the state capital in 1912). Even more important in economic terms, starting in 1903 more than 200,000 acres in the city and surrounding Maricopa County were organized into the Salt River Valley Water Users' Association (often called the Salt River Project). With this, the city became the beneficiary of Roosevelt Dam—one the most prominent early dams built by the U.S. Reclamation Service. Completed in 1911, the Roosevelt Dam (along with associated work on the Granite Reef Diversion Dam) cost more than $10 million. Although there was general accord that the water stored behind Roosevelt Dam would be of great value to the community, there was widespread concern about the huge federal debt incurred by local landowners. In 1917, the water users' association wrested operational control over the Salt River Project from the Reclamation Service and signaled that—at least until the 1930s and the New Deal—the federal government would play a limited financial role in the region's hydraulic development.[33] In the 1920s, this was reflected in the work of the water users' association's general manager, C. C. Cragin, who implemented construction of three large hydroelectric power dams on the Salt River below Roosevelt Dam without any federal financing or subsidies.[34] It was also reflected in the construction of the Cave Creek flood control dam.

Usually a dry arroyo, during heavy rainstorms Cave Creek becomes a troublesome torrent. As reported in the engineering journal *Southwest Builder and Contractor:*

> [Cave Creek] arises some fifty miles north of the city in a steep and mountainous country . . . [it] has an average gradient of about fifty feet to the mile and a drainage area of 210 square miles. Electric storms passing over the Cave Creek watershed often result in cloudbursts and great volumes of water gather and rush down the steep gradient, sometimes exceeding 30,000 [cubic feet per second], to the flat slopes of the Salt River Valley. Here Cave Creek loses its channel and floods the surrounding country.[35]

Concern over Cave Creek floods did not arise suddenly in the early 1920s, but dated back to the nineteenth century. In particular, completion of the Arizona Canal in the mid-1880s opened up land north and west of downtown Phoenix to irrigation; the water supply for these lands could be affected by washouts and breaks in the canal walls caused by uncontrolled flood surges. Thus, Cave Creek floods engendered significant maintenance expenses for the Arizona Canal that, after 1903, became the responsibility of the water users' association. Similarly, roads and bridges administered by the City of Phoenix and Maricopa County—as well as the Santa Fe trackbed—were susceptible to washouts. Although myriad industrial facilities and residences were potentially at risk, most embarrassingly the Arizona State Capitol building lay in the lower Cave Creek flood plain.[36]

Cave Creek floods were not particularly deadly or horrendously devastating, but the social and economic havoc they incurred grew increasingly burdensome. The issue came to a head in August 1921, when a large rainstorm struck the region and brought Cave Creek well above flood stage. The floods received national attention, with the *Engineering News-Record* reporting:

> On the night of the 21st of August, a strip of the western portion of Phoenix, Ariz., from 2 to 10 blocks, was flooded by water originating in Cave Creek. No lives were lost and the principal damage to the city was to the state capitol where water stood about 2–1/2 feet deep on the first floor. . . . The flood waters broke over the Arizona Canal, which has a flood capacity of 800 [cubic feet per second], 8 miles from Phoenix . . . and reached their peak [at] 3 p.m. of the 21st with a flow estimated at 25,000 [cubic feet per second]. . . . It is estimated that the property damage in Phoenix is $300,000.[37]

Even before the August 1921 flood, there existed widespread consensus that a dam across Cave Creek could hold back flood waters and allow their release over the period of a few days (rather than a few hours). Earlier that summer the State of Arizona, the City of Phoenix, Maricopa County, and the Salt River Water Users' Association had joined together to form the Cave Creek Flood Control Board, but as *Engineering News-Record* noted in September, "the work had not yet begun pending completion of financial arrangements."[38] After August, interest in the flood control initiative accelerated, with responsibility for engineering plans falling to the water users' association's general manager C. C. Cragin and Arizona State Engineer Thomas Maddock. At this time, a dam site was located "where Cave Creek leaves Paradise Valley and passes through a low ridge of mountains at a point of confluence of several tributaries."[39]

Eastwood first learned of the proposed Cave Creek Dam in October 1921, when George Davenport, the same Santa Fe Railway engineer who had endorsed the Lake Hodges design three years earlier, met with him in Los Angeles. Soon thereafter, Eastwood provided Davenport with a preliminary description and estimate that the Santa Fe engineer forwarded to Cragin.[40] In making the formal connection between Eastwood and the flood control board, Davenport counseled Cragin:

> In accordance with our discussion in regard to the proposed Cave Creek dam for the protection of Phoenix from flood waters, I have gone somewhat further in regard to the possibilities of constructing a multiple arch dam and have taken the liberty of asking John S. Eastwood, Consulting Hydraulic Engineer, who has built fourteen of this type, for a preliminary estimate of the cost of a multiple arch dam . . . it seems to me that if there is any possibility of constructing a multiple arch dam for the same money that an earthfill dam would cost, or even somewhat more money, it should be seriously considered, due to the fact that the arch dam is able to carry a flood over its top if the spillway capacity turns out to be inadequate, while the earthfill dam would fail under such a flood. . . . If you agree with me in this, I would suggest that you send to Mr. Eastwood a set of plans showing the dam as now proposed so that on his return from the north he can make up more definite figures.[41]

In his letter to Cragin, Davenport's reference to the frailties of an earthen embankment design reflected a fundamental truth regarding the ability (or lack thereof) of earthen and rockfill embankment structures to survive overtopping. Prior to this time, it was assumed that the flood control board would build Cave Creek Dam as an earthen embankment and Davenport voiced his lack of enthusiasm for such a proposal. Although unstated in his letter, Davenport could have made reference to such disasters as the nationally famous Johnstown flood (Pennsylvania, 1889) or collapses of the Lower Otay Dam (southern California, 1916) and Walnut Grove Dam (Arizona, 1890), all of which involved failures of embankment dams caused by overtopping.[42] In particular, the Walnut Grove disaster held special meaning because it had occurred less than one hundred miles north of Cave Creek and caused scores of deaths within an Arizona mining settlement along the Hassayampa River. In contrast, Davenport could speak authoritatively about the potential strength of Eastwood's designs in withstanding overtopping because, as described in *Engineering News-Record*, Eastwood's Big Bear Valley Dam had safely passed water over the length of its crest during the same storm that destroyed the Lower Otay Dam.[43]

Eastwood appreciated that Davenport's initiative and Fletcher's close relationship with the Santa Fe were indispensable in bringing his services to the attention of the Cave Creek Flood Control Board. In November 1921, Eastwood advised Fletcher that "[State Engineer] Maddock has requested me through the Santa Fe engineers to get up plans for the Cave Creek Dam, so you can count that your boosting has been doing a great deal of good and I thank you."[44] Taking advantage of the entree offered by Davenport and Fletcher, in December Eastwood sent Maddock estimates for his "latest design" featuring a "curved front slope" and forty-four-foot span arches.

In the time since his work at Lake Hodges, Eastwood had devised curved face designs as a way to reduce concrete costs.[45] Based upon laborious studies calculating the amount of concrete quantities required for various buttress-arch configurations, his curved face dams did not feature a sharp angle between the inclined and vertical sections of each arch. Instead, these new designs used a gradual curve. Thus, at the top of Cave Creek Dam the arches are vertical; as they extend downward they slowly flatten out until reaching a point where they maintain a constant upstream inclination of about fifty-six degrees. With this refinement, Eastwood could reduce the amount of concrete in, and thus the cost of, the structure even beyond that achieved at Lake Hodges and Murray.[46]

In December 1921, the flood control board requested bids for an earthen dam across Cave Creek, while advising contractors that they also could submit bids on alternative designs. Previous to this, Eastwood had given his multiple arch design to Lynn S. Atkinson, a Sacramento-based contractor who subsequently used it as the basis of a bid.[47] Apparently only one bid for the earthfill design was offered to the flood control board and, although no evidence is available documenting the exact amount of the embankment dam bid, it engendered less than enthusiastic support. As an alternative, the board moved to accept Atkinson's bid on the multiple arch design so that construction could begin in early 1922. As reported in the Salt River Valley Water Users' Association's annual history for 1921–1922: "It was found that the details of construction of an earthfill dam . . . could not be agreed upon by the Cave Creek Flood Control Board and a compromise design submitted by John S. Eastwood December 17, 1921 was accepted by all members of the Board."[48]

"All Lines Are Curves"

Given the scale of the project, things moved quickly from planning to construction in early 1922.[49] The only stumbling block emanated from S. M. Cotten, an assistant engineer with the City of Phoenix, who apparently maintained close ties with a contractor who had anticipated building the earthen dam across Cave Creek. Cotten had no experience with multiple arch dam technology, but boldly proclaimed that Eastwood's dam would fail during a major flood.[50] In the end, Cotten's objections proved unconvincing, with his complaints perhaps best characterized by a thirteen-page, single-spaced letter where—rather amazingly—he readily acknowledged his "unfamiliarity with the subject of multiple arch dam design and practice." Referring to the design accepted by the flood control board as "this Eastwood monstrosity," Cotten also confided that "I have in mind a dam design for Cave Creek which appears feasible."[51]

In the spring of 1922, Cotten took his campaign public through a letter published in *Engineering News-Record* where his criticism involved seemingly esoteric technical issues. Specifically, they centered on whether it was appropriate to consider—as Eastwood did—that the buttresses would absorb the entire weight of the adjacent arches and water load.[52] Eastwood understood this assumption was not absolutely accurate, but he knew that it erred on the side of safety and (of no small consequence) helped facilitate computations. As Eastwood explained to State Engineer Maddock: "Of all the methods used to determine these stresses, there is none that meets fully the conditions of mathematical certainty, but the method [I have] adopted finally is the one . . . most nearly of determining the safety under the most severe conditions."[53] Cotten's letter in *Engineering News-Record* sought support for his assault on Eastwood's method of analysis.[54] But the gambit soured when Gardiner S. Williams, an engineering professor at the University of Michigan and a designer of several small multiple arch dams in Michigan, dismissed Cotten's contentions and averred that:

> There can be little question that the method of computation employed by the designer of the dam [Eastwood] . . . was the better one. . . . In solving problems of unusual type it is well to remember that rules and formulas derived from certain definite assumptions will not give exact values when mechanically applied to cases where the conditions do not fit those assumptions . . . the best we can do with a foundation is to make reasonable assumptions and find approximate stresses.[55]

Eastwood reacted by advising Maddock that "the method of treating the subject . . . is well explained by Professor Gardiner S. Williams . . . [who offers] a proper explanation of why there is such a wide variation of opinion, and such wild statements made."[56]

Although initially troublesome, Cotten's attack dissipated after a few months and ultimately did nothing to impede the project or force changes to the Eastwood design. Nonetheless, it is significant that Maddock, Cragin, and Santa Fe officials did not back down from their support of Eastwood's design because, without their steadfast backing, Eastwood would have been powerless to retain the design commission. As further evidence of the strength shown by the Santa Fe and the flood control board, it is worth noting complaints made by A. G. Wishon, general manager of the San Joaquin Light and Power Corporation in central California, who commented upon the legacy of John Freeman's opposition to Eastwood's work in March 1922. At the time Wishon made the following comments, Freeman was starting a term as president of the American Society of Civil Engineers and held great authority within America's engineering community: "It is a tremendous undertaking to sell the idea [of multiple arch and single thin arch dams] to the financier who slips behind your back and consults a man like John R. Freeman . . . the prominence of John R. Freeman in the engineering world and his influence with men of capital [has] delayed the storing of water for a period of ten or fifteen years."[57]

No evidence is available as to how the flood control board specifically reacted to Cotten's complaints—except that construction proceeded without interruption—but Wishon's comments reveal that any decision to abandon Eastwood's design would have found support among a significant segment of the engineering establishment. For Eastwood, the controversy (or appearance thereof) proved exasperating and prompted him to consider philosophical issues underlying the practice of structural design. In an essay entitled "Limitations to Formulae," Eastwood mused upon the potential misuse of mathematical analyses and eloquently assessed how engineering represents much more than simplistic application of mathematical equations to reach answers of supposed absolute validity; as he expressed it: "Einstein is Right! All lines are curves and all things mundane are relative."[58]

"Already More than Paid for Itself"

On 29 July 1922, the board's faith in Eastwood's design was publicly affirmed in a major article published in *The Arizona Gazette*. No mention was made of Cotten's complaints, as the article proclaimed that "When the last cubic yard of concrete is poured into the arch and buttress [structure] . . . Phoenix will have a concrete barrier capable of withstanding the fury of a flood twice the volume of any which Cave Creek run-offs have developed."[59] Responsibility for construction fell to Atkinson and his crew, who labored under the supervision of the prominent Arizona hydraulic engineer James B. Girand (appointed by the flood control board as its on-site representative).

Sand and aggregate used in the concrete were excavated near the site and concrete from a central mixing plant was delivered to various parts of the structure via gasoline-powered Ford trucks. Ironically, the biggest physical impediment hindering completion comprised a lack of water for mixing the concrete. In fact, *Southwest Builder and Contractor* reported that "for several weeks the production [of concrete] was . . . somewhat curtailed" until excavation of a special well that was more than 100 feet deep.[60] As built, the dam consists of thirty-eight arches each with a 44-foot span; its aggregate length is 1,692 feet. Although most of the structure is less than 90 feet high, it has a maximum height of 120 feet in the central section, where solid foundations proved to lie much deeper than originally envisaged. When finished, the structure could impound 14,000 acre-feet of water.

On 4 March 1923, the structure was declared complete. Eastwood came to the ceremony marking the project's conclusion and—as transcribed by *The Arizona Gazette*—he heard Cragin declare "that contractor Atkinson had done his work well . . . [and] the dam has met fully every specification and regulation laid down by the board."[61] Even more significantly, just a few days before the official completion ceremony, the dam had demonstrated its integrity by holding back a sudden flood. As evidence of this, a photo showing water "backed up against the dam for a height of 20 feet" appeared in *The Arizona Gazette* under the headline "Cave Creek Dam Stops a Flood."[62] As Cragin expressed it in his ceremonial remarks: "During the floods of last week [the dam] had demonstrated its worth, while the [gradual] release of waters stored during the rain had proved that the Arizona Canal, crossing the path of Cave Creek, can take care of the run-off from the dam without damage."[63] *The Arizona Gazette* was probably overly effusive in reporting that the dam had "already more than paid for itself," in holding back this initial flood. But there is little question that completion of Eastwood's dam marked a dramatic transition point in the hydraulic history of western Maricopa County. For the next fifty years (until supplanted by a larger dam built by the Army Corps of Engineers in the late 1970s), Cave Creek Dam did its job in holding back potentially devastating floods.

Significance of Cave Creek Dam

In assessing the historical significance of the Cave Creek Dam, it is useful to move beyond events in western Maricopa County and adopt a reference frame of wide scope. First, it is important to appreciate that the dam stands as one of America's premier examples of efficient structural design. This can be illustrated by relating the structure's overall size to the amount of concrete it required to build. Almost 1,700 feet long with an average height of about 75 feet, the Cave Creek Dam incorporated about 18,750 cubic yards of concrete into the completed structure.[64] On average, this means that each lineal foot of the structure, with an average height exceeding 70 feet, required only 11 cubic yards of concrete. In contrast, a gravity dam of comparable size would require almost 60 cubic yards of concrete per lineal foot. In other words, Eastwood's design offered savings of more than 80 percent in the material necessary to erect a similar-sized concrete gravity dam. This allowed the entire structure to be built for less than $560,000.[65] Although this chapter has not attempted to offer a detailed technical treatise on Cave Creek Dam vis-à-vis more conventional dam design, the structural and economic attributes of Eastwood's curved face design necessarily undergird its distinctive status within hydraulic history.

Within less than two years after completion of the Cave Creek Dam, Eastwood was dead. However, even within this short time period, significant extensions of his work at Cave Creek were underway. One of these involved construction of the Bluewater Dam in New Mexico.[66] Lying adjacent to the Santa Fe Railway's main-line route about one hundred miles west of Albuquerque, the Bluewater-Toltec Irrigation District was closely allied to the railroad because of the relationship between agriculture production and freight revenue. Beginning in mid-1923, Eastwood prepared a series of designs for the Bluewater Dam; at the time of his death, plans for the structure were essentially complete and the start of construction lay only a few months in the future.[67] As it turned out, in 1925 the district erected an arch dam at the site using plans prepared in part by W. S. Post, an engineer previously associated with Ed Fletcher on many irrigation projects in San Diego County.[68] Thus, although an Eastwood design was not used at Bluewater, the influence of Fletcher and the Santa Fe on regional hydraulic development was nonetheless evident in a manner akin to Phoenix's Cave Creek project.

A more provocative—and instructive—connection forged at Cave Creek involved plans spearheaded by James B. Girand (the flood control board's supervisory engineer at Cave Creek) to build a privately financed hydroelectric power dam across the Colorado River in northwestern Arizona. Known as Diamond Creek Dam (the site was later referred to as Bridge Canyon), this structure would have inundated the lower end of the Grand Canyon. Girand first proposed this power development in 1916, but political issues impeded its immediate implementation.[69] After work concluded at Cave Creek, in 1923 Girand engaged Eastwood to develop designs for a structure more than three hundred feet high to impound a major reservoir at Diamond Creek.[70] The ultimate fate of the Diamond Creek Dam is particularly intriguing because it relates to the overall development of the Colorado River and the role of the federal government in controlling this process. In his 1941 book *The Boulder Canyon Project: Historical and Economic Aspects,* Paul Kleinsorge describes Girand's proposal:

> By 1921, eleven applications had been filed with the Federal Power Commission for sites on the Colorado and its chief tributaries. One of the most important of these was the application of Mr. James B. Girand of Phoenix, Arizona . . . [whose proposal] had been recognized by the Federal Power Commission when a preliminary permit was granted for a period of one year for a power project on the Colorado River near the outlet of Diamond Creek. . . . As time went on, the complicated character of the development of the Colorado River was realized more and more clearly. The Girand permit was extended, but no permanent grant was made."[71]

In early 1927, the proposed Boulder Canyon project (focused around what became Hoover Dam) fostered a congressional resolution blocking "the establishments of any rights on the river that might interfere with the proposed dam and power plant at Boulder Canyon."[72] After passage of the Boulder Canyon Project Act, the Colorado River power permit embargo was lifted, but the effect of the federal government's involvement in building Hoover Dam had far-reaching effects. As Kleinsorge described it:

> [In 1932] it was decided that the [Federal Power] commission should withhold authorization for those projects which might affect interstate water allocations . . . pending applications found not to be adapted to existing conditions were to be rejected. Under these classifications, the Girand application, after many postponements was finally rejected. . . . Thus with the rejection of one of the oldest applications for power development on the Colorado River, the Federal Power Commission indicated definitely that all power development would be under its close supervision.[73]

The struggle between California and Arizona over construction of Hoover Dam and its effect on the use and distribution of Colorado River water provides a provocative background for considering the reactions to Eastwood's designs by the respective state engineers in the period following completion of Lake Hodges Dam. Whereas California State Engineer

Wilbur F. McClure worked to obstruct Eastwood's dam design innovations, Arizona State Engineer Thomas Maddock offered strong support, even in the face of S. M. Cotten's vitriolic objections. After completion of the Cave Creek Dam, Maddock became one of Arizona's most outspoken critics of the Boulder Canyon Project (which existed first and foremost to serve the economic interest of southern California). Maddock published treatises that spelled out why Arizona opposed the project and the Colorado River Compact that accompanied it. Maddock's opposition did not stop with passage of the Boulder Canyon Project Act in 1928 and extended through the 1950s as he steadfastly resisted California's claims to the Colorado River.[74]

The conjoining of California's and the federal government's interests over Hoover Dam and the control of the Colorado River represents a major touchstone in the evolution of western water policy. It also represents a distinctively different mode of water project development than that responsible for creation of Cave Creek Dam. Although federal agencies such as the Bureau of Reclamation and the Army Corps of Engineers have, on occasion, exhibited interest in materially conservant dam design, the great majority of federally sponsored dams—especially those built since the New Deal—use massive gravity designs. The process of technological transfer and the political mechanisms that brought the Cave Creek Dam into existence as a public-works structure involved no federal participation; as such, the network of information exchange responsible for fostering Cave Creek Dam stands in significant counterpoint to what is often considered a central factor in western water development.

In this light, it is important to appreciate that the possibilities signified by Eastwood's work were not pie-in-the-sky "dreams on a drawing board," but found expression in concrete structures that operated in the real, physical world. The political economy—based upon an interaction of private and public actors—that brought the Cave Creek Dam into being represents a meaningful alternative to the federally dominated system of water project development that held sway over much of the West by mid-century. This does not mean that environmental costs associated with dam construction could somehow have been avoided had Eastwood's vision of dam design been more widely embraced. Nor does it mean that control over the region's limited water supplies may have been radically redistributed among less advantaged social groups. But it suggests that the choices made and the paths followed in the wake of the New Deal were hardly preordained. They were no more—and no less—inevitable than the forces that fostered construction of an Eastwood dam to protect Phoenix from floods.

There are many ways to study history, and it is conceivable that a future historian could offer a revealing analysis of the Cave Creek Dam that would foreground the internal politics of Phoenix that fostered the formation and operation of the Cave Creek Flood Control Board. In such a history, the place of John Eastwood and his dam design might well assume a supportive, almost secondary role. However, in this chapter the emphasis has been on the relation of technology to business networking and to the larger political economy of western water development. Such relationships—and the manner in which people and personal interactions influence its evolution—are central to every western water project of any size or scale ever built.

Notes

1. Hoover Dam (also called Boulder Dam) was authorized long before the 1929 stock market crash; nonetheless, President Franklin Roosevelt enthusiastically embraced the construction of the dam as part of his administration's New Deal. In contrast, dams such as Grand Coulee, Bonneville, Shasta, Friant, Fort Peck, Norris, and a host of others were the subject of studies and proposals prior to Roosevelt's inauguration, but their actual authorization and financing did not occur until after he took office.

2. *Salt River Valley Water Users' Association Annual History, 1922–1923*, (Tempe, Ariz.: Salt River Project, 1923), 105.

3. For more on this, see Donald C. Jackson, *Building the Ultimate Dam: John S. Eastwood and the Control of Water in the West* (Lawrence: University Press of Kansas, 1995).

4. James Marshall, *Santa Fe: The Railroad that Built an Empire* (New York: Arbor House, 1945); L. L. Waters, *Steel Trails to Santa Fe* (Lawrence: University Press of Kansas, 1950); Donald Duke, *Santa Fe: The Railway Gateway to the American West*, vol. 1, *Chicago–Los Angeles–San Diego* (San Marino, Calif.: Golden West Books, 1995). In 1866, federal legislation authorized a three-million-acre land grant if trackage reached Colorado by the end of 1872. After meeting this deadline, the Santa Fe line did not proceed much further west until obtaining the Atlantic & Pacific Railroad franchise. In 1880, the city of San Diego and the Kimball Brothers (who controlled the Rancho de la Nacion, a large real estate tract renamed National City) signed an agreement granting the railroad almost 20,000 acres of land in return for making San Diego the Pacific terminus of the railroad. The con-

nection with the Santa Fe mainline was made in 1884, but the right-of-way in northeast San Diego County proved difficult to maintain. It was soon abandoned in favor of a "coast line" extending through Los Angeles. This subordinated San Diego to Los Angeles in terms of transcontinental commerce, but maintained the distinctive relationship between the railway and San Diego that fostered Ed Fletcher's subsequent association with the company. See Robert Mayer, comp., *San Diego: A Chronological and Documentary History, 1535–1976* (Dobbs Ferry, N.Y.: Oceana Publications, 1978), 108–10; Glenn S. Dumke, *The Boom of the Eighties in Southern California* (San Marino, Calif.: Huntington Library, 1944), 22–23.

5. Phoenix's early history is recounted in Bradford Luckingham, *Phoenix: The History of a Southwestern Metropolis* (Tucson: University of Arizona Press, 1989).

6. The Santa Fe's connection between Ash Fork and Prescott was completed in 1885. Rough terrain, in combination with the depression of 1893–1894, slowed its extension into Phoenix for several years.

7. James D. Schuyler, "The Construction of Sweetwater Dam," *Transactions of the American Society of Civil Engineers* 19 (1888): 201–32.

8. For more on the technology of arch dams and its relationship to massive gravity dams, see Jackson, *Building the Ultimate Dam*, 14–27; F. E. Brown, "The Bear Valley Dam," *Engineering News* 19 (23 June 1888): 513–14.

9. The overtopping of Sweetwater Dam is described in James D. Schuyler, *Reservoirs for Irrigation, Water Power, and Domestic Water Supply* (New York: John Wiley, 1909), 213–37.

10. "Steel Dam at Ash Fork, Arizona, A. T. and S. F. Ry," *Engineering News* 39 (12 May 1898): 299–300.

11. Terry S. Reynolds, "A Narrow Window of Opportunity: The Rise and Fall of the Fixed Steel Dam," *IA: The Journal of the Society for Industrial Archeology* 15 (1989): 1–20.

12. John S. Eastwood Papers, Water Resources Center Archives (WRCA), University of California, Berkeley. The Eastwood papers are divided into numbered folders, hereafter referenced as JSE.

13. George Low, "The Fresno Transmission Plant," *Journal of Electricity* 2 (April 1896): 79–89.

14. Jackson, *Building the Ultimate Dam*, 47–58.

15. The early history of the Big Creek power system is discussed in David Redinger, *The Story of Big Creek* (1949; reprint, Los Angeles, Calif: Angeles Press, 1986); William Myers, *Iron Men and Copper Wires: A Centennial History of the Southern California Edison Company* (Glendale, Calif.: Trans-Anglo Press, 1983); Jackson, *Building the Ultimate Dam*, 59–83.

16. John S. Eastwood, "Hume Lake Dam," *Journal of Electricity, Power, and Gas* 23 (30 October 1909): 398–404; also see Hank Johnston, *They Felled the Redwoods* (Glendale, Calif.: Trans-Anglo Press 1966); JSE 12, WRCA.

17. John S. Eastwood, "The New Big Bear Valley Dam," *Western Engineering* 3 (December 1913): 458–470; JSE 1, WRCA.

18. "John Ripley Freeman," *Transactions of the American Society of Civil Engineers* 98 (1933): 1471–76.

19. Jackson, *Building the Ultimate Dam*, 109–33, documents the conflict between Freeman and Eastwood over Big Meadows Dam.

20. John S. Eastwood, "Syllabus of an Illustrated Lecture Proposed for the Internal Water Ways Congress, January 15–18, 1914," JSE 3, WRCA.

21. Jackson, *Building the Ultimate Dam*, 146–52; Ed Fletcher, *Memoirs of Ed Fletcher* (San Diego, Calif.: Privately printed, 1952). In 1888, sixteen-year-old Ed Fletcher sailed from Massachusetts to San Diego, where he lived until his death in the 1950s. In 1896, he established his own grocer/dry goods business and from this financial base explored regional investment opportunities. He also assisted others interested in real estate development. For example, in 1905 Fletcher acted as an agent for Henry Huntington's South Coast Land Company in assembling land options in northern San Diego County. See "Agreement between C. W. Gates and W. G. Kerckhoff and Frank Salmons and Ed Fletcher," November 1905, box 35, drawer 1, Henry W. O'Melveny Papers, Henry E. Huntington Library, San Marino, California; Fletcher, *Memoirs*, 205–10.

22. The Riverside Cement Company was a prominent cement manufacturer and publisher of *Good Concrete* (Riverside, 1914), a sophisticated reference work on concrete construction methods.

23. Fletcher, *Memoirs*, 231–54. Eastwood's involvement with Fletcher and Henshaw on this project in 1914 are documented in JSE 56, WRCA.

24. John S. Eastwood to W. S. Post, 12 March 1917, box 8, Ed Fletcher Papers, University of California San Diego (UCSD). In 1914–1915, Fletcher first explored the idea of building a multiple arch dam at the Carroll site (the initial name for the site of Lake Hodges Dam) using designs prepared by W. S. Post (a San Diego–based engineer associated with Fletcher). Evidence of Eastwood's early work on Carroll/Lake Hodges Dam also appears in John S. Eastwood to W. G. Henshaw, 2 June 1916; and John S. Eastwood to W. G. Henshaw, 5 February 1917, JSE 31, WRCA.

25. The Santa Fe's review of Eastwood's design is discussed in W. S. Post to E. O. Faulkner, 26 March 1917; John S. Eastwood to W. S. Post, 27 March 1917; and George L. Davenport to G. C. Millett, 7 April 1917, all in box 8, Ed Fletcher Papers, UCSD.

26. See the construction contract "Arthur S. Bent and H. Stanley Bent, Doing Business as a Copartnership under the Firm Name of Bent Brothers and the San Dieguito Mutual Water Company," 15 May 1917, box 8, Ed Fletcher Papers, UCSD.

27. The San Dieguito Dam is documented in JSE 51, WRCA.

28. Jackson, *Building the Ultimate Dam*, 163, 195–96.

29. Ibid., 197–208.

30. In 1910, Fletcher joined with James Murray, a Montana-based businessman, to take control of the San Diego Flume Company and its thirty-seven-mile-long wooden flume that served the eastern edge of San Diego. In January 1916, heavy floods came within a foot of overtopping the earthfill La Mesa Dam that stored water delivered by the flume. Because of this near calamity, Fletcher and Murray sought to build a larger, safer structure for their

system. In 1917, they commissioned Eastwood to design a replacement for La Mesa Dam (subsequently named Murray Dam). See John S. Eastwood, "Recent Multiple Arch Dams," *Journal of Electricity* 42 (15 March 1919): 263–64; Fletcher, *Memoirs*, 149–85; JSE 54, WRCA.

31. Fletcher, *Memoirs*, 98; JSE 43, WRCA.

32. Construction costs for the Murray and Lake Hodges Dams are delineated in Ed Fletcher to A. P. Davis, 25 April 1919, "Discussion Related to Dams," box 289, entry 3, U.S. Reclamation Service, RG 115, Denver, National Archives.

33. Karen Smith, *The Magnificent Experiment: Building the Salt River Reclamation Project, 1890–1917* (Tucson: University of Arizona Press, 1986); Donald C. Jackson, "Engineering in the Progressive Era: A New Look at Frederick Haynes Newell and the Early U.S. Reclamation Service," *Technology and Culture* 34 (July 1993): 539–74.

34. The development of hydroelectric power plants on the Salt River is described in David M. Introcaso, "A History of Mormon Flat Dam," file AZ-14, Historic American Engineering Record, Prints and Photographs Collection, Library of Congress.

35. W. H. Peterson, "Cave Creek Flood Control Dam Multiple Arch Structure Built along New Lines," *Southwest Builder and Contractor* 62 (16 February 1923): 30–33.

36. Photographs in the Hayden Library, Arizona State University, Tempe, show water from a 1905 Cave Creek Dam flood inundating the first floor of the Territorial Capitol.

37. "Cave Creek Flood Does Damage at Phoenix," *Engineering News-Record* 87 (15 September 1921): 464.

38. Ibid.

39. Peterson, "Cave Creek Flood Control Dam."

40. John S. Eastwood to George L. Davenport, 15 October 1921, JSE 45, WRCA. This letter refers to a recent "meeting at your office" in which the two engineers discussed the Cave Creek project.

41. George L. Davenport to C. C. Cragin, 18 October 1921, Salt River Project Archives, Tempe, Arizona. This letter included Eastwood's 15 October letter to Davenport as an attachment. Copies of the letter to Cragin were also sent to State Engineer Maddock.

42. David McCullough, *The Johnstown Flood* (New York: Simon and Schuster, 1968); "The Walnut Grove Dam Disaster," *Engineering News* 22 (8 March 1890): 229–30; "The Walnut Grove Dam Disaster," *Engineering News* 22 (26 April 1890): 389–90; "Otay Rock-Fill Dam Failure," *Engineering News* 75 (3 February 1916): 236–39.

43. "Bear Valley Multiple Arch Dam Overflows for First Time," *Engineering News* 75 (18 May 1916): 961.

44. John S. Eastwood to Ed Fletcher, 16 November 1921, JSE 6, WRCA.

45. Eastwood claimed to have first developed a curved-face design while working on a project for the San Joaquin Light & Power Corporation. See John S. Eastwood to J. T. Crabbs, 12 February 1922, JSE 9, WRCA.

46. A demonstration that the buttress of minimum weight should be built with a curved upstream face appears in Herman Schorer, "The Buttressed Dam of Uniform Strength," *Transactions of the American Society of Civil Engineers* 96 (1932): 681–83.

47. Thomas Maddock to John S. Eastwood, 5 December 1921; and George L. Davenport to C. C. Cragin, 18 October 1921, JSE 45, WRCA.

48. *Salt River Project Annual History, 1921–1922*, 63.

49. "Progress on Flood Prevention at Phoenix, Ariz.," *Engineering News-Record* 88 (26 January 1922): 162.

50. S. M. Cotton to L. B. Hitchcock, 6 January 1922, JSE 45, WRCA.

51. S. M. Cotten to W. C. Foster, 1 June 1922, JSE 45, WRCA

52. See S. M. Cotten to L. B. Hitchcock, 30 December 1921; S. M. Cotten to L. B. Hitchcock, 6 January 1922; S. M. Cotten to W. C. Foster, 1 June 1922; S. M. Cotten to Thomas Maddock, 3 July 1922, all in JSE 45, WRCA. Cotten proposed a method of analysis in which the arches and buttresses are considered to share the arch weight and the water load at all elevations and, at least superficially, this technique did not appear unreasonable. However, Cotten's approach predicted that the pressure exerted by a full reservoir would cause the upper portions of the Cave Creek design to suffer from tensile stresses at the *downstream* edge of the buttresses. This meant that—rather illogically—the dam would somehow act to tip over *upstream* into the reservoir under full hydrostatic load. See S. M. Cotten's discussion accompanying Fred Noetzli, "An Improved Type of Multiple Arch Dam," *Transactions of the American Society of Civil Engineers* 87 (1924): 399–403.

53. John S. Eastwood to Thomas Maddock, 9 August 1922, JSE 45, WRCA.

54. S. M. Cotten [writing as "Enquirer"], "How Should We Figure Foundation Pressures in a Multiple Arch Dam?" *Engineering News-Record* 88 (13 April 1922): 623–24; Cotten identifies himself as "Enquirer" in *Engineering News-Record* 89 (13 July 1922), 78–79.

55. Gardiner S. Williams and Albert S. Greene, "Foundation Pressures in Multiple Arch Dams," *Engineering News-Record* 89 (13 July 1922): 79.

56. John S. Eastwood to Thomas Maddock, 9 August 1922, JSE 45, WRCA.

57. A. G. Wishon to John S. Eastwood, 29 March 1922, JSE 48, WRCA.

58. John S. Eastwood, "Limitations to Formulae" (unpublished manuscript, circa 1922), JSE 45, WRCA.

59. "Control Dam at Cave Creek to Banish All Menace of Floods," *The Arizona Gazette*, 29 July 1922.

60. Construction is best described in Peterson, "Cave Creek Flood Control Dam."

61. "Cave Creek Supervisors Accept Dam," *The Arizona Gazette*, 6 March 1923.

62. "Cave Creek Dam Stops a Flood," *The Arizona Gazette*, 8 March 1923.

63. "Cave Creek Supervisors Accept Dam," *The Arizona Gazette*, 6 March 1923.

64. Peterson, "Cave Creek Flood Control Dam," indicates that 19,000 cubic yards of concrete were used in the dam. However, the chart "Cave Creek Dam: Total Quantities" indicates that 18,774.47 cubic yards of concrete went into the structure; JSE 45, WRCA.

65. *Salt River Valley Water Users' Association Annual History, 1922–1923*, 105–6. Because the dam proved to be much taller than the flood control board initially believed necessary, the structure significantly exceeded preconstruction estimates of approximately $350,000. (Originally, the maximum height was stipulated at less than 85 feet—as built, the

maximum height reached 120 feet.) Of course, additional expenditures of this type would have been incurred for any type of dam built at the site. In total, $556,982.39 was spent to complete the dam including "$18,300 paid to John S. Eastwood on contract." Payment responsibilities were as follows:

$169,494.43, City of Phoenix
$169,494.43, Maricopa County
$119,494.43, Salt River Valley Water Users' Association
$50,000.00, State of Arizona
$25,000.00, Santa Fe Railway Company
$10,000.00, Standard Oil Company
$5,000.00, Arizona and Eastern Railway
$5,000.00, Union Oil Company
$1,750.00, Greenwood Cemetery
$1,749.10, interest on bank account

$556,982.39, total cost

66. Ed Fletcher to John S. Eastwood, 23 June 1923; Ed Fletcher to John S. Eastwood, 14 August 1923; and John S. Eastwood to George L. Davenport, 26 January 1924, JSE 37, WRCA; also see John S. Eastwood to Ed Fletcher, 3 April 1924, JSE 4, WRCA.

67. The advanced state of Eastwood's involvement with the Bluewater Dam project at the time of his death is evident in Ed Fletcher to Ella Eastwood, 24 September 1924; and Ella Eastwood to Ed Fletcher, 30 September 1924, box 5, Ed Fletcher Papers, UCSD.

68. T. Mermel, *Register of Dams in the United States* (New York: McGraw-Hill, 1958), 44.

69. "Surveying Dam Sites in the Grand Canyon," *Engineering News* 76 (17 August 1916): 306–7. Girard's participation in the project is noted in "Personal Notes," *Engineering News-Record* 80 (14 February 1918): 335.

70. See Diamond Creek Dam folder, JSE 28, WRCA. Also see "Hunt Reelection Enlivens Colorado River Problems," *Engineering News-Record* 93 (13 November 1924), 805.

71. Paul L. Kleinsorge, *The Boulder Canyon Project: Historical and Economic Aspects* (Stanford, Calif.: Stanford University Press, 1941), 52–54.

72. Ibid.

73. Ibid.

74. Thomas Maddock, "Reasons for Arizona's Opposition to the Swing-Johnson Bill and Arizona Compact with Tentative Tri-state Compact Submitted to California and Nevada by Arizona Commission on February 7, 1927," and Thomas Maddock, "An Argument Presented to the U.S. Senate Committee on Irrigation and Reclamation at Phoenix, Arizona, November 3, 1925," Thomas Maddock Papers, Arizona Historical Foundation, Hayden Library, Arizona State University, Tempe.

14

The Changing Fortunes of the Big Dam Era in the American West

Mark Harvey

A few summers ago while vacationing in the West, my family and I stopped off for a short visit at Buffalo Bill Dam on the north fork of the Shoshone River east of Yellowstone National Park in northwestern Wyoming. Built by the Bureau of Reclamation in 1910, Buffalo Bill is one of the oldest of the agency's big dams.

The bureau had recently opened a new visitor center, perched on the cliff on the north side of the dam. Strolling through this modest building, I quickly became engaged with well-trodden paths of western history: books on cowboys, homesteaders, fur traders, and Indians; a pictorial history of homesteading and irrigation in northwestern Wyoming; and a video on the Bureau of Reclamation. This history celebrated the transformation of this corner of Wyoming from a desert to an oasis made possible by the foresight of water developers.

Holding good thoughts about hardy pioneers and their irrigation ditches, I moved across the building to large glass windows and found a terrifying view down the sheer face of Buffalo Bill Dam. In the next instant, my mind's journey into history vanished. The dam itself seemed to transcend history; its sheer walls, dizzying heights, and precise placement inside the canyon inspired a sense of timelessness. I was experiencing what David Nye has called the "technological sublime."[1] The size and domination of the dam seemed to surpass all human endeavors. Large, impersonal forces dominated here: the laws of physics and gravity and the anonymity of a technological wonder. Dams are tricky things; they stand like huge stone monuments virtually outside of time, nearly silent, unchanging, as if disconnected from the human world, from history.[2]

In fact, the opposite is true. Dams are human creations and products of history; more to the point, dams can teach us much about the history of the American West. A growing body of literature about dams has illuminated many aspects of the region's past: its politics, economic and social aspirations, and environmental history.[3] For most of the last one hundred years, dams have represented technological triumph and "American know-how." The large dams, such as Hoover, Grand Coulee, Shasta, Friant, and a host of others, have symbolized technological might and the power of humans over nature. Dams, the traditional dictum went, harnessed "wild" and "untamed" rivers and transformed them into calm, docile waterways. This taming of rivers always inspired noble thoughts of the conquerors. Here, for instance, is a poem by H. C. Wones, which caught the sense of triumph and celebration that marked the opening of American Falls Dam on Idaho's Snake River during the 1920s:

> To-day across the mountain peaks
> A glorious light appears,
> And men with Gladdened hearts arise,
> To greet their dreams of years.
> Deserts remote, and sage-clad plains,
> In symphonies sublime,
> Shall herald to the ends of earth
> This monument of time.[4]

Twenty years later Grand Coulee Dam on the Columbia River won similar praise. The dam was "the biggest thing on Earth" Richard Neuberger wrote, while other accounts labeled it "the greatest structure built by man," "the world's most monstrous dam," and "the eighth wonder of the world."[5] Joseph Stevens, author of a 1988 history of Hoover Dam, demonstrates that such reverence continues: "Confronting this spectacle in the midst of emptiness and desolation first provokes fear, then wonderment, and finally a sense of awe and pride in man's skill in bending the forces of nature to his purpose. In the shadow of Hoover Dam one feels that the future is limitless, that no obstacle is insurmountable, that we have in our grasp the power to achieve anything if we can but summon the will."[6]

Stevens also wrote that the soaring spectacle and "brute strength" of Hoover Dam brings to mind "the romance of the engineer."[7] In a recent essay, Helen Ingram points out that the engineering profession has been central in examining water problems and devising solutions to them, and engineers have been at the forefront of the dam business.[8] We are apt to think of engineers as faceless individuals who perform in the service of large corporations or government agencies. Their work is pure science, it is generally assumed, insulated from the political, economic, and even aesthetic aspects of dam construction.

In *Building the Ultimate Dam,* Donald Jackson has greatly revised this benign image. Focusing on the design of dams erected in early-twentieth-century California, such as Big Meadows Dam on the Sacramento River north of the state capital, Jackson argues that the big dam era, which the Hoover Dam ushered in, had more complex and contingent origins than has been assumed. Large dams, upward of one hundred feet high, were being built on rivers and streams in California's gold mining country in the 1860s. Later in the nineteenth century, water and power companies in California erected arched concrete gravity dams, which, although expensive, proved to be a superior design. Still, in the late nineteenth and early twentieth centuries no single design structure predominated, and the overwhelming majority of dams in the West were erected with private capital.[9]

That would change in the 1930s, when massive gravity dams arose on western rivers, built with federal dollars. These new dams' curved gravity design—a choice determined by "the social nature of the technological decision-making process"—came to symbolize Americans' desire to conquer nature, which gained strength from the anxieties of the Great Depression. As Jackson puts it, "during the New Deal the 'celebration of mass' became the dominant ideology associated with dam construction; the more material a dam required, the more acclaim and adulation it received. In an era of limits and diminished expectations, American culture apparently derived psychological satisfaction from creating something big in the face of adversity."[10] The nation's economy and its psyche would be rebuilt, in part, through the construction of these massive projects.

Much of the recent literature on the West's large dams reveals how the political and economic context of the Depression and New Deal forged the crucible of the big dam era. During the 1930s, Boulder, Bonneville, Grand Coulee, Fort Peck, and other mammoth dams symbolized the emergence of federal responsibility for water development as well as the determination of the New Dealers to lift the nation from the worst depression in its history. Boulder Dam on the Lower Colorado River was constructed between 1931 and 1935 and was renamed for Herbert Hoover in 1947. The dam tightly regulated the Lower Colorado and protected agricultural land in the Imperial Valley devastated by the river's floodwaters early in the century, while creating a massive reservoir, Lake Mead. Hoover Dam also insured a ready supply of hydropower to southern California, thus contributing to substantial industrial growth in Los Angeles during World War II. Finally, work on the dam employed more than 5,200 construction workers in the depths of the Depression, making it a major employer in southern Nevada.[11]

The Six Companies of San Francisco, a consortium of several firms, won the contract for building Hoover Dam with a bid of nearly $49 million (at that time "the largest single contract ever let by the United States government") and helped build Grand Coulee Dam on the Columbia.[12] These contracts enriched Henry Kaiser, Warren Bechtel, Frank Crowe, and other construction giants and engineers in the West and reinforced the continued success of the firm in winning lucrative federal contracts during and after World War II.[13]

Meanwhile, the big dams offered President Franklin Roosevelt a superb opportunity to remake the West's regional economy, while also relieving widespread unemployment. Dam construction put thousands of men and women to work, thus providing much needed employment in California, Nevada, Arizona, Washington, Oregon, Montana, and Texas. As Hoover Dam was a great showcase of unemployment relief in the Southwest, Grand Coulee Dam in Washington played a similar role in the Pacific Northwest. President Roosevelt, eager to expand New Deal work relief programs into the region, initiated construction of Grand Coulee Dam with an executive order in 1933 (Congress later approved funding of the dam in 1935).[14] More than 7,000 workers labored on Grand Coulee; 10,500 built Fort Peck Dam in Montana, the first major dam along the Missouri River; and about 2,000 built dams for the Lower Colorado River Authority in Texas.[15]

Recent histories provide extensive coverage of those who labored on the big dams. They highlight the numbers engaged in the work and profile their age, gender, and race. The most engaging book on this subject focuses on Roosevelt and other dams on Arizona's Salt River, built by white, black, Hispanic, and Native American workers.[16] Paul Pitzer's book on Grand Coulee, Joseph Stevens's on Hoover, and Russell Martin's on Glen Canyon also remind us of the dangers attending the work. Highscalers at Boulder Dam risked their lives blasting rock from the cliffs while dangling from ropes; 110 workers lost their lives building the dam. Eighteen died building Glen Canyon, while 365 suffered serious injuries.[17] The devotion of these workers to various dams was such that legends arose about the numbers of them supposedly entombed beneath millions of tons of concrete.[18]

New Deal dams not only symbolized the dignity of work and full stomachs, they promised sweeping transformation of people's lives by the still relatively new miracle of electricity. As Paul Pitzer puts it: "electrical power preoccupied the twentieth century. It could unburden our lives and improve our standard of living both physically and spiritually."[19] Low-cost electricity generated at the dams promised to bring power to rural and urban areas, attract industry, diversify regional economies, and launch the West on a new course of opportunity. Lewis Mumford, a major social critic whose works probed the adverse effects of industry, technology, and urbanization, lauded this change. In his book *Technics and Civilization,* he contended that hydropower offered a way out of the Paleotechnic age, a time dominated by coal and iron and the resulting air pollution and endless toil for masses of workers. Hydropower offered a new beginning: a clean source of power, expansion of wealth for all, and liberation from burdensome work.[20] While Mumford provided the intellectual rationale for hydropower, singer Woody Guthrie's folk songs supplied the words and images that galvanized public interest in and support for dams on the Columbia. "Grand Coulee Dam," "Roll on, Columbia," and other songs captured the soaring hopes of dam supporters in the 1930s.[21]

By the end of the 1930s, the foundations of the big dam industry had been established. The decade had seen a burst of federal dam-building that elevated the Bureau of Reclamation's profile throughout the region. Thousands of workers, building contractors, and numerous related industries and banks all benefited from the big projects. A coalition of interests now helped to sustain the dam industry, including major construction companies, natural resource extraction industries, banks, and labor unions. This coalition compelled the attention of western lawmakers from both parties, thereby exerting a powerful influence in Congress well into the 1970s.

Nevertheless, if the West's newly minted dams during the New Deal created the institutional and political foundation of the big dam era, they also fueled grand expectations that were never entirely realized. Unmet expectations and unexpected results were to become another theme of the big dam era. In Washington, for instance, the early hopes for Grand Coulee Dam were dashed. Rufus Woods, editor of the *Wenatchee World*, who had been at the forefront of a considerable effort of boosters and lawmakers in the 1920s and 1930s to gain federal approval of Grand Coulee Dam, understood that eastern Washington was sparsely populated, reliant on outside capital, and prone to boom and bust in its mainly extractive economy.[22] Lacking crop diversification as well as industry, the Columbia Basin lagged behind the urban west coast of Washington and suffered from the colonial dependence that characterized nearby Montana and Idaho. Woods contended that Grand Coulee Dam would transform the regional economy. Irrigation would promote crop diversification (especially sugar beets), whereas hydropower would encourage manufacturing, population growth, and an enlarged tax base. The dam promised a short-term boom and long-term economic stability; it would remake the regional economy.[23]

This lofty vision for Grand Coulee never quite materialized. Over time, the electric power the dam generated was sold to urban markets far from the basin—Portland, Spokane, and across the Cascades in Puget Sound. Because local communities like Wenatchee and Ephrata in the Big Bend region received no preferential treatment from the Bonneville Power Administration on power rates, industry had no incentive to locate there; during World War II, the bulk of manufacturing went to Portland and Seattle. Hydroelectric power clearly improved the lives of thousands in the Northwest, but not in the way that Rufus Woods and the early boosters had once hoped. "The history of Columbia River development," historian Robert Ficken concluded, "is replete with irony and unanticipated result."[24]

In similar fashion, some of the New Deal dams privileged certain Westerners at the expense of others. Engineers from the Bureau of Reclamation feared that silt from the Colorado River could rapidly build up behind Hoover Dam and reduce the dam's effectiveness as a storage reservoir and power generator. Accordingly, they pushed their colleagues in other federal agencies to mount an all out assault on soil erosion in the Lower Colorado Basin. Soon, the large Navajo Reservation in northeastern Arizona, where sheep herding had sustained the economy and cultural practices of one of the West's largest Native American communities, became a major target. In 1933, a federally imposed livestock reduction program began, with sheep and goats either purchased or shot. Reduction hit the subsistence base of the Navajo enormously hard. Yet the goal of protecting Hoover Dam from a premature death overrode such concerns. According to Richard White, "the dam was the catalyst that prompted drastic stock reduction. The government saw itself not only as saving the Navajos from themselves, but also as saving much of California, and indeed the entire Southwest, from Navajo herds."[25] Benefits from Hoover Dam flowed south to the Imperial Valley and southwest toward Los Angeles, while the Navajo economy was devastated.

World War II and the Cold War

As the United States entered World War II, the West's big dams became major contributors to the war effort. Here was the start of the "American Century," as Henry Luce had framed it, when the United States fought the good war against fascism and became an international power. The West's big dams proved themselves during the war, supplying hydroelectric power for defense industries, especially in Washington, Oregon, and California. Grand Coulee Dam, once an object of sharp criticism because of the inadequate market for its abundant hydroelectricity, now justified itself by supplying power to the aluminum industry, aircraft manufacturers, and the shipbuilders in Washington and Oregon.[26] Meanwhile, workers and contractors finished off Shasta Dam on the Sacramento River, a key edifice in the Central Valley Project in California. Near the war's end, Congress enacted the Flood Control Act of 1944, which authorized several large dams along the Missouri to control floods, generate power, and irrigate thousands of acres in the dry northern Plains states.[27] In the postwar years, thousands of workers built huge earth-rolled dams (a design made necessary by the flat topography of the Plains) including Oahe Dam near Pierre, South Dakota, and Garrison Dam in central North Dakota. The large reservoir behind Garrison Dam was named Lake Sakakawea, after the Shoshone woman who had accompanied Lewis and Clark on their epic journey west.[28]

Champions of big dams found an even more favorable atmosphere during the first stages of the Cold War, when the United States and Soviet Union became locked in an ideological struggle for international dominance. President Harry Truman, who presided over the early expansion of nuclear weapons, approved development of the hydrogen bomb in 1950, thus opening a new era in the arms race. His decision, we know now, persuaded his Interior Secretary Oscar Chapman to approve the controversial Echo Park Dam in Colorado (later defeated in Congress), which the Atomic Energy Commission sought for its large power reserves.[29] Historian Keith Petersen claims that the Truman administration gave initial approval to Ice Harbor Dam on the Lower Snake River due to demands for power by the commission at the Hanford nuclear site in Washington.[30]

The Cold War placed a heavy premium on extraction of the nation's natural resources, especially certain minerals in the West; national preparedness was taken to mean full-scale development of timber, rangelands, soils, and western rivers to maximize food and hydroelectric power production. This too underpinned the big dam era. In short, the huge structures dotting the West's rivers were now signs of national preparedness, of the United States' will and determination to confront Soviet power. Some conservationists in the 1950s tried to argue that too many dams offered a convenient target to Russian bombs and should not be built, but the argument carried little weight.[31]

The Cold War and the expanding national economy sustained the big dam era for more than two decades. The huge dams arising from the West's rivers from the 1940s through the 1960s reflected American international power and national confidence. To many, they furnished proof that the world could be remade through science, technology, and American strength. In the American Century dams symbolized the nation's technological might, rapidly expanding capitalist economy, and democratic system. As Donald Worster has argued, "to friends and foes alike, [the dams] said that we were a people who had risen, through destiny and virtue, to pre-eminent leadership over the entire planet, that we were now ready to push the world around."[32]

For the engineers, designers, building contractors, and federal agency officials who oversaw these enormous projects, big dams controlled and improved rivers. Whether the problem was too much water or too little, builders and designers understood that dams resolved the problem by evening out river flows and creating streamlined waterways punctuated with huge reservoirs, which perfectly balanced multiple demands on that particular river. Those who built the Hoover, Glen Canyon, Flaming Gorge, and other dams along the Colorado, took satisfaction that the river now served millions of residents from Wyoming to the Mexican border who required electrical power, storage, flood control, and irrigation. Left alone, rivers "went to waste," but regulated by dams, they served numerous interests in cities and rural communities throughout the seven-state Colorado Basin.

Moreover, in the minds of many engineers—and undoubtedly countless other observers—the great dams had actually enhanced nature's wonders. For them, Lake Powell, the enormous reservoir behind Glen Canyon Dam, turned a once-desolate and seldom-visited area in the Southwest into a magnificent lake. While environmentalists decried the flooding of Glen Canyon, Bureau of Reclamation commissioner Floyd Dominy insisted that the new lake in the desert had a rare beauty of its own. As for the mammoth concrete structure that created Lake Powell, Dominy likened it to an eternal object of beauty: "The manmade rock of the dam has become as one with the living rock of the canyon. It will endure as long as time endures."[33]

Demise of the Big Dam Era

The remarkable affluence of the United States in the two decades following World War II had been another major factor sustaining the big dams. Economic expansion from the late 1940s through the late 1960s was fueled by government spending; advances in chemistry, physics, and electronics; the GI bill; and other factors. Besides raising living standards and enlarging the middle class, this affluence contributed to a period of liberal reform as well as social movements on behalf of blacks, women, and the environment.[34] Eventually, though, the economy turned downward. By the middle 1970s, inflation and foreign competition ended the great postwar expansion and, coupled with an energized environmental movement, put the brakes on the big dam era.[35]

It became clear to many critics that the dam industry had developed a powerful life of its own. Too many dams, they argued, resulted from political horse trading rather than from genuine need. Although on Capital Hill dams had to be justified with cost-benefit analyses, western politicians had learned that the Bureau of Reclamation and Army Corps of Engineers could easily be persuaded that costs and benefits need not necessarily be perfectly balanced, as bargaining with eastern and southern members of Congress helped guarantee dams for the West. After World War II, congressional authorization of new dams for the West became part of the annual bargain, as Bernard DeVoto once observed, in which party lines meant nothing. While Republicans in Congress during the Truman administration fought to roll back elements of the New Deal, few Republicans from the West questioned the need for more dams. Neither, of course, did Democrats. By the 1970s, however, big dams and multimillion-dollar water projects lost their broad base of support.

A rising chorus of dam critics aided the process. In the early 1970s, a Ralph Nader study group issued a sharp critique of the Bureau of Reclamation, pointing especially to its hidden subsidies, cost overruns, and insensitivity toward American Indian water needs and the environment.[36] The best known of these critics was Marc Reisner, formerly a writer with the Natural Resources Defense Council, whose book *Cadillac Desert* became a kind of cult classic with opponents of big dams.[37] This fast-paced, hard-nosed exposé read like an early-twentieth-century muckraking tract. Quick to denounce most of the West's dams as the result of grasping politicians and greedy boosters, the book laid out a history of dams in the West as an unbroken tale of corruption, bureaucratic rivalry, and power-hungry politicians. Reisner's biting prose took aim at such men as Floyd Dominy, the cigar-chomping commissioner of the Bureau of Reclamation in the 1960s, whose determination to dam the West's rivers was unparalleled. Dominy wanted a dam wherever he could get one, to squeeze every drop of water for irrigation and power.[38] Reisner's scorn for Dominy and other eager dam-builders also arose from his conviction—grounded in the intellectual assumptions of Walter Prescott Webb, DeVoto, and Wallace Stegner—that dams could not fundamentally change the West's environment: "Desert, semidesert, call it what you will, the point is that despite heroic efforts and many billions of dollars, all we have managed to do in the arid

West is turn a Missouri-sized section green—and that conversion has been wrought mainly with nonrenewable groundwater."[39]

Reisner's book reflected changing political and economic currents in the 1980s as federal budget deficits soared and the environmental movement matured. *Cadillac Desert,* which was also the basis of a four-part PBS television series in 1997, was not so much a scholarly history as it was an impassioned argument against the economic and especially environmental follies of the big dam era. Filled with indignation, scorn, and not a small amount of sarcasm, Reisner looked upon the big dam era as a world gone mad.[40]

The altered political and social climate of the 1980s also inspired the work of Donald Worster, whose *Rivers of Empire* dissected the twentieth-century West's water history with an even more powerful scalpel. Worster drew on the work of German historian Karl Wittwogel and argued that the West's rivers had fallen victim to a powerful, centralized, capitalistic, and technological elite. Dams, along with irrigation canals and other water control structures, had arisen out of the enormous growth and power of the American West born of World War II, when, as Worster argued, the West threw off its colonial bonds and underdeveloped status and attained the rank of empire. As dams arose on western rivers in the postwar years, they marked the emergence of the region as a new center of political power in the United States. Such power manifested itself in total control and dominance over nature as well as through the class-based dominance of growers and business leaders over migrant workers and small farmers, most evident in California's Central Valley.[41] Worster's description of the massive Central Valley Project, anchored by Shasta Dam in the north and Friant Dam to the south near Fresno, along with the Delta-Mendota Canal and Friant-Kern Canal, led him to conclude that "that was how the Bureau of Reclamation defined salvage in Depression America—and how it made more rational an irrational nature."[42] Dams, then, meant the conquest of nature and the expanding hegemony of the federal government, agribusiness, urban power, and capitalism in the modern West. Dams and waterworks were at the heart of what Worster called the hydraulic society, the western empire that came into its own after 1945.

For Worster, the most important development in the twentieth-century West has been a centralization and concentration of power over nature and people.[43] "The chief political lesson of Hoover Dam . . . is that a new concentration of economic, social, and political power is the outcome of the domination of nature." This has had social implications, too: "[W]hen the tourist looks into the flat water backing up behind a dam like Hoover, he is in fact seeing his own life reflected. What has been done to the Colorado has been done to him as well; he too has, in a sense, been conquered and manipulated, made to run here and there, made to serve as an instrument of production."[44]

Worster's analysis can certainly be challenged, but it gains credence by remembering the devastating consequences of big dams on Native American peoples in the West. Garrison and Oahe Dams on the Missouri River in North and South Dakota inundated thousands of acres of reservation land held by the Three Tribes at Fort Berthold and the Sioux at Standing Rock, Cheyenne River, Lower Brule, and Yankton Reservations. Tribal governments challenged the dams, but were admonished to back off so as not to impede economic progress. Although compensated for their losses, the tribes considered the payments wholly inadequate for replacing lands where they had buried their dead, gathered sacred plants, and cultivated gardens on the Missouri's bottomlands. Other Native American communities experienced similar treatment. In the Northwest, Indian fishing sites along the Columbia River, such as the Celilo Falls beneath the reservoir at Dalles Dam, were eliminated.[45]

Worster's denunciation of hegemonic power invites comparisons with other periods in American history such as the early republic, when much political wrangling sprang from different conceptions of how a republican government should work. The literature on republicanism analyzes public fears of corruption and tyranny in a young nation still testing the strength of self-government. Thus, in 1832 Andrew Jackson destroyed the Second Bank of the United States, regarding it as a monster, a force of corruption and tyranny, an institution whose excessive power directly threatened survival of the young republic. What Jackson said about the bank is not unlike what Donald Worster has said about the Bureau of Reclamation and other power elites in the modern American West.[46] Yet if the bureau and its allies are monsters—akin to the Second Bank of the United States—why haven't Westerners revolted against them?

The answer, in part, is that many Westerners see little reason to criticize an agency that has greatly benefited their lives. Many of the big dams provide substantial security from the terrible floods that had devastated the Imperial Valley in California in 1905 and portions of the Lower Missouri River Basin during World War II. (Hoover Dam and the main stem dams on the Missouri from Montana through the Dakotas have ended that danger.) Hydroelectric power helped attract industry to the Pacific Northwest, southern California, Arizona, Texas, and the Mountain West in the middle of the century when many states began to emerge from their colonial status and develop more mature and stable regional economies.[47]

Another obvious benefit has been the dams' giant reservoirs, which have added substantially to the recreation and tourist sectors in many regions. Lake Sakakawea in north-central North Dakota generates $60 million annually for the state from boaters, fishing enthusiasts, and other recreationists. Lake Powell in southern Utah has long been a major tourist attraction in the Southwest and contributes a whopping $500 million annually to the surrounding region.[48] Artificial lakes across the West have added significantly to the recreation and tourist industries in many states across the region. The added income from boating, fishing, and associated activities has been a welcome stabilizing development in a region accustomed to an economic cycle of boom and bust.

Despite these benefits, few could deny that critics of dams had become a powerful force by the 1970s and 1980s and that the great dam-building era had been all but halted. In part, the demise of the big dam era is attributable to its stunning successes: Virtually all of the good dam sites were gone, and nearly all western rivers had been dammed. Remaining dam sites were often much more marginal in terms of their cost efficiency. Furthermore, rising federal budget deficits in the 1970s and early 1980s curtailed or eliminated numerous proposed projects. Deficits hamstrung new dams because federal dollars had become scarcer and because they fueled an alarmingly high inflation rate in the 1970s that culminated during President Jimmy Carter's administration. Numerous water projects authorized one or two decades earlier (and still being built) had suddenly become frightfully more expensive. President Carter tried to eliminate numerous dams in the West by drawing up a "hit list" of expensive projects he regarded as "pork barrel." However, the West's power in the Congress was formidable, and Carter was attacked and forced to back down. If the economics of dams had been dramatically reversed, few in the West were ready to admit it.

But the years ahead proved even more devastating to the dam industry. Since the 1930s, government policy had subsidized farmers to reduce production to elevate farm prices; this too worked against additional dams as members of Congress from other regions increasingly opposed dams for the West that would simply add to food surpluses. Also, costs continued to escalate. After the 1976 collapse of Teton Dam on the Teton River in Idaho, which caused a flood that killed eleven people, 13,000 cattle, and devastated Rexburg, Wilford, and other towns, subsequent federal legislation required improvements in safety, thus adding to construction costs. Because of such improvements, the cost of Red Fleet Dam in Utah soared from $12 million to $80 million.[49] As Marc Reisner has observed, "inflation, diminishing returns, and environmentalism have made new dams almost impossible to build."[50]

Dams and Environmentalism

Reisner's remark offers a reminder that the most persistent and longstanding criticisms of dams have come from conservationists and environmentalists. Since the early twentieth century, many concerned with wildlife, national parks, and wilderness have questioned the need for so many dams and denounced their adverse environmental effects, including the destruction of fish migratory routes, the inundation of national park lands and monuments, and the submersion of white-water rapids.

Controversies surrounding these threats have focused public attention on scenic areas in the American West, contributing to the emergence of the preservationist and environmental movements. The earliest such fight, from 1907 to 1913, centered on the Hetch Hetchy Valley inside Yosemite National Park, which San Francisco wished to turn into a water supply reservoir by damming the Tuolumne River at its mouth. Hetch Hetchy became the center of conflict between advocates of the dam and opponents led by the spirited John Muir, leader of the Sierra Club, and J. Horace McFarland, president of the American Civic Association. In the end the dam was built, but not before Muir and McFarland awakened the nation to the beauties of Hetch Hetchy.[51]

Forty years later, a remarkably similar controversy surrounded the proposed Echo Park Dam in Dinosaur National Monument. This dam, one of several within the Colorado River Storage Project introduced in Congress early in the 1950s, threatened to inundate miles of the deep and curvaceous Lodore and Yampa Canyons and submerge dramatically scenic Echo Park at the confluence of the Yampa and Green Rivers. The Echo Park Dam controversy pitted the Upper Colorado Basin states and the Bureau of Reclamation against the Sierra Club, National Parks Association, and the Wilderness Society.[52]

The fight to save Echo Park proved especially challenging, because few Americans knew about this remote national monument in an isolated part of the West. Only a few hundred tourists had ever seen Echo Park or the vast canyons of Dinosaur National Monument. Stirred by the threat of the dam, in 1953–1954 the Sierra Club arranged river trips down the Green and Yampa Rivers that enabled hundreds of club members to experience the thrilling rapids and priceless scenery in Echo Park, Whirlpool Canyon, Island Park, and Split Mountain Canyon. In 1956, conservationists defeated the dam in Congress. The outcome was a major triumph for national park and wilderness activists, although it also proved to be merely the first of many controversies over dams on the Colorado River.[53]

Glen Canyon Dam, authorized by the Colorado River Storage Project Act of 1956, caused great anguish within the wilderness movement. Although not a part of the national park system, the tremendous beauties of this stretch of the Colorado River became clear to wilderness activists in the late 1950s, after Congress authorized the dam in the Colorado River Storage Project. In 1955, photographer Charles Eggert wrote to David Brower that "Glen Canyon reservoir is going to flood some of the most outstanding God-created places on Earth—Hidden Passage, Music Temple, Twilight Amphitheater, Guardian Pool to name only a very few. Glen—from the San Juan down has the most extraordinary scenery I have ever seen."[54] Eight years later, in 1964, Richard Bradley wrote that "Glen Canyon is gone, and with it some of the most remarkable canyon country on the continent. It died its needless death just as people were beginning to learn of its wonders."[55] Sierra Club river trips through Glen Canyon gave birth to what Michael Cohen calls perhaps "the greatest of the Exhibit Format books [and] certainly . . . the saddest." *The Place No One Knew,* with photographs by Eliot Porter and a gamut of wilderness writers and poets, became a sign of the vigilance of the rapidly growing environmental movement's position toward the Colorado River. To conservation activists in the 1960s, Glen Canyon Dam itself became a powerful symbol of environmental ruin.[56]

Lake Powell also threatened Rainbow Bridge, a small but spectacular arch inside a national monument in southern Utah. Conservationists fought mightily to secure a protective dam from the Bureau of Reclamation to safeguard the bridge from the rising waters of Lake Powell, but their effort failed.[57] While the bridge itself was not threatened by the water, the Sierra Club and its friends were dismayed that the reservoir crossed into the national monument boundary, and this added to their bitterness at the imminent inundation of the wondrous inner gorge of Glen Canyon beneath Lake Powell.

Closing of the gates of Glen Canyon Dam in 1963 proved a critical point in the long history of the big dam era. Glen Canyon Dam was one of the last of the great giants in the West, a descendant of Hoover Dam downriver. While the size of Glen Canyon Dam and the scope of the construction project recalled the mighty dams of the 1930s and 1940s, it also signaled the end of this era. At the same time, laments over the loss of Glen Canyon represented not only the vigilance of conservation and wilderness devotees who had defended Echo Park a decade earlier, but they also revealed the blossoming of a broader-based environmental movement. Dam-builders no longer faced national park and wilderness devotees alone, but a larger, more potent movement of environmentalists.

Those concerns emerged in the climatic battle for the Colorado River between 1964 and 1967, when the Bureau of Reclamation proposed two dams at Marble and Bridge Canyons inside the Grand Canyon, threatening national park and monument boundaries once again. The struggle over the Grand Canyon dams occupied the Sierra Club, National Parks Association, and other organizations for several years in the mid-1960s. David Brower of the Sierra Club galvanized the environmental movement and the public to defend the Grand Canyon with full page advertisements in the *New York Times* and *Washington Post*. The most famous of these asked rhetorically, "Should we also flood the Sistine Chapel so tourists can get nearer to the ceiling?"[58] Publication of the ads led to the nullification of the Sierra Club's tax-deductible status by the Internal Revenue Service, a move that generated a massive public outcry and brought countless numbers of Americans into the environmental camp.[59]

By 1966, Interior Secretary Udall decided to scrap Marble Canyon Dam, and the following year he did the same to the Bridge Canyon Dam. The end of the controversy marked a significant victory for environmentalists. In blocking dams from Grand Canyon, they emerged triumphant and armed with a broader definition of wilderness than during the earlier fight over Echo Park. Then, wilderness had been equated with protection of national parks. Now, preservation had come to include free-flowing rivers, an argument that was embodied in the Wild and Scenic Rivers Act that Congress passed that same year.[60]

The Colorado River battles awakened the wilderness and environmental movements to a distinctive region in the West that thereafter became nearly a sacred place on their agenda. Protection of the Colorado could never be perfect, given the presence of Hoover, Davis, Parker, Glen Canyon, and other dams along its course, but the wilderness groups now stoutly resisted any more dams and fought fiercely in the following years to challenge other threats to the surrounding Colorado Plateau.[61] Furthermore, the increasingly powerful environmental movement had helped to provide the tools for challenging dams. In 1970, Congress enacted the National Environmental Policy Act (NEPA), which required environmental impact statements for all federally funded projects. NEPA proved to be a major weapon in combating dams.

In the 1970s, environmentalists used NEPA to challenge the multimillion-dollar Central Utah Project (CUP) and its various dams, canals, dikes, and other structures. Besides questioning the economics of the vast plan, environmentalists focused their attack on how the CUP would eliminate rare trout streams and wildlife habitats the streams supported in the High Uintas Range in northeastern Utah. In 1991, the CUP was reformulated to include provisions requiring wildlife habitat restoration.[62] Another major battle in the 1970s ended with elimination of proposed dams in Hells Canyon along the Oregon-Idaho border. Environmentalists considered this a forceful affirmation of the value Americans now placed on free-flowing rivers.[63]

All of these contentious battles laid a sharp divide between dam supporters and environmentalists, between those who championed "improved" landscapes and those who preferred ones that are pristine. When it comes to dams, there has not been any middle ground, as the actions of George Hay-duke, one of the protagonists in Edward Abbey's 1975 novel *The Monkey Wrench Gang*, makes clear. His plot to blow up Glen Canyon Dam later inspired members of Earth First! to unfurl a huge black banner that resembled a giant crack on the face of the dam as a sign of their dedication to the same end.

In recent years, several dams in the West have been targeted for removal, including two on the Elwha River, Condit Dam on the White Salmon River in Washington, several dams on Butte Creek in California, and, not surprisingly, Glen Canyon. Removals have been slow in coming, however, nor is it clear whether many will actually occur. Taking down Glen Canyon Dam, some maintain, could create as much environmental havoc as it would undo.[64]

These proposals to decommission dams have sprung from the knowledge of how much damage many dams have caused. Across the West, dams have dramatically altered rivers and riparian habitats. By eliminating seasonal flooding, they have altered sandbars and biotic zones supporting plants and wildlife. Instream flows, once corridors for fish as well as for the movement of wildlife in securing their feeding, shelter, and reproduction, have been drastically changed or blocked. Replacement of seasonally flooding rivers with huge, fluctuating reservoirs has dramatically altered riparian habitats of birds and mammals as well as transformed riverine habitats of fish.[65] Along the Sacramento River, which has been massively altered by dams, levees, and diked-off islands, "the bird populations have declined to a small fraction of their original numbers, and they crowd together on the few remaining wetlands protected in national wildlife refuges. The sky no longer grows dark with geese."[66] Lost Creek Dam on the Rogue River in Oregon has degraded the ecosystem downstream. According to one source, "gravel bars are becoming grown over, and native plants that could withstand floods are being replaced with noxious invaders such as purple loosestrife, jimsonweed, and star thistle. These are nearly useless to wildlife."[67]

The Columbia River, along with its biggest tributary, the Snake, best illustrates the adverse environmental consequences of big dams. Bonneville, John Day, Priest Rapids, Grand Coulee, Ice Harbor, and other large dams have devastated salmon habitat and the fisheries, although other factors including water pollution from mining, overgrazing, and clearcutting also have contributed to the salmons' demise. Yet the dams have been key in reducing their once flourishing numbers, so that tremendous energy and money had to be spent on fish hatcheries, fish ladders, and on transporting fish around the dams in trucks. Saving the salmon also meant aiding juvenile fish, because the turbines in the huge dams killed thousands of smolts trying to swim downstream. There have been no easy answers; one technique, flushing juvenile fish down the spillways to bypass the deadly turbines, also killed the fish by saturating the water with nitrogen gas bubbles.[68] On the whole, dams on the Columbia River intruded into a major ecosystem and vastly complicated survival of fish and economic livelihoods of those who rely on them for subsistence and sport.[69] Dams, as environmentalists like to say, kill rivers.[70]

Here is the crux of much of the debate about dams on western rivers. On one side are a majority of environmentalists who maintain that dams are, for the most part, disasters. Philip Fradkin titled his history of the Colorado *A River No More,* suggesting that the dams are the last chapter of that river's environmental history.[71] This notion suggests that the West's dammed rivers have been made wholly subservient to mankind's use. Resembling the national forests, rivers have been controlled, sliced up into manageable "units," and regulated to produce saleable commodities such as hydropower and profits for barge companies and a host of businesses that depend on them. Rivers have been tamed and placed in the market economy.

Or have they? Donald Worster maintains that nature cannot so easily be destroyed. In his provocative essay, "Hoover Dam: A Study in Domination," Worster contends that human mastery of nature has never been complete. Irrigation of dry lands in the Colorado Basin has produced enormous volumes of saline water that threatens farms in the United States and Mexico. Salinity has become a major economic and political problem, but for Worster it also demonstrates the folly of efforts to control the Colorado—or any other river—forever:

> What an utter foolishness it is to believe that any lasting mastery over nature has really been won. Even at this very moment the Colorado River is busy preparing to saw through the Hoover Dam, laying down its silt behind the wall,

gathering force to remove this new obstruction just as it has removed everything else ever put in its way by the forces of geology. Human domination over nature is an illusion, a passing dream by a naive species.[72]

In his provocative environmental history of the Columbia, *The Organic Machine,* Richard White concurs. Although he does not question the severe effects of the Columbia's dams on salmon or their habitat, neither does he accept the notion that the river has been completely destroyed: "No matter how much we have created many of its spaces and altered its behavior, it is still tied to larger organic cycles beyond our control."[73] Humans have not conquered nature completely; the energy of the river is still present, the salmon still swim upriver and try to return to the ocean, while agencies and interest groups still wrestle over how to manage the river to accommodate conflicting demands. White argues that neither nature nor culture has triumphed, but are intermingled in an evolving relationship.

White maintains that although the dams have enabled humans to control and master the Columbia River, they have clearly not eliminated the constant struggles over just who or what the river should serve. The river is still a point of conflict among commercial fisheries, Indian fishing communities, navigation traffic, and many interests who rely on hydroelectric power. Huge dams mean the river is intensively managed, but this has only solidified divisions between different groups over the Columbia's flow. Barge operators, aluminum companies, cities and rural electrical cooperatives, farmers and ranchers, sporting and commercial fishing interests, and Native Americans continue to struggle over the river, and none are able to transcend their self-interest, which often leads to gridlock. While environmentalists clamor for the reservoirs to be drawn down to help the salmon, draw-downs threaten boat navigation as well as industries and millions of customers who rely on cheap hydropower. Trying to make the river give way in one place threatens someone in another: As White asserts, "the Columbia runs through the heart of the Northwest in ways we have never imagined. It flows along the borders of the numerous divisions in our fractured society. To come to terms with the Columbia we need to come to terms with it as a whole, as an organic machine, not only as a reflection of our own social divisions but as the site in which these divisions play out."[74]

Yet despite the gridlock, White insists that the common view of environmentalists that the river is dead and that efforts should be made to restore the once mighty Columbia to its "original" state is unrealistic, self-serving, and incapable of addressing the challenges that the dams present. For him, there can be no going back to some pristine condition of nature, a time of harmony of humans and the landscape. Humans have always been manipulators of the river and the landscape; the line between nature and humanity has never been solid. Human labor, whether exercised to catch fish or build dams, has always intruded into nature, and nature has always placed limits on what humans can accomplish. "To call for a return to nature is posturing," White contends. "It is a religious ritual in which the recantation of our sins and a pledge to sin no more promises to restore purity. Some people believe sins go away. History does not go away."[75]

White reminds us that dams belong to the West's multifaceted past. Far from being removed from history, as the casual observer might think, they take up an imposing place in the twentieth-century West. The big dams offer reminders of the grand visions and hopes born of the Great Depression, of the glories of the American Century, and of the postmodern era, which considers earlier notions of progress and growth naive and damaging. Today, Westerners with a sense of the past appreciate that many of their dams are surrounded by layers of history. They will understand that the engineers, politicians, boosters, water district managers, and others who promoted the great structures earlier in the twentieth century succeeded in implementing their vision. Dams across the West are powerful reminders of the desire for a tamed nature inasmuch as they have controlled floods, generated power, and established recreational lakes throughout the region. In this sense, the dams represent the classical vision of early- and mid-twentieth-century conservationists: to control nature so as to benefit society.

At the same time, the history of dams is by no means a one-dimensional narrative in which the conquest of nature has been complete or communities throughout the West evenly affected. Instead of simply altering ecosystems, dams have brought into being distinctively new kinds of riverine and riparian environments marked by the introduction of exotic species of plants and major changes in the habitats of fish, birds, and other animals. Some dams, like Glen Canyon, instigated such massive environmental effects as to give rise to a host of scientific studies focusing on rivers below the dams and reservoirs behind them.[76] The environmental changes set into motion by dams have been monumental and often surprising, and they starkly reveal that conquering nature is hardly a simple task.

Just as nature has not been entirely conquered, neither have the human communities surrounding the dams and giant reservoirs always welcomed their presence or looked kindly upon the agencies that operate them. Although countless residents of the West gain economic benefits from dams and reservoirs, others are mindful of the high cost they exacted and of the heavy-handed manner in which the Bureau of Reclamation or Army Corps of Engineers built the dams and manage them today to suit their particular purposes. In the twenty-first century, scholarly histories of dams should graduate to a higher stage of sophistication. The newer works will be multilayered histories, mindful not only of the visions and expectations held by the dam boosters and builders, but also of the ways that dams transformed landscapes and human communities. These histories will acknowledge the undeniable benefits that the dams wrought, but will also indicate the limitations of those who dreamed of and designed them. These histories will chronicle environmental and social changes, especially those that were unanticipated.

Perhaps most important, these histories will reveal that the past is often difficult to separate from the present. To look at a dam is to discover a substantial amount of history and to recognize that the past constantly affects the present. Consider, for example, Garrison Dam and the enormous Lake Sakakawea, constructed along the Missouri River in North Dakota during the late 1940s and early 1950s. In 1948, Montana Sen. James Murray received an eloquent letter from a resident of Butte, who expressed deep sorrow and sympathy for members of the Three Tribes of North Dakota who lost more than 150,000 acres of land behind Garrison Dam. Murray responded by saying he believed the Indians were compensated adequately, and while "all of us feel sympathetic to the Indians and the loss of their tribal lands . . . this is just another example of what is known as progress in these days."[77]

Of course, "progress" can be defined in a variety of ways. For the Three Tribes, Garrison Dam destroyed a self-sufficient economy on the Missouri's bottomlands, and progress was hard to detect for many years following its construction. Now, the Three Tribes at New Town, North Dakota, have gained economic benefits from the reservoir and the Four Bears Casino that sits on the edge of the artificial lake.

Still, many have not forgotten the dark days when "progress" took away their lands. In June 1999, members of the Three Tribes held a reenactment of a historic meeting between their members and representatives of the Army Corps of Engineers at Elbowoods, North Dakota. As the grandsons and granddaughters read the original speeches made on that day in 1946, their words recalled the lives and hopes—as well as the anger—of residents of a small town on the Fort Berthold Reservation more than fifty years ago. Garrison Dam may have brought "progress" to North Dakota in some respects, but the meaning of that progress was defined in a particular way by those who held power and authority in the United States at the time. For all of what Garrison Dam has brought about in North Dakota, it has not erased the memories of those whose way of life was displaced a half century ago. The tribes have not forgotten the past. They have not forgotten their history. Such are the multiple meanings attached to one of many of the big dams, and by extension, many dams throughout the American West.

Notes

For their careful reading and thoughtful comments on earlier drafts of this chapter, I wish to thank Mark Fiege, Donald Jackson, Donald Pisani, Jeffrey Stine, and Richard White.

1. *American Technological Sublime* (Cambridge, Mass.: MIT Press, 1994).

2. Richard White, *The Organic Machine: The Remaking of the Columbia River* (New York: Hill and Wang, 1995), x.

3. Joseph E. Stevens, *Hoover Dam: An American Adventure* (Norman: University of Oklahoma Press, 1988); John E. Thorson, *River of Promise, River of Peril: The Politics of Managing the Missouri River* (Lawrence: University Press of Kansas, 1994); John R. Ferrell, *Big Dam Era: A Legislative and Institutional History of the Pick-Sloan Missouri Basin Program* (Omaha, Nebr.: Missouri River Division, U.S. Army Corps of Engineers, 1993); Keith C. Petersen, *River of Life, Channel of Death: Fish and Dams on the Lower Snake* (Lewiston, Idaho: Confluence Press, 1995); John A. Adams Jr., *Damming the Colorado: The Rise of the Lower Colorado River Authority, 1933–1939* (College Station: Texas A&M University Press, 1990); Paul C. Pitzer, *Grand Coulee: Harnessing a Dream* (Pullman: Washington State University Press, 1994); Russell Martin, *A Story That Stands Like*

a Dam: Glen Canyon and the Struggle for the Soul of the West (New York: Henry Holt and Co., 1989).

4. Quoted in Mark Fiege, *Irrigated Eden: The Making of an Agricultural Landscape in the American West* (Seattle: University of Washington Press, 1999), 180.

5. These quotes are attributed to various sources and appear in Pitzer, *Grand Coulee*, 164–65.

6. Stevens, *Hoover Dam*, 266–67.

7. Ibid., 267.

8. "Place Humanists at the Headgates," in *Reopening the American West*, ed. Hal K. Rothman (Tucson: University of Arizona Press, 1998), 158.

9. Donald C. Jackson, *Building the Ultimate Dam: John S. Eastwood and the Control of Water in the West* (Lawrence: University Press of Kansas, 1995), 34–40.

10. Ibid., 246–47.

11. Along with Steven's book, see also Marc Reisner, *Cadillac Desert: The American West and Its Disappearing Water* (New York: Viking Penguin, 1986), 126–36; Andrew J. Dunar and Dennis McBride, eds., *Building Hoover Dam: An Oral History of the Great Depression* (New York: Twayne Publishers, 1993), xvii.

12. Stevens, *Hoover Dam*, 47.

13. Donald E. Wolf, *Big Dams and Other Dreams: The Six Companies Story* (Norman: University of Oklahoma Press, 1996), 51–57.

14. Pitzer, *Grand Coulee*, 79–80; Richard Lowitt, *The New Deal and the West* (Bloomington: Indiana University Press, 1984), 167.

15. Michael Malone, Richard B. Roeder, and William Lang, *Montana: A History of Two Centuries*, rev. ed. (Seattle: University of Washington Press, 1991), 300; Adams, *Damming the Colorado*, 62.

16. A. E. Rogge, D. Lorne McWatters, Melissa Keane, and Richard Emanuel, *Raising Arizona's Dams: Daily Life, Danger, and Discrimination in the Dam Construction Camps of Central Arizona, 1890s–1940s* (Tucson: University of Arizona Press, 1995).

17. Philip L. Fradkin, *A River No More: The Colorado River and the West* (Tucson: University of Arizona Press, 1984; reprint, Berkeley: University of California Press, 1996), 239; Martin, *Story That Stands Like a Dam*, 205.

18. Pitzer, *Grand Coulee*, 136.

19. Ibid., xiv; White, *Organic Machine*, 55–58.

20. White, *Organic Machine*, 55–56.

21. William Dietrich, *Northwest Passage: The Great Columbia River* (New York: Simon and Schuster, 1995; reprint, Seattle: University of Washington Press, 1996), 297–304; see also Wesley Arden Dick, "When Dams Weren't Damned: The Public Power Crusade and Visions of the Good Life in the Pacific Northwest in the 1930s," *Environmental Review* 13 (Fall/Winter 1989): 113–15.

22. Robert E. Ficken, *Rufus Woods, The Columbia River, and the Building of Modern Washington* (Pullman: Washington State University Press, 1995), 29–32.

23. Ibid., 95–96; see also Ficken's essay, "Grand Coulee Dam, the Columbia River, and

the Generation of Modern Washington," in *Politics in the Postwar American West*, ed. Richard Lowitt, (Norman: University of Oklahoma Press, 1995), 280–82; Pitzer, *Grand Coulee*, 56–57; Lowitt, *New Deal and the West*, 167.

24. "Grand Coulee Dam," in Lowitt, *Politics in the Postwar American West*, 294; White, *Organic Machine*, 72–73; Pitzer, *Grand Coulee*, 361–70.

25. Richard White, *The Roots of Dependency: Subsistence, Environment, and Social Change among the Choctaws, Pawnees, and Navajos* (Lincoln: University of Nebraska, 1983), 252.

26. Pitzer, *Grand Coulee*, 250–51.

27. Ferrell, *Big Dam Era*, 73–86.

28. Ibid., passim; Robert Kelley Schneiders, *Unruly River: Two Centuries of Change along the Missouri* (Lawrence: University Press of Kansas, 1999).

29. Mark W. T. Harvey, *A Symbol of Wilderness: Echo Park and the American Conservation Movement* (Seattle: University of Washington Press, 2000), 90.

30. Petersen, *River of Life, Channel of Death*, 122.

31. Harvey, *A Symbol of Wilderness*, 270.

32. Donald Worster, "An End to Ecstasy," in *The Wealth of Nature: Environmental History and the Ecological Imagination*, (New York: Oxford University Press, 1993), 136.

33. Floyd Dominy, *Lake Powell: Jewel of the Colorado* (Washington, D.C.: U.S. Department of the Interior, 1965), 9.

34. On the latter, see Samuel Hays, *Beauty, Health, and Permanence: Environmental Politics in the United States, 1955–1985* (New York: Cambridge University Press, 1987); see also James T. Patterson, *Grand Expectations: The United States, 1945–1974* (New York: Oxford University Press, 1996), 61–68.

35. Patterson, *Grand Expectations*, 783.

36. Richard L. Berman and W. Kip Viscusi, *Damming the West* (New York: Grossman Publishers, 1973).

37. Reisner, *Cadillac Desert.*

38. Dominy could not have attained the power he had without help from the West. Throughout the region, powerful interests supported the big dams, including the aluminum and aircraft firms in the Pacific Northwest and urban boomers and the agribusiness interests in Los Angeles and in the Central Valley of California. Stewart Udall, telephone interview by author, 10 January 1997; Reisner, *Cadillac Desert*, 222–63; for another sketch of Dominy and his views, see John McPhee, *Encounters with the Archdruid* (New York: Farrar, Straus, Giroux, 1971), 153–245.

39. Reisner, *Cadillac Desert*, 5.

40. For excellent commentary on Reisner's book, see Donald J. Pisani, "Deep and Troubled Waters: A New Field of Western History?" *New Mexico Historical Review* 63 (October 1988): 322–25.

41. Donald R. Worster, *Rivers of Empire: Water, Aridity, and the Growth of the American West* (New York: Pantheon Books, 1985; reprint, New York: Oxford University Press, 1992), 233–56.

42. Ibid., 240–41.

43. See Worster's essays, "The Legacy of John Wesley Powell" and "Water as a Tool of Empire," in *An Unsettled Country: Changing Landscapes of the American West* (Albuquerque: University of New Mexico Press, 1994), 1–53.

44. "Hoover Dam: A Study in Domination," in *Under Western Skies: Nature and History in the American West,* ed. Donald Worster (New York: Oxford University Press, 1992), 73, 71; Worster, "Thinking Like a River," in *The Wealth of Nature,* 123–31.

45. The standard treatment of this topic is Michael L. Lawson, *Dammed Indians: The Pick-Sloan Plan and the Missouri River Sioux, 1944–1980* (Norman: University of Oklahoma Press, 1982); see also Peter Iverson, *"We Are Still Here": American Indians in the Twentieth Century* (Wheeling, Ill.: Harland Davidson, 1998), 131.

46. Harry L. Watson, *Liberty and Power: The Politics of Jacksonian America* (New York: Hill and Wang, 1990), 132–56; Theodore Steinberg, " 'That World's Fair Feeling': Control of Water in 20th Century America," *Technology and Culture* 34 (April 1993): 401–9.

47. Norris Hundley, "The Great American Desert Transformed: Aridity, Exploitation, and Imperialism in the Making of the Modern American West," in *Water and Arid Lands of the Western United States,* ed. Mohamed T. El-Ashry and Diana C. Gibbons (New York: Cambridge University Press, 1988), 51–52.

48. Tom Kenworthy, "A Voice in the Wilderness," *Washington Post Magazine,* 1 June 1997, 15.

49. Worster, *Rivers of Empire,* 308; Reisner, *Cadillac Desert,* 398–425; Steve Hinchman, "Why Utah Wants 'The Bureau' Out," *High Country News,* 15 July 1991; this article appears in a *High Country News* compilation, *Water in the West: A Collection of Reprints* (Paonia, Colo.: High Country News, 1998), IV-b-15.

50. Reisner, "The Fight for Reclamation," in *High Country News, Water in the West,* I-24.

51. Stephen Fox, *John Muir and His Legacy: The American Conservation Movement* (Boston: Little, Brown and Co., 1981), 139–47; Ernest Morrison, *J. Horace McFarland: A Thorn for Beauty* (Harrisburg: Pennsylvania Historical and Museum Commission, 1995), 153–72.

52. Harvey, *A Symbol of Wilderness.*

53. Ibid., 287–93.

54. Charles Eggert to David Brower, 31 January 1956, box 320, Otis Marston Papers, Huntington Library, San Marino, California.

55. "Requiem for a Canyon," *Pacific Discovery* 17 (May/June 1964): 9.

56. Michael Cohen, *The History of the Sierra Club, 1892–1970* (San Francisco: Sierra Club Books, 1988), 318–19; Eliot Porter, *The Place No One Knew: Glen Canyon on the Colorado* (San Francisco: Sierra Club Books, 1963); Martin, *Story That Stands Like a Dam,* 6.

57. Mark W. T. Harvey, "Defending the Park System: The Controversy over Rainbow Bridge," *New Mexico Historical Review* 73 (January 1998): 45–67.

58. For further analysis of the club's ads on the controversy, see Cohen, *History of the Sierra Club*, 352–65.

59. Roderick Nash, *Wilderness and the American Mind*, 3rd ed. (New Haven, Conn.: Yale University Press, 1982), 230–31.

60. Cohen, *History of the Sierra Club*, 352–67, 388.

61. In this regard, the writings of Edward Abbey became popular among preservationists. His books *Desert Solitaire* (New York: Simon and Schuster, 1968) and *The Monkey Wrench Gang* (Philadelphia, Pa.: J. P. Lippincott, 1975) focused on the Colorado Plateau; see also Roderick Nash, "Wilderness Values and the Colorado River," in *New Courses for the Colorado River: Major Issues for the Next Century*, ed. Gary D. Weatherford and F. Lee Brown (Albuquerque: University of New Mexico Press, 1986), 206–9.

62. Analysis of the CUP is found in Daniel McCool, ed., *Waters of Zion: The Politics of Water in Utah* (Salt Lake City: University of Utah Press, 1995).

63. Hays, *Beauty, Health, and Permanence*, 106.

64. Marc Reisner, "Deconstructing the Age of Dams," and "Dam Deconstruction—What's Next?" in *High Country News, Water in the West*, XI-25–XI-32.

65. Dietrich, *Northwest Passage*, 359–60; Schneiders, *Unruly River*, 227–43.

66. Sarah F. Bates, David H. Getches, Lawrence J. MacDonnell, and Charles F. Wilkinson, *Searching out the Headwaters: Change and Rediscovery in Western Water Policy* (Washington, D.C.: Island Press, 1993), 111.

67. Jim Leffmann of the Bureau of Land Management, quoted in Tim Palmer, *America by Rivers* (Washington, D.C.: Island Press, 1996), 264.

68. Petersen, *River of Life, Channel of Death*, 138–41.

69. Pitzer, *Grand Coulee*, 223–30; Dietrich, *Northwest Passage*, 323–53; Lisa Mighetto and Wesley J. Ebel, *Saving the Salmon: A History of the U.S. Army Corps of Engineers' Efforts to Protect Anadromous Fish on the Columbia and Snake Rivers* (Seattle, Wash.: Historical Research Associates, Inc., 1994); Joseph E. Taylor III, *Making Salmon: An Environmental History of the Northwest Fisheries Crisis* (Seattle: University of Washington Press, 1999).

70. Wallace Stegner, "Striking the Rock" in *The American West as Living Space* (Ann Arbor: University of Michigan Press, 1987), 50; Petersen, *River of Life, Channel of Death*, 229–50.

71. See full citation in note 17.

72. "Hoover Dam: A Study in Domination," in Worster, *Under Western Skies*, 77–78.

73. White, *Organic Machine*, 112.

74. Ibid., 113.

75. Ibid., 112.

76. See Steven W. Carothers and Bryan T. Brown, *The Colorado River through the Grand Canyon: Natural History and Human Change* (Tucson: University of Arizona Press, 1991).

77. James Murray to Lee Nye, 19 June 1948, series I, box 239, James Murray Papers, K. Ross Toole Archives, University of Montana, Missoula, Montana.

15

Building Dams and Damning People in the Texas-Mexico Border Region

Mexico's El Cuchillo Dam Project

Raúl M. Sánchez

Overview

The governments of the United States and Mexico ignore, and even encourage, high levels of environmental degradation in the U.S.–Mexico border region for the sake of increased trade, economic development, and industrial growth. U.S. and Mexican governmental authorities also are willing to tolerate many negative impacts on the environment ("environmental harms") that originate on the opposite side of the Rio Grande ("transboundary harms") and the many violations of international environmental law that result from such transboundary harms.[1] For residents on either side of the international border, obtaining domestic legal remedies in the courts of the neighboring country for transboundary harms is only a theoretical possibility. In fact, U.S. and Mexican border residents have virtually no access to immediate or effective legal remedies for the transboundary environmental harms that are common in the region.

A transboundary environmental harm usually begins as a domestic harm in violation of domestic laws on one side of the international border. Upon crossing the international border, the environmental harm often results in violations of international legal norms, especially in the area of international environmental law. The willingness of the governments of the United States and Mexico to overlook many transboundary environmental harms, related violations of international law, and their respective domestic environmental harms reveals their complicity in fouling the environment in the border region, irrespective of the human costs and ecological impacts.

How can residents secure state accountability? Until the signing of the Universal Declaration of Human Rights in 1948, which marked the advent of the modern human-rights era, international law had been solely the province of states. Today, it is universally recognized that all human beings possess certain inalienable rights, which are recognized as "human rights" by international law. Under international human-rights law, individuals whose human rights have been violated may seek redress in various international fora in their own voice without having to rely on the intercession of their own government; although a minority of states, like the Peoples' Republic of China, consider violations a state matter and not actionable in any international forum. A human-rights approach would label environmental harms that interfere with the human enjoyment of a healthy environment and/or any other human right, such as civil and political liberties, as an actionable and sanctionable human-rights violation.

Human-rights violations most frequently are perpetrated by agents of the victim's home government; therefore, it would be unrealistic to expect the violator to champion the victim's cause in international legal fora. With respect to environmental concerns in the U.S.–Mexico border region, it is similarly unrealistic to expect the governments to champion the causes of their citizens, including transboundary environmental harms, because the two neighboring governments are coconspirators in contaminating the environment. A legal approach to environmental harms, particularly transboundary environmental harms, based on actionable human-rights claims presented to accessible legal tribunals could help impede such governmental complicity.

An extensive literature supports the view that a right to a healthy environment is a fundamental human right, whether as an independent right or as a necessary corollary right for the enjoyment of other human rights.[2] Human-rights doctrines surrounding such a right have strong roots that date from the U.N. Conference on the Human Environment convened in 1972.[3] Many countries and U.S. states already have proclaimed environmental protection guarantees in their constitutions.[4]

Governments bear full responsibility for protecting human rights. If governments permit and even promote activities that negatively affect the environment and human life, such actions should be labeled as human-rights violations and the responsible governments called to account as human-rights violators. Even so, appropriate judicial fora do not currently exist to entertain relevant complaints.

The mandates and procedures of existing human-rights fora could be interpreted liberally or modified to permit environmental victims to bring legal actions against offending governments. For example, in the Americas, the Inter-American Commission for Human Rights and the Inter-American Court for Human Rights arguably have the authority to begin hearing such actions. New independent legal fora also could be developed, such as an international environmental law court or, in the North American context, a North American court of environmental justice. In the border region, one might imagine the creation of a U.S.–Mexico border region environmental law tribunal.

The purpose of this chapter is to present a case study of Mexico's El Cuchillo Dam Project as an example of why national governments, domestic law, and international environmental law cannot be expected to remedy environmental harms in the U.S.–Mexico border region. The case study also serves as a basis for advocacy for the recognition of an enforceable human right to a healthy environment. The El Cuchillo Dam Project, located primarily in the Mexican state of Nuevo León, has caused severe social and environmental harms in the Mexican state of Tamaulipas and the U.S. state of Texas. A human right to a healthy environment, enforceable through effective, independent legal fora, could help provide remedies for such harms. Rational binational efforts for managing transborder resources according to principles of sustainable development should be accompanied by regional and international efforts to develop legal concepts, instruments, and independent judicial fora that can enable residents in the U.S.–Mexico border region to vigorously protect their health, environment, and future.

The El Cuchillo Dam Project

In water volume, the San Juan River is the second largest tributary of the Rio Grande on the Mexican side of the international border and usually represents approximately one-fourth of the total flow of water into the Rio Grande from all Mexican tributaries.[5] The San Juan River springs to life in the mountains to the southeast of the city of Monterrey, the capital of the Mexican state of Nuevo León. The river, which is wholly within the territory of Mexico, flows principally from south to north out of Nuevo León and crosses into the Mexican state of Tamaulipas. Tamaulipas borders the U.S. state of Texas across the Rio Grande, starting a few miles upstream of the twin cities of Laredo, Texas, and Nuevo Laredo, Tamaulipas, down to the Gulf of Mexico.

Until the Marte R. Gómez Dam began operating in 1943, approximately five miles from the U.S.–Mexico border and forty miles west of Reynosa, the San Juan River had flowed unimpeded into the Rio Grande. Sugar Lake, the reservoir behind the Gómez Dam, was created to provide a relatively clean source of drinking water for the inhabitants of Reynosa and irrigation water for the Twenty-Sixth Irrigation District, which surrounds Sugar Lake and includes farms north to the banks of the Rio Grande. The entire region surrounding the San Juan River varies from arid to semiarid conditions.

The 1944 treaty, negotiated long before the current population and water-use boom in the border region, labels the waters of the San Juan River as "Mexican."[6] Accordingly, the government of Mexico may consume all of the water of the San Juan River without violating Mexico's treaty obligations with the United States. This situation ignores the ecological and social damage such consumption could cause elsewhere in the Rio Grande Basin. Such action is sanctioned by the 1944 treaty, but is contrary to the modern trend in international river law toward a basinwide approach to international watercourse issues.[7]

Today, coriparian states, like the United States and Mexico, which share the Rio Grande Basin, are expected to respect the following three fundamental principles of general international law on shared water resources development, conservation, and use. Each state possesses: (1) a duty not to cause substantial injury; (2) a right to an equitable and reasonable share in the use of the waters of the international watercourse or basin; and (3) a duty to inform, consult, and engage in good-faith negotiations.[8] These three principles were codified, in part, in the new Convention on the Law of the Non-Navigational Uses of International Watercourses, adopted by the U.N. General Assembly and opened for signatures in 1997. Mexico and the United States voted for the treaty, but have not yet signed the agreement.

The El Cuchillo Dam Project consists of the El Cuchillo Dam, a forty-three-meter-high dam on the San Juan River and related infrastructure, including pumping stations, more than sixty miles of enclosed aqueduct, sewage collection and treatment facilities, and water metering installations.[9] The El Cuchillo Dam, located approximately forty-eight miles south of the Rio Grande near the town of China, Nuevo León, possesses a maximum storage capacity of 784 million cubic meters. Five pumping stations convey the water from the El Cuchillo Dam to Monterrey, a distance of approximately sixty-three miles. The project also includes additional infrastructure that was built to facilitate water delivery and metering to the end users. The project's stated purpose is to increase water availability for residential and industrial purposes in the Monterrey area.

Additional major elements of the El Cuchillo Dam Project include a sewage and wastewater collection system in an area northeast of Monterrey. Effluent is to be treated in three new treatment plants and pumped into the Pesquería River, which originates in and flows out of Nuevo León. In Tamaulipas, the Pesquería River flows into the San Juan River approximately twenty miles upstream of Sugar Lake. The project's design appears to have anticipated substantially reduced water levels in Sugar Lake, because it also provided for the building of a new canal for Reynosa to pump the city's drinking water directly from the Rio Grande. Apparently, project planners expected from the beginning that the inhabitants of Reynosa no longer would drink the relatively clean water from Sugar Lake, which previously had been pumped to the city.

The El Cuchillo Dam Project was completed at breakneck speed during the six-year administration of President Carlos Salinas de Gortari, who assumed office in January 1988. During Salinas's presidency, the project was designed, an agreement was signed between the federal government and the state government of Nuevo León for the construction of the project, environmental impact studies were written, financing was requested and received from the Inter-American Development Bank (IDB), and construction was completed.[10] (In the United States, a comparable project would have been in litigation for at least ten years just to define the content of the environmental impact statement.)

Construction of the El Cuchillo Dam Project cost approximately $650 million (U.S. dollars), but exact figures may never be revealed to the public.

The IDB initially had intended to finance the Monterrey sewage treatment plants; however, sources at the IDB indicate that the Japanese Overseas Economic Cooperation Fund provided last-minute financing. The IDB contribution consisted of approximately $325 million (U.S. dollars) in loans to agencies of the government of Mexico.

In October 1994, key portions of the project remained unfinished, including the sewage treatment plants that were to send treated effluent down the Pesquería and San Juan Rivers to Sugar Lake; nevertheless, with considerable pomp and fanfare, President Salinas inaugurated the El Cuchillo Dam. Shortly thereafter, the floodgates were closed, and the reservoir behind the dam began to fill. Simultaneously, a slow-motion disaster began to unfold downstream as the life-sustaining waters of the San Juan River stopped flowing into Tamaulipas.

The building of the project was not expected to cause any problems downstream of the El Cuchillo Dam, according to Mexican government documents. Environmental impact assessments written by Mexican agencies and delivered to the IDB did not envision any significant environmental problems.[11] Possible impacts in Tamaulipas or Texas were not addressed in the assessments, and IDB experts did not raise any red flags.

In 1990, Sócrates Rizzo and Américo Villareal Guerra, the governors of Nuevo León and Tamaulipas, respectively, signed an agreement, witnessed by several federal representatives including now-President Ernesto Zedillo, that purported to guarantee the future effective management of the entire San Juan River Basin (the 1990 San Juan River Basin Agreement).[12] The two governors specifically pledged to maintain the level of Sugar Lake and guarantee a water supply to the more than three thousand farmers of the Twenty-Sixth Irrigation District from the Pesquería River. The 1990 agreement is written in vague, general language; does not establish an effective mechanism for carrying out the agreement; and contains no enforcement language; therefore, the signatories' precise intentions are open to speculation.

The governors of Nuevo León and Tamaulipas could not realistically have expected that the water needs of everyone who had depended on the San Juan River would continue to be met after the El Cuchillo Dam Project was completed. In signing the 1990 San Juan River Basin Agreement, two loyal governors of President Salinas's ruling party, the Institutional Revolutionary Party (Partido Revolucionario Institucionalizado, PRI) probably were seeking to deflect criticism for a project favored by their then-popular president, at least temporarily until the project had become a fait accompli. In 1990, two PRI governors dared not challenge their powerful chief executive and party leader.

President Salinas also was a favorite son of Nuevo León. Born in Mexico City, he claimed the state as his home, and the family hacienda is in Agualeguas, Nuevo León. Salinas owed many favors to the industrial elite of Monterrey.[13] During his six years as president, Salinas had three major infrastructure projects built in Nuevo León: the El Cuchillo Dam Project; the Colombia-Solidarity Bridge over the Rio Grande, which connects Nuevo León directly to the United States; and an airport in Agualeguas suitable for landing the Boeing 747s of the U.S. presidents during negotiations for the North American Free Trade Agreement (NAFTA).[14]

In the fall of 1993, the people and ecosystems downstream of the El Cuchillo Dam, particularly in Tamaulipas, began to suffer the inevitable devastating impacts of the closing of the dam's floodgates. Monterrey began siphoning away water almost as quickly as the new reservoir filled up, and the downstream lands simply dried up. Unless the flow of treated effluent down the Pesquería River matched the volume of water that had flowed down the San Juan River, the game had become zero-sum. The water that had belonged to Tamaulipas now belonged to Nuevo León. Of course, the Nuevo León perspective was that the waters of the San Juan River had always belonged to that state.

The shortfall of water for end users in Tamaulipas immediately became obvious. Reynosa began drawing its drinking water from the highly contaminated Rio Grande, and the water level in Sugar Lake began to drop rapidly as farmers from the Twenty-Sixth Irrigation District continued to water their fields. By the summer of 1996, Sugar Lake was below 20 percent of normal capacity, and the reservoir was becoming increasingly contaminated with untreated effluent because the new sewage treatment plants in the Monterrey area were not fully operational.

A devastating drought that has laid siege to northern Mexico, from 1993 to the present day, has severely aggravated the downstream water shortage. The coincidence of the start-up of the El Cuchillo Dam and the regional drought provided Mexican authorities with a convenient excuse: The severe water shortage in Tamaulipas was caused by nature. Undoubtedly, the drought has made matters much worse; however, but for the El Cuchillo Dam, the water level of Sugar Lake would be higher and not as contaminated.

IDB officials also adhere to the same "natural" explanation about the water shortages in Tamaulipas; however, as late as the spring of 1996, they acknowledged that the new sewage treatment plants were not completely operational and that untreated effluent had been pumped into Sugar Lake via the Pesquería River since 1993.[15] IDB officials also refuse to acknowledge or assume any liability for having funded a large infrastructure project that failed to take into account substantial negative impacts on the environment and large segments of the population. IDB representatives insist that Mexican authorities effectively consulted affected populations, but have not provided any proof of such consultations. Considering that the environmental impact statements provided to the IDB by the Mexican government did not appraise any negative impact outside of Nuevo León, it is highly unlikely that any consultations took place in Tamaulipas.

Consultations of affected populations were not undertaken in Texas. According to the official views of the government of Mexico, and probably of the U.S. government as well, no such consultations were needed because the San Juan River is a Mexican river under the 1944 treaty. Mexico may be entitled to consume the waters of the San Juan River under the 1944 treaty, however, such a right is not accompanied by unlimited discretion to cause transboundary harms.

The cessation of a downstream flow of the San Juan River, except for those contaminated waters that enter from the Pesquería River and the drastic drop in the water level in Sugar Lake, has devastated northern Tamaulipas. Approximately three hundred families who earned a living from fishing in Sugar Lake have lost their livelihood. The farmers of the Twenty-Sixth Irrigation District have virtually no irrigation water and have lost their crops over several growing seasons. Local merchants and owners of rustic motels who earned a living from recreational fishing, especially from U.S. tourists, were ruined financially.[16]

Many Tamaulipan residents who had been dependent on Sugar Lake for their livelihoods became environmental refugees; they sold their instruments of trade, boarded up their homes, and moved in search of a better life.

Some moved to other Mexican cities, while others slipped over the international border to work as undocumented laborers in the United States. Residents who have remained in the Sugar Lake area report that colonies of friends and relatives are living in San Antonio and Houston, Texas.[17] Some individuals may have died trying to walk north, either by drowning in the Rio Grande or expiring in the harsh countryside of South Texas.[18]

Residents in the Sugar Lake area also report confidentially that a substantial number of local residents who can no longer fish or farm have turned to a more lucrative trade: drug trafficking.[19] Area residents are familiar with local shortcuts that lead to the U.S. border and fording points across the Rio Grande. Such knowledge is invaluable to drug dealers who need local contacts to move drug shipments northward to the United States. These reports are consistent with press reports and police announcements of increased drug seizures along the Texas-Tamaulipas border in recent years.[20]

Other residents who remained around Sugar Lake turned to alternative water sources. Some drilled authorized and/or unauthorized wells, which have lowered the water table on both sides of the Rio Grande. The United States and Mexico share numerous underground aquifers, but these countries have not signed a bilateral treaty to regulate groundwater use. Some farmers of the Twenty-Sixth Irrigation District resorted to unauthorized pumping of water directly from the Rio Grande.

In 1995, drought conditions worsened in Mexico. As of mid-summer, newspaper reports estimated that Mexican ranchers had lost 300,000 cattle and farmers had lost one million acres of crops. Mexico's share of water in Falcon and Amistad Reservoirs, the two binational reservoirs on the Rio Grande, had dropped to less than 10 percent of the overall allotment. The remaining U.S. allotment of water was below 50 percent. In October 1995, the U.S. government granted Mexico's request for an emergency water loan. The United States agreed to lend Mexico up to 81,000 acre-feet of U.S. water stored in the binational reservoirs for drinking and home use only.[21] The loan expired eighteen months later, but Mexico never drew on it, possibly to avoid political criticism.[22]

By April 1996, severe drought conditions also had spread to Texas. The U.S. allotment of Rio Grande water in the binational reservoirs was below 20 percent and Mexico's was approximately 5 percent. Unauthorized pumping, known as "diversions," on the part of Mexican farmers, particularly in the Twenty-Sixth Irrigation District, became more widespread, and patience and tolerance on the part of U.S. water users evaporated. Growers and ranchers in South Texas became extremely vocal about what they considered to be the theft of U.S. water—their water—by Mexican farmers. The Texans complained to their federal and state government representatives and directly to the U.S. section of the binational International Boundary and Water Commission (IBWC; the Mexican section is known as the Comisión Internacional de Límites y Aguas, CILA), which is responsible for addressing binational water and boundary issues between the United States and Mexico. The IBWC was powerless to take police action, but began to log carefully all diversions that it detected on the Mexican side of the Rio Grande. Texas state government officials sought to intercede and complained directly to the Mexican federal government and to the CILA.

Through the CILA, the Mexican government responded that it would honor all treaty obligations concerning the Rio Grande and would take steps to control diversions. Mexican diversions continued, but were less frequent. Mexican crop yields were devastated. Anger in Tamaulipas against Nuevo León, the PRI, and the federal government quickly mounted.

Historically, diversions from either bank of the Rio Grande have not presented a serious binational problem. The IBWC and the CILA usually agree to subtract diverted amounts from the relevant national water allocation in the Falcon or Amistad Reservoirs. However, by early 1996, Mexico's water allocation in the binational reservoirs had dropped to 20 percent and the drought showed no signs of abating. By August 1996, Mexico's allocation was below 10 percent. All Mexican diversions documented by the IBWC were eventually settled through negotiations with the CILA, but in the future an amicable settlement may be impossible if Mexican reserves drop to zero. Mexico may have to seek additional loans of U.S. water for human consumption in its northern border cities, which undoubtedly will anger U.S. water users.

Discontent in northern Tamaulipas had begun to build as soon as the floodgates of the El Cuchillo Dam were closed. The central issue was not quality of life or the cost of water, but survival itself. Area farmers and merchants were having difficulty feeding their families. Unemployment was spreading and the drought was worsening. Protests erupted throughout Tamaulipas in January 1996.

Gov. Manuel Cavazos Lerma of Tamaulipas began a legal action that demanded that a presidential decree of 1952, which gave legal rights to Tamaulipas for the use of the San Juan River, be honored. Gov. Rizzo of Nuevo León insisted that the presidential decree had been abrogated by the 1990 San Juan River Basin Agreement; the latter mentions the 1952 presidential decree, but does not specifically abrogate it. The National Water Commission (Comisión Nacional del Agua, CNA), which operates the nation's water infrastructure, including dams, sided with Tamaulipas and ordered that the floodgates of the El Cuchillo Dam be opened to send water downstream to Tamaulipas.

Following vigorous demonstrations in Nuevo León, a march on the El Cuchillo Dam, and threats of violence against the director of the CNA, the floodgates were closed ten hours after they had been opened when Nuevo León's governor convinced a federal judge to overrule the CNA. Additional legal actions, counter-actions, and a public exchange of threats between the governors of Nuevo León and Tamaulipas followed. In the end, the two governors agreed to amicably resolve the problem; in late January 1996, they reached an agreement that required Nuevo León to periodically transfer certain volumes of water to Tamaulipas.

Gov. Rizzo's local critics in Nuevo León now added the "giveaway" of Nuevo León's water to their allegations of fiscal mismanagement and improprieties, corruption, and nepotism. A movement began to build that called for the governor's resignation. In April 1996, Rizzo resigned. Later, his critics also would allege corruption in the construction of the El Cuchillo Dam.

President Salinas named Benjamín Clariond Reyes Retama, a PRI member, to complete Rizzo's term in office. Clariond Reyes's government failed to meet the target amounts of water promised to Tamaulipas under the previous governor's agreement, and the pressure from Tamaulipas continued. The new Nuevo León governor was compelled to sign yet another agreement with his counterpart in Tamaulipas, which provided even less water for the Twenty-Sixth Irrigation District.

The first member of the conservative National Action Party (Partido de Acción Nacional, PAN) to occupy the statehouse, Fernando Canales Clariond was elected governor of Nuevo León in July 1997. This election reflected the growing power of the PAN in both the state and nation. The party also made inroads in neighboring Tamaulipas, due, in part, to widespread resentment against the PRI, which was viewed as responsible for creating the water problems surrounding the El Cuchillo Dam Project.

In November 1997, Gov. Canales Clariond penned yet a third comprehensive "final" agreement with the governor of Tamaulipas. Each "final" agreement provided for partial indemnity payments to the farmers of the Twenty-Sixth Irrigation District for lost crops and a ration of water to be sent downstream once or twice per year; however, the amounts of water stipulated in the agreements have never been sent downstream by Nuevo León. Apparently, each agreement was merely a hasty response to substantial public pressure and street demonstrations, but did not accurately express the intentions of the governor of Nuevo León.

The agreements between Nuevo León and Tamaulipas have not and will not solve the water problem between the two states. The one thing that is most needed, and that the two governors cannot create, is more water. It is apparent that under current consumption rates, the San Juan River cannot meet water demands on the part of users in Tamaulipas *and* Nuevo León, particularly under drought conditions.

The drought in northern Mexico abated somewhat in 1997, but resumed with a vengeance in 1998. As of 4 July 1998, the U.S. share of water stored in Amistad and Falcon Reservoirs totaled 22 percent of the U.S. capacity and the Mexican share totaled 14 percent of capacity. The capacity of the El Cuchillo Dam was down to 15 percent, and Sugar Lake was down to a mere 6 percent.[23] The flow of the Pesquería River into the San Juan River was virtually nil. In response to the renewed drought, the governor of Nuevo León all but abrogated the latest agreement he had signed, which required Nuevo León to transfer water from the El Cuchillo Dam to Tamaulipas. The governor stated publicly that the needs of Nuevo León must be met first, and that "Nuevo León will not grant even one millimeter of water to Tamaulipas."[24]

The political "fixes" to the Nuevo León–Tamaulipas water problem have neither ended the problem nor fully satisfied anyone, and efforts to obtain legal justice in the matter have not fared any better. Legal complaints were filed before the governor and Congress of Tamaulipas, the State Human Rights Commission of Tamaulipas (Comisión Estatal de Derechos Humanos de Tamaulipas), the National Human Rights Commission (Comisión Nacional de Derechos Humanos), several federal courts, and the Office of the Federal Attorney General for Environmental Protection (Procuraduría Federal de Protección Ambiental) on behalf of the fishing families of Sugar Lake, the farmers of the Twenty-Sixth Irrigation District, several municipalities, and the entire state of Tamaulipas. The bases for the complaints included alleged violations of the 1952 presidential decree that gave the use of the San Juan River to Tamaulipas, violations of the 1990 San Juan River Basin Agreement, constitutional violations regarding permissible water uses, and destruction of property. Most actions were dismissed on procedural grounds or on findings of no harm. A few remain stalled in the fora in which they were filed with no immediate prospects of a resolution. A dismissal on procedural grounds is convenient for any court that may not want to reach the substance of the controversy, especially in cases such as this one, which are wrought with political controversy. The disingenuousness of a finding of no harm is obvious in the dry, desolate fields of Tamaulipas.

Additional widespread, negative economic and environmental impacts of the El Cuchillo Dam Project can be gleaned from press reports, field inspections, and field interviews. The precise extent of such impacts must remain largely anecdotal and speculative, because neither the Mexican government nor the U.S. government acknowledges that the El Cuchillo Dam Project has caused such impacts. Therefore, badly needed environmental assessments have not been undertaken. Mexican authorities insist that the ongoing drought, which has plagued northern Mexico for several years, is solely to blame. U.S. officials are, or pretend to be, ignorant of the consequences north of the Rio Grande. Admittedly, harm in the United States has not yet been as severe as in Tamaulipas, but it has been significant and will likely grow in the future.

Certain direct environmental impacts may be occurring only south of the Rio Grande; however, even in such cases, the indirect stressing of neighboring ecosystems across the international border also may be inferred.[25] The creation of the El Cuchillo Reservoir, the diminution of Sugar Lake, the contamination of the Pesquería River and Sugar Lake, and the disappearance of a forty-mile stretch of the San Juan River probably affected the habitats of many plant and animal species in the United States and Mexico. Erosion of dry river and lake beds and previously irrigated crop lands has increased. The ranges of some animals are found only in the area affected by the El Cuchillo Dam Project. Migratory fowl are affected, notwithstanding international treaties that are supposed to protect them. Groundwater levels in the United States and Mexico have dropped. The inhabitants of Reynosa, Tamaulipas, now must drink water from the Rio Grande, which is highly contaminated with raw sewage, heavy metals, and other industrial wastes. Disease vectors, which affect residents of both countries, have been enhanced. For example, increasing incidences of gastrointestinal diseases, cholera, hepatitis, and dengue fever have been reported in northern Tamaulipas and South Texas.[26] Other social impacts in rural areas include increased incidence of crime, alcoholism, and family distress.[27]

In November 1997, in addition to the new Nuevo León–Tamaulipas agreement, the CNA and the governor of Tamaulipas announced the construction of a new dam, the Las Blancas Dam, on the Alamo River, another Mexican tributary of the Rio Grande. The new dam, a response to the political pressure generated by the farmers of the Twenty-Sixth Irrigation District and other water users in Tamaulipas, is to be built in the vicinity of Ciudad Mier, Tamaulipas, which is located only a few miles from the U.S.–Mexico border. The Las Blancas Dam and a new aqueduct will convey water to Sugar Lake for the benefit of the Twenty-Sixth Irrigation District; however, once again, local populations and elected officials have not been consulted, and it remains to be seen whether appropriate environmental impact studies were, or will be, conducted.[28] The problems of the El Cuchillo Dam Project may be repeated very soon, but this time even closer to the U.S.–Mexico border.[29]

International Environmental Law

The field of international environmental law, which is only decades old, continues to grow and develop as international understanding of the precarious nature of the global environment matures.[30] As the global impact of what used to be understood as "local" environmental harms become more obvious, the international impetus to address environmental concerns from a global perspective becomes greater. Through global environmental treaties and state practice, nations are slowly relinquishing small slices of state sovereignty, ostensibly for the sake of the greater global good. Whether, in the end, the current effort will be sufficient to save the environment for future, or even present generations, is an open question.

Traditional norms of international law offer no meaningful protection or remedy to the victims of the El Cuchillo Dam Project, especially those within Mexico's borders. Generally recognized norms address responsibility for environmental harm only in cases of transboundary pollution. In such a context, an international duty arises from the principle that a state should not use its territory in a manner that could harm other states. This principle, commonly called the principle of "good neighborliness," is supported by the decisions of international tribunals, government declarations, and the work of international legal scholars.[31]

Individuals harmed by environmental mismanagement, such as those affected by the El Cuchillo Dam Project, cannot rely on the principle of good neighborliness to remedy their injuries for several reasons.[32] As a threshold matter, such individuals would encounter great difficulty in establishing that the Mexican state is directly or indirectly responsible for the alleged harms. Although the government of Mexico is unquestionably responsible for building the El Cuchillo Dam Project, the responsibility for the sequence of consequences that followed is legally unclear. Furthermore, Mexico undoubtedly will continue to plead an act of nature, the ongoing drought, as an intervening cause for many of the environmental harms caused by the El Cuchillo Dam Project.

A list of the elements to be proved to establish Mexico's responsibility for a transboundary harm in the United States would include the following: successful attribution of the offending act or omission to the government of Mexico, a showing that the government of Mexico breached an international duty, a demonstration that a casual relationship existed between Mexico's conduct and the injury claimed, and proof of material damage. Even if complainants could establish the foregoing elements, no effective legal fora or mechanism exists to adjudicate or settle any claim, and access to affordable legal counsel is extremely limited. In addition, international legal tribunals, as currently constituted, are severely limited in their ability to advance the development of an international liability regime.[33] Furthermore, only states have standing to appear as parties to proceedings before the International Court of Justice; therefore, individuals and/or associations of individuals would have to convince the U.S. government to espouse their claims. States are usually reluctant to pursue such a claim, in part for fear of relinquishing any portion of their sovereignty to binding third-party arbitration and of inviting complaints about their own environmentally degrading conduct. In the case of the El Cuchillo Dam Project and related transboundary environmental harms that have affected U.S. territory, the U.S. government has not manifested any detectable concern about such impacts and is not likely interested in upsetting bilateral relations by filing a complaint against Mexico before the International Court of Justice. In addition, the precise source and content of relevant international law to be applied by international tribunals remains unclear.[34]

Human Rights

Clarity may be brought through the application of human-rights concepts and doctrines. Until the signing of the Universal Declaration of Human Rights in 1948 and subsequent developments in the human-rights arena, the field of international law stood exclusively as the province of nation-states. Today, the subfield of human rights stands as an important exception in the larger field. With respect to human rights, the individual is now a subject as well as an object of international law. As such, the individual may seek to assert and vindicate directly individual universal rights in several international legal fora and, where possible, in domestic tribunals.

The Stockholm Declaration of the U.N. Conference on the Human Environment (the Stockholm Declaration), which was signed in 1972, recognized the connection between the exercise of human rights and a need for environmental protection.[35] The declaration states, in part: "Man has the fundamental right to freedom, equality and adequate conditions of life, in an environment of a quality that permits a life of dignity and well-being."[36] This reference to a healthy environment invokes fundamental human rights that are guaranteed by the Universal Declaration of Human Rights and other international human-rights instruments, but does not unequivocally proclaim a right to a healthy environment.[37] The human rights directly threatened by environmental degradation often include rights to life, health, privacy, adequate working conditions, political participation, and information. U.N. documents since the Stockholm Declaration have used such an indirect approach and have avoided an explicit claim of a right to a healthy environment.[38]

A right to a healthy environment may be understood as a "third-generation," or "solidarity," right. The "first generation" is usually understood as referring to civil and political rights. The "second generation" concerns rights to economic and social welfare.[39] The second- and third-generation rights are neither universally nor uniformly accepted as affirmative and binding on the state. For example, the U.S. government views such rights as merely aspirational: "A right to enjoy and use a healthy environment, one that is clean, ecologically balanced and protected, and whose physical, social, and cultural elements are adequate for both individual well-being and dignity and collective development, can be seen as necessarily underlying all other rights."[40] Nevertheless, the lack of an explicit international proclamation of the right to a healthy environment has fueled debates in an extensive human-rights legal literature as to whether and/or to what extent a right to a healthy environment actually exists and how such a right might be enforced and further developed. Some legal scholars insist that such a right does not exist or ought not exist.[41] Others take the opposite position, but debate whether such a right exists as a human right, as a prerequisite to the enjoyment of human rights, or as a component of environmental protection.[42] Even among those scholars who agree that a right to a healthy environment is a human right, debates persist as to the scope of the substantive and procedural aspects of such a right.[43]

It is not the purpose of this chapter to add to these debates in such a limited space. Rather, this study of the El Cuchillo Dam Project is intended to convince the reader that if a right to a healthy environment does not exist, resources should be marshaled and efforts should be directed at recognizing such a right. Alternatively, if such a right already exists, existing legal instruments, mechanisms, and legal fora should be used or new ones developed to enforce this right.

Numerous human rights have been trampled in connection with the El Cuchillo Dam Project's environmental degradation including rights to life, health, substantive and procedural justice, a livelihood, adequate working conditions, political participation, and information. On the Mexican side of the Rio Grande, remedies under international environmental law for such violations are unavailable and domestic remedies have proved inadequate. On the U.S. side of the border, the U.S. government prefers to ignore any problems; therefore, remedies under international environmental law also are unavailable. An enforceable right to a healthy environment demands recognition and protection on the south *and* north banks of the Rio Grande. In the absence of recognition of such a right and effective legal mechanisms through which to protect that right, advocates for victims of environmental harms are left to pursue claims under international law for the violation of other human rights that have been violated by the El Cuchillo Dam Project.

The governments of Mexico and the United States claim extreme concern about environmental degradation on their common border. As evidence, the two governments can point to a number of bilateral environmental agreements and a binational governmental program for cooperative environmental efforts in the U.S.–Mexico border region. In fact, all such agreements are ineffective because their respective scope is limited, enforcement provisions are lacking, and sanctions are few and weak.[44] Moreover, none of these agreements creates individual rights enforceable in a court of law in the United States or Mexico. In addition, the political will for strict and rigorous execution of these environmental agreements is lacking in the United States and Mexico.

For example, the author provided a public comment to the U.S. Environmental Protection Agency (EPA) regarding the Framework Document of the Border XXI Program (the binational environmental cooperation program of the United States and Mexico), which requested that the EPA insert into the document the need to study the ecological impact of the El Cuchillo Dam Project. In response, the EPA modified the text of the framework document to state that "future" environmental impact of "additional" infrastructure, such as dams, in the Texas-Tamaulipas regions should be given "maximum consideration." The meaning of "maximum consideration" is extremely vague. What is not vague is that, in the context of the Border XXI Program, the environmental impacts of the El Cuchillo Dam Project will *not* be studied. Whether the impacts of the Las Blancas Dam will be studied also is unlikely.

The governments of the United States and Mexico would oppose the recognition of a legally enforceable international human right to a healthy environment. The current legal framework and particularly the lack of available remedies with respect to transboundary environmental harms, as demonstrated by the El Cuchillo Dam Project, give maximum discretion to the two governments regarding environmental degradation. The United States and Mexico probably would view additional enforceable individual rights as an impermissible restraint on trade, industrial development policies, and sovereignty. Private industry would lobby as hard against such rights as it has against global warming agreements.

The most direct and immediate avenue to secure an enforceable human right to a healthy environment in the entire territories of the United States and Mexico would be for the governments to sign and ratify the American Convention on Human Rights (the American Convention) and the related Additional Protocol of the American Convention in the Area of Economic, Social, and Cultural Rights (the Protocol of San Salvador). The latter specifically recognizes a right to a healthy environment.[45] In fact, the government of Mexico already signed the American Convention and the Protocol of San Salvador and ratified the former instrument. President Jimmy Carter signed the American Convention on behalf of the United States, but twenty years later, the treaty still awaits approval of the U.S. Senate. The United States is not a signatory of the Protocol of San Salvador.

Mexico and the United States also could permit actionable claims to be brought in domestic courts under these treaties. The Inter-American Commission for Human Right and the Inter-American Court of Human Rights also could entertain complaints under the two agreements. In the end, determination of the content and justiciability of a right to healthy environment would rest with the courts.

The United States and Mexico also could amend existing binational agreements to include a human right to a healthy environment. The La Paz Agreement could be amended to narrowly provide for such a right only in the U.S.–Mexico border region, covering one hundred kilometers on either side of the international border, an area already designated by the agreement as environmentally significant. More ambitiously, the United States, Mexico, and Canada could seek to amend the North American Agreement for Environmental Cooperation.

The United States and Mexico could pursue new instruments and/or new institutions. For example, either country could take the initiative to move the international community, or at least the NAFTA signatories, toward a convention on human rights and the environment. A Draft Declaration of Principles on Human Rights and the Environment currently exists to provide a guide.[46] In the United States, efforts to approve a constitutional amendment for a right to a healthy environment could be advanced on the state and federal level. As a hemispheric free trade agreement moves closer to reality, so too could a provision for a right to a healthy environment. The North American Agreement for Environmental Cooperation could be amended to provide for an environmental court. The international community recently created an International Criminal Court. A similar court for the environment certainly warrants serious consideration.[47]

Conclusions

The wisdom of government policy in building a dam may be subject to cost-benefit analyses, but the value of human rights should not. Human-rights violations associated with the construction of a dam usually are discernable by examining how the dam was built, who benefits and who does not, and who has to pay the price of environmental and other negative impacts. Former President Salinas and his subordinates; the governments of Mexico, Tamaulipas, and Nuevo León; the CNA; the IDB; and others share the blame for the violations of the El Cuchillo Dam Project. The U.S. government and other U.S. governmental subdivisions also share the blame for transboundary environmental harms in the United States as a result of their silence, inaction, and/or complicity.

Such blame may extend even further still. To the extent that Monterrey industry is benefiting from the El Cuchillo Dam Project by enhancing industrial production and exports abroad, so too are First-World consumers benefiting. The El Cuchillo Dam Project is a case study of the costs of economic globalization. In particular, the project shows how the forces of globalization sometimes coincide with traditional pork-barrel and special-interest politics in the distribution of an increasingly precious commodity in the U.S.–Mexico border region: water.

In recent years, World Bank officials and others have commented that wars in the next millennia will be fought over water.[48] It is difficult to imagine, or preferable not to imagine, the United States and Mexico ever going to war again; however, binational water issues undoubtedly will add serious strain to their relationship. The two governments are making some efforts toward binational cooperation in the area of environmental protection in the U.S.–Mexico border region; however, both governments also are willing to play fast and loose with human-rights and environmental protections when such actions suit them and the powerful business interests in both countries—with the complicity of the neighboring government. As a result, residents on both sides of the international border are condemned to live in an increasingly polluted environment.

The United States and Mexico are hemispheric leaders, and the United States is the only remaining global superpower. Both nations could take aggressive roles in the hemisphere and in the world in advancing the cause of sustainable development and human rights. However, border residents cannot wait for their respective governments to do the right thing. Legal protection of an individual human right to a healthy environment is desperately needed *now* in the U.S.–Mexico border region.

Once in a great while, the proverbial line is crossed by either the U.S. or Mexican government with respect to the environmental degradation in the border region, and the neighboring government complains of an actual or potential transboundary environmental harm. Recently, an effort to establish a low-level nuclear waste dump in a small Texas community, only sixteen miles from the U.S.–Mexico border, led the government of Mexico to protest to the United States.[49] Unfortunately, many government actions or inactions, which fall far short of the potential of a nuclear accident, also lead to serious environmental harms in the border region; however, in these cases neither the U.S. nor the Mexican government complains. The border residents in the United States and Mexico cannot afford to wait for their respective governments to defend their interests only when a threatened or existent harm is as serious as a potential nuclear accident.

The environmental harms caused by the El Cuchillo Dam Project in Mexico and the United States are not immediately lethal, like the accident of the Chernobyl nuclear power plant in the former Soviet Union; however, the damage caused by dams can be quite extensive.[50] Governments that build dams and the multilateral lending institutions that finance them often are blind to the damage such structures cause and deaf to the complaints from the affected populations. By necessity, a vigorous, nongovernmental, international anti-dam movement has grown around the world, largely organized one dam at a time by groups of affected populations.[51] Such pressure has led to the establishment of internal review mechanisms within some of the multilateral lending institutions, the cancellation of a few dam projects, and the creation of an independent World Commission on Dams, which will undertake a comprehensive two-year review of large dam projects.[52] At the very least, the governments of the United States and Mexico should undertake a similar review of dams in the U.S.–Mexico border region, especially before another dam is built. In addition to the Las Blancas Dam, dams also have been proposed on the Rio Grande near Laredo/Nuevo Laredo and Brownsville/Matamoros.

The people of the U.S.–Mexico border region already are acutely aware that the actions and inactions of their respective governments speak for themselves. Ultimately, the entire population of each country should strive to ensure that their respective government represents their fundamental interests without also condemning any human beings, at home or abroad, to an environmental hell. Demanding and struggling for an enforceable human right to a healthy environment is only the beginning.

Notes

The author gratefully acknowledges receipt of support and/or editorial comments from Professor Barbara Bader Aldave (formerly dean of St. Mary's University School of Law), Emily Hartigan, John Teeter, Reynaldo Valencia, and Monica Schurtman. The author also acknowledges the singular bravery of Professor Char Miller in inviting a lawyer to a conference of historians and social scientists.

1. The term "environmental harm" is used throughout this paper to mean detriment, or "harm," arising from negative impacts on the environment.

2. Dinah Shelton, "Human Rights, Environmental Rights, and the Right to Environment," *Stanford Journal of International Law* 28, no. 1 (Fall 1991): 103–38.

3. Neil A. F. Popovic, "Pursuing Environmental Justice with International Human Rights and State Constitutions," *Stanford Environmental Law Journal* 15 (June 1996): 348.

4. Ibid.; see Megan Brynhildsen, "Human Rights and the Environment," *Colorado Journal of International Environmental Law and Policy 1996 Yearbook* 97 (1997); see also Kevin C. Papp, "Environmental Constitutional Protection, Human Rights and the Eighth

Draft of the Temporary Constitution for the Palestinian National Authority in the Transitional Period," *Transnational Law and Contemporary Problems* 7, no. 2 (Fall 1997): 529–76.

5. Portions of this section appeared in the author's article "Mexico's El Cuchillo Dam Project: A Case-Study of Nonsustainable Development and Transboundary Environmental Harms," *University of Miami Inter-American Law Review* 28 (Winter 1996/1997): 425–36. See also the author's article "To the World Commission on Dams: Don't Forget the Law, and Don't Forget Human Rights—Lessons from the U.S.–Mexico Border," *University of Miami Inter-American Law Review* 30 (Winter/Spring 1999): 629–657. The author also acknowledges the contributions of the Center for Border Studies and the Promotion of Human Rights (Centro de Estudios Fronterizos y de Promoción de los Derechos Humanos, A. C., CEFPRODHAC) in Reynosa, Tamaulipas, and its president, Arturo Solis, in helping to piece together the story of the El Cuchillo Dam project and fighting tirelessly for the human rights of the people of Tamaulipas. See CEFPRODHAC, "La construcción de la Presa 'El Cuchillo' y el impacto en el medio ambiente en Tamaulipas," October 1996; see also Ismael Aguilar Barajas, "Inter-Regional Transfer of Water in Northeastern Mexico—The Dispute over El Cuchillo Dam: An Overview of Its Economic, Engineering, Political and Legal Aspects," Departamento de Economía, Instituto Tecnológico y de Estudios Superiores de Monterrey (1998; unpaginated, English-language translation of original, unpublished Spanish-language manuscript, on file with author). For a general discussion of U.S.–Mexico water issues, see Carlos A. Rincón Valdes, "Disponibilidad de Aguas Superficiales y Su Demanda Futuro a lo Largo de la Franja Fronteriza Desde Ciudad Juárez, Chihuahua, Hasta Matamoros, Tamaulipas," in *The U.S.–Mexico Border Region: Anticipating Resource Needs and Issues to the Year 2000*, ed. César Sepúlveda and Albert E. Utton, (El Paso: Texas Western Press, 1992), 119–25.

6. The Convention of 21 May 1906 apportions waters of the Upper Rio Grande and the Treaty Relating to the Utilization of Waters of the Colorado and Tijuana Rivers and the Rio Grande ("the 1944 treaty") apportions waters of the Lower Rio Grande.

7. The term "watercourse" refers to all related or connected rivers, lakes, other surface waters, and groundwater in an entire river basin.

8. Dante A. Caponera, "The Role of Customary International Water Law," in *Water Resources Policy for Asia: Proceedings of the Regional Symposium on Water Resources Policy in Agro-Socio-Economic Development*, ed. M. Ali, G. Radosevich, and A. Ali Khan (Brookfield, Vt.: A. A. Balkema Publishers, 1985), 365, 380–81.

9. The description that follows was taken from the following environmental impact statements filed with the Inter-American Development Bank by agencies of the Mexican government agencies, as indicated (on file with the author): Servicios de Agua y Drenaje de Monterrey, Comisión de Agua Potable y Drenaje de Monterrey, "Manifiesto de Impacto Ambiental (Modalidad General)—Proyecto de Abasto de Agua del Sistema Regional China-Monterrey," October 1989; Comisión Nacional del Agua, "Proyecto de Abasto de Agua del Distrito Regional China-Monterrey, Nuevo León—Manifestación de Impacto Ambiental Modalidad Intermedia," October 1990.

10. Signed in Monterrey, on 9 October 1989, such agreement did not include Tamaulipas. Barajas, "Water in Northeastern Mexico," section entitled "The Conflict between Tamaulipas and Nuevo León."

11. Ibid.

12. "Acuerdo de Coordinación que Celebran el Ejecutivo Federal Atraves de las Secretarías de Programación y Presupuesto, Contraloría General de la Federación, de Agricultura y Recursos Hidraúlicos, de Desarrollo Urbano y Ecología, la Comisión Nacional del Agua y los Ejecutivos de los Estados Libres y Soberanos de Nuevo León y Tamaulipas, Para la Realización de un Programa de Coordinación Especial Para el Aprovechamiento de la Cuenca del Río San Juan, con el Objeto de Satisfacer Demandas de Agua Para Usos Urbanos e Industriales de la Ciudad de Monterrey y Preservar las de Usos Múltiples del Distrito de Riego no. 026, en el Estado de Tamaulipas," May 1990, on file with author.

13. Antonio Jáquez, "Política, negocios, créditos y favores, en el toma y daca de los Salinas con los magnates de Monterrey," *Proceso* (5 February 1996).

14. Ralph Schusler, " 'Solidarity Bridge': Symbol of a New Era" *Business Mexico* (August 1991); see Andrew Downie, "Loyal Supporters Defend Salinas's Scorned Name," *The Houston Chronicle,* 3 March 1996.

15. Nan Borroughs, interviewed by author, Washington, D.C., spring 1996.

16. See CEFPRODHAC, "El Cuchillo."

17. Anonymous local peasants, interviewed by author, fall 1995.

18. See John Ward Anderson, "Death on the Frontier: U.S. Border Crackdown Pushes Illegal Migrants onto Treacherous Ground," *The Washington Post,* 27 May 1998.

19. Anonymous local peasants, interviewed by author, fall 1995.

20. S. Lynne Walker, "Drug Traffic a Way of Life along Border: Money, Narcotics Flow Freely in Many Towns," *The San Diego Union-Tribune,* 28 February 1998 (documenting the pervasiveness of drug trafficking in border towns); Jodi Bizar, "Mexican Border Mayors Seek Help in Drug Battle," *San Antonio Express-News,* 1 October 1996 (documenting increased drug flows at border).

21. Wayne Slater, "U.S. Agrees to Loan Water to Drought-Stricken Mexico," *The Dallas Morning News,* 5 October 1995.

22. Telephone conversation between a representative of the International Boundary Water Commission (IBWC) and the author, 21 June 1998.

23. See the IBWC website (http://www.ibwc.state.gov/storage.htm).

24. Reynaldo Escalante, "Este año no cederá Nuevo León agua a Tamaulipas," *El Nacional* (13 May 1998).

25. For a comprehensive discussion of the environmental impacts of large dams, see Patrick McCully, *Silenced Rivers: The Ecology and Politics of Large Dams* (London: Zed Books, 1996).

26. Susan Duerksen, "Tuberculosis: When Disease Knows No Boundary," *The San Diego Union-Tribune,* 28 December 1997; U.S. Department of Health and Human Services, "Dengue Fever at the U.S.–Mexico Border, 1995–1996," *Morbidity and Mortality Weekly Report* (4 October 1996).

27. Barajas, "Water in Northwestern Mexico," section entitled "The Conflict between Tamaulipas and Nuevo León."

28. Public statements made by a representative of a local ranchers' association and the mayor of Ciudad Mier in the presence of the author, 27 June 1998.

29. The position of the IBWC is that Alamo River is a Mexican river; therefore, Mexico need not consult with the U.S. government. E-mail exchanges on file with the author.

30. Editors of the Harvard Law Review, *Trends in International Environmental Law* (Washington, D.C.: American Bar Association, Section of International Law and Practice, 1992).

31. Michelle Leighton Swartz, "International Legal Protection for Victims of Environmental Abuse," *Yale Journal of International Law* 18, no. 1 (Winter 1993): 357–58. See also Harvard Law Review, *Trends in International Environmental Law,* 15–46.

32. The discussion below consists substantially of an application of the legal analysis in Swartz, "Protection for Victims," 358–59, to the facts of the El Cuchillo Dam project.

33. Harvard Law Review, *Trends in International Environmental Law,* 21–28.

34. But see also sources in note 8, and accompanying text concerning aspects of international river law that Mexico may have violated with the El Cuchillo Dam project.

35. Adopted by the U.N. Conference on the Environment at Stockholm, 16 June 1972, U.N. Doc. A/CONF.48/14/Rev. 1 at 3 (1973), U.N. Coc. A/CONF.48/14 at 2–65, and Corr. 1 (1972), 11 I. L. M. 1416 (1972).

36. Ibid.

37. Shelton, "Right to Environment," 112.

38. Ibid.

39. Burns H. Weston, "Human Rights," in *Human Rights in the World Community,* 2d ed., ed. Richard Pierre Claude and Burns H. Weston (Philadelphia: University of Pennsylvania Press, 1992), 18–21.

40. Mark Allan Gray, "The International Crime of Ecocide," *California Western International Law Journal* 26, no. 2 (Spring 1996): 223–24; see also Janusz Symonides, "The Human Right to a Clean, Balanced and Protected Environment," *International Journal of Legal Information* 20, no. 1 (Spring 1992): 28–29.

41. For example, see Gunther Handl, "Human Rights and Protection of the Environment: A Mildly 'Revisionist' View," in *Human Rights, Sustainable Development and the Environment* ed. Cançado Trindade (San Jose, Costa Rica: Instituto Inter-Americano de Derechos Humanos, 1992), reprinted in Anthony D'Amato and Kirsten Engel, *International Environmental Law Anthology* (Cincinnati, Ohio: Anderson Publishing Co., 1996), 68–69.

42. Shelton, "Right to Environment," 112–16.

43. Ibid.

44. Joseph F. DiMento and Pamela M. Doughman, "Soft Teeth in the Back of the Mouth: The NAFTA Environmental Side Agreement Implemented," *Georgetown International Environmental Law Review* 10, no. 3 (Spring 1998): 651–752; Christopher N. Bolinger, "Assessing the CEC on it Record to Date," *Law and Policy in International Busi-*

ness 28, no. 4 (22 June 1997): 1107; Public Citizen, *NAFTA's Broken Promises: The Border Betrayed* (Washington, D.C.: Public Citizen, 1996), available via http://www.citizen.-org/pctrade/nafta/reports/enviro96.htm

45. "Everyone shall have the right to live in a healthy environment," Article 11, Additional Protocol to the American Convention on Human Rights in the Area of Economic, Social, and Cultural Rights.

46. See Neil A. F. Popovic, "In Pursuit of Environmental Human Rights: Commentary on the Draft Declaration of Principles on Human Rights and the Environment," *Columbia Human Rights Law Review* 27, no. 3 (Spring 1996): 487–603.

47. For a discussion of several proposals, see Joshua P. Eaton, "The Nigerian Tragedy, Environmental Regulation of Transnational Corporations, and the Human Right to a Healthy Environment," *Boston University International Law Journal* 15 (1997): 305–7.

48. Barbara Crossette, "Severe Water Crisis Ahead for Poorest Nations in Next Two Decades," *The New York Times*, 10 August 1995.

49. Howard La Franchi, "Mexico on Nuclear Dump: Not on Our Border," *Christian Science Monitor*, 18 June 1998.

50. McCully, *Silenced Rivers;* Robert S. Devine, "The Trouble with Dams: Environmental Problems, High Cost of Operation," *Atlantic Monthly*, August 1995, 64.

51. For relevant information, documents, and web links, see the following websites: RiverNet by the European Rivers Network (http://www.rivernet.org); International Rivers Network (http://www.urn.org).

52. For example, Janet Bell, "Protest: Small Is Powerful," *The Guardian (London)* (16 August 1995): T5; see the website of the World Commission on Dams (http://www.dams.org).

The Coming Fight

16

Water and the Western Service Economy

A New Challenge

Hal K. Rothman

A few years ago, after a long delay at Chicago's O'Hare Airport, I arrived around 3:00 A.M. in New Haven to attend a professional meeting at Yale University. Making a breakfast meeting four hours later seemed like a stretch when I fell into bed, but I managed to haul myself down to the hotel dining facilities, only to find a room full of eager historians from around the country who had enjoyed closer to a full night's sleep. Among them was Louis Warren, a native of southern Nevada finishing his graduate work at Yale. We sat at the same table and simultaneously reached for the water jug. "Leave it to the Nevadans to fight over the water," Warren quipped.

The quip was instructive. Nevada is the most arid state in the union and also the most federally owned—I have always suspected a correlation. It is also the most colonized, the most dominated from outside the state. This crystallizes on a trip to the Hoover Dam, where vehicles inscribed with the insignia of the Los Angeles Municipal Utility District sport Nevada license plates, proof positive that resources held in Nevada, especially the always-precious water stored in Lake Mead, do not belong to the state. The control of Nevada from outside its borders could not be more clearly defined. No state holds more resources targeted for use in other places than Nevada. It is also the place people gamble and frequently leave their financial resources, with stunning results. Las Vegas household per capita income is nearly double that of San Antonio and other large southwestern cities.

But if Nevada is unique in many respects, it is also an archetype for western states, different only—and not for long—in its embrace of a booming service economy. The state's realities are regionwide realities, especially when it comes to water. In California in 1990, 19.4 percent of the water consumed went to *all* nonagricultural purposes. The more than 80 percent of rural water generated about 2 percent of the state's gross income. In Nevada, the 18.7 percent of the water used for nonagricultural purposes generated greater than 99 percent of the state's income. Even though California produces much of the nation's vegetables and arid Nevada adds $1 billion in gross agricultural product, the amount of state revenue generated by more than 80 percent of the water in the two states pale in comparison to the jobs, tax revenues, and income of the much smaller nonagricultural share.

Something is clearly amiss here and maybe it is our perception of desert life and aridity. Profligate Las Vegas, where the average household uses more gallons per day than any other desert city in the United States; dubious Phoenix, supported by water from federal projects; tyrannical Los Angeles, draining first the Owens Valley and then the Colorado River to support its ever-expanding growth. Maybe we should reassess the way we regard the issue of water in the West.

The global economic realignment of the past two decades has had a profound effect on the social value of natural resources. It has devalued their worth by conventional measures, while adding new and different kinds of valuation to the economic totals on which we base so much of our decision-making. The so-called Information Revolution, the consequence of the tiny microchip, has spawned a transformation as great as the Industrial Revolution: Knowledge and the ability to manipulate it have become formidable forms of power and dramatically increased the range and significance of the service economy. Without the microchip, the revolutions of the past twenty years in everything from banking to health technologies simply could not have taken place.

The Microchip Revolution has also been responsible for a constellation of factors that have changed much about human life worldwide. Cyber transactions now move capital around the globe, creating what Christopher Lasch recognized as a transnational class of monied individuals who are essentially stateless in their sense of national obligations. This boundary-free wealth has few obligations to people and places and its incredible mobility can be a threat to the stability of individual countries and national economies. The tremendous increase of exporting manufacturing to developing countries, from the *maquiladora* corridor in Mexico to the Nike factories of the Far East, has profound implications for anyone dependent on a paycheck.

In the end, this revolution has changed the character of the American economy. We do not work as we did even fifteen years ago; personal computers and the Internet create access and information transfer in ways that seemed like science fiction in the 1970s. We no longer rely on the sources of capital that once drove the American economy. The junk bonds of the 1970s, themselves an innovation, became the derivatives, a computer-generated sort of splitting and matching, of the 1990s. Americans do not do the same tasks in the same ways, and we barely experience the constraints that once placed limits on the way business was conducted.

The transformation has been powerful and comprehensive. It has diminished not only the market value of American resources, but a significant piece of the national mythology as well. The United States was built on two premises: cheap labor, often provided by immigrants and family members, and cheap natural resources. Labor has not been cheap in the United States since before World War II and the cheap natural resources are both more expensive and less necessary in a global economy. Although it may be premature to declare the end of the natural resources–based world economy, it is clear that ways in which resources are organized, used, and the purposes to which they are put have dramatically changed, especially in the First World.

Like all its predecessors, this new revolution picks winners and losers. Some sectors of the economy fare marvelously. The educated, the upper end of the service sector—everyone from attorneys to consultants of all stripes—find opportunities galore at fees that reward their choice of career. The technically proficient—and here skilled health-care professionals stand out, along with the denizens of Silicon Valley—also benefit greatly. These workers *are* their skills; they are able to sell their knowledge, which is transferable from the equipment of one employer to another. In contrast, wage workers of all kinds as well as agricultural, semiskilled, and unskilled workers bear the brunt of this change.

The Information Age has also fomented a progressively more-dense urbanism. San Jose, California, once a parochial, agricultural town, has expanded into the megalopolis of San Francisco and the Bay Area. The patterns of life, the ways of making a living have been overcome by space-based commuter culture, dependent in many ways on the proportionately outrageous number of telephone and fiber-optic lines and the cellular phones of the region. This transformation has been devastating, not only to economies, but if you follow writers such as James Howard Kunstler, to the human spirit as well.

The Microchip Revolution also, and at long last, calls into question the allocation of federally subsidized water in the West. For most of this century, western water development has been funded by federal dollars and centralized in the hands of a very few, increasingly large corporations. Many of these are agribusiness concerns or conglomerate holding companies, often located in the very cities that they deprive of water. This process has created huge operations, often staffed by immigrant pickers *sin papeles* (without papers) who travel from place to place as the crops ripen, that exert enormous political influence in local, county, and especially state and national elections. This oligarchy of control maintains its power by flexing its political muscle.

The oligarchy's power is now under attack. A revolution in water distribution is taking place. Under the name of reallocation, water is being shifted from traditional uses to new ones. Water is going from rural to urban use, from agriculture and ranching to household, factory, or recreational use. A mythic capsule version of this transformation would be that water that used to irrigate cotton fields around the Gila River in southern Arizona is now entertaining children at the Schlitterbahn, a water park in New Braunfels, Texas. Although not a one-to-one exchange, this example dramatizes the emerging, fluid relationship between water distribution and consumption.

For those in the rural West who depend on large quantities of federally subsidized water, this is a prospect fraught with peril. It threatens more than mere livelihood. Reallocation places limits on individual prerogative, often curbs a generations-old albeit heavily subsidized way of life, and applies the mythic free market—what Richard White has become fond of calling "the weirdness of late capitalism"—to a smug cocoon, the fictive water markets of the West.

The sole defense against this radical change is the culture-and-custom argument—"we've always done it this way"—but even that does not hold water. To interventionists, those who see the government as the best regulating force, accepting a culture-and-custom argument means embracing the Sagebrush Rebellion and the County Movement, local-rights doctrines that interpose—in the tradition of opposition to the civil rights—local rules over national ones. This is a squatter's argument, the idea that local use conveys a proprietary right that is first among rights. For free marketers, culture-and-custom means embracing blatant inefficiency as a form of government subsidy of unessential economies; for these bottom-liners who count only dollars, such arguments are hard to swallow.

So the region faces a conundrum in dealing with an issue the Microchip Revolution has posed: how to regard and treat the people and institutions its opportunities damage. Are the advocates of the County Movement mere Luddites, destroying the machinery of the future that threatens their way of life, or perhaps Don Quixotes, tilting at the windmill of technology that has already displaced them? Are they resisting change by destroying symbols or are they more dangerous, along with agribusiness, hoarding a finite and precious commodity? Are these advocates determining patterns of use—and growth and development besides—by oligarchic control of an essential and irreplaceable commodity? Or are they simply slow-moving dinosaurs, unable to respond to changing scenarios?

No matter what the response, the revolution is clearly underway. In the eleven interior western states in 1994, only sixteen counties employed more than 35 percent of their workforce in once-traditional economic endeavors such as mining, logging, ranching, and farming. Of these, four—Pershing, Lander, Eureka, and Esmerelda Counties—are in Nevada. Together they total less than 50,000 people. Who cares about them? Why are they relevant to the future of the their state or even the western region?

Except as curiosities, they are not. Indeed, they serve as examples of the three things in American society that have not changed since the turn of the twentieth century. The first is the internal combustion engine, essentially the same as it was, despite all the bells and whistles. The second is the flush toilet, largely as it was invented by Thomas Crapper in the mid-nineteenth century. The third is the way western ranching and agriculture use water. All three stem from a time when resources—whether gasoline or water—were cheap and the people who competed for them were few in number. In all three cases, the technology and the ideology that support the practice embody archaic principles that clash with the new western economy. Its development is today driven by a transfer-payment economy. Much of the western economy's funding is derived from retirement income, welfare, inherited wealth, and what Henry George would have called "unearned increment," the return on investment. In the greater Yellowstone area, retirement contributes more to the regional economy than mining. In Clark County, Nevada, home to Las Vegas, 22 percent of the population is retired. The expenditures of that segment of the population comprise a far more important part of the regional economy than any extractive industry, agriculture, or ranching. Transfer payment economies demand service.

The shift to a service economy is so pronounced in the West because of the region's colonial characteristics. Western economic growth is shaped from outside the region and increasingly by newcomers, sometimes called "amenity migrants," who seek in the West a landscape in which to replicate the worlds they left behind. They play many roles—from political candidates correctly labeled as carpetbaggers to buyers of vast ranches in rural Montana—but their impact and the reasons they are drawn to region are colonial in character. One of the most significant industries in the West, for instance, is also one of its most colonial: tourism. Moreover, with the exception of the coastal and industrial belt from Seattle to San Diego, the West remains an economic colony, supported by federal and outside dollars that are subject to both extra- and intraregional influences. The structure of these communities and their economic and political evolution, the way they use transient and semipermanent labor, and how they constantly become reinvented as new forms of themselves highlight the problems of tourist-based economies. Their identity is malleable, as national, resort-based chains have replaced local businesses. As these chains become ubiquitous, they obscure local economies and culture so that the traveling public sees just what it saw at home, only in a different setting. This homogenization of the landscape reflects rather than foreshadows transformation. Although the arrival of such businesses illustrates the increased economic importance of tourist communities, it also spells the end of existing cultures. Often this arrival amounts to "killing the goose that laid the golden egg." The inherent problem of communities that succeed in attracting so many tourists is that their very presence destroys the cultural and environmental amenities that once made the place special.

But from another perspective, tourism offers a great benefit: It uses a lot less water and generates a lot more revenue than agriculture, ranching, or most extractive industries. Consider that it takes more than six thousand gallons of water to produce one six-inch silicon wafer and producing one ton of grain requires one thousand tons of water, whereas San Antonio's Riverwalk uses a one-mile-long, four-foot channel of water as a foundation of a $3 billion tourist economy. The use of water for tourism starts to look like a pretty good deal.

In this context, it is no wonder that the Hyatt Hotel along San Antonio's Riverwalk diverted a little stream of water into the hotel lobby as a connection between the faux sacred space of tourism and the profane commercial recreational space of the hotel. The water, stagnant as it often is, has great meaning. It connotes recreation, pleasure, good times, and its presence in the lobby is symptomatic of how water can create wealth in a way that agriculture and ranching no longer do—and in fact probably never did.

Tourism as a strategy is not without problems. Outside capital wields great power in local settings and it is transformative by its very nature. Tourism requires the marketing of identity, which over time can cause near-schizophrenia in community identity. Work in tourism is overwhelmingly low skill and poorly paid; it is difficult to raise a family on seasonal, minimum-wage work. And it is work that has little status: Those who do "real work"—in factories and fields—regard tourism with disdain. Its workers produce little but smiles; there is nothing people can hold in their hands to feel a sense of accomplishment at the end of the day.

But as a transitional strategy or a stopgap to tide a place or region over until new forms of economy coalesce, tourism has promise. It uses fewer resources than other economic activities, it can become an important regional baseline, and it can provide structure while planning a response to the conditions of the postindustrial era. As a bridge to something else, instead of an end in itself, tourism works pretty well.

Ask any Nevadan. Tourism is at the heart of the state's economy. The volcano on the Las Vegas Strip that explodes every hour and the battle between the HMS *Brittania* and the *Jolly Roger* at the Treasure Island every ninety minutes are integral parts of the most important economy in the state, in some ways, the only genuine source of sustenance in an otherwise poor, arid colony. When Nevadans look at the fact that more than 80 percent of their water produces only $1 billion in revenue and realize that 18 percent accounts for more than three hundred times that amount in gross revenue, they cast their eyes on the rural parts of the state with wonder. With tongue in cheek, urban Nevada can easily envision this bargain with its rural counterparts: Take the absolutely best year that the farmers and ranchers have ever had, call it their "parity year," and add 20 percent to it; then promise them a cost-of-living raise *and* an annual 5 percent increase in revenues. In exchange for this munificent deal, all the farmers and ranchers have to do is to cede their water rights to the urban regions of the state. While they remain in their fields, perhaps tossing fictitious bales of hay or watching center-pivot irrigation devices roll across the land without spraying water—imagine the tourist attraction this might make—the urban economy will create considerably more opportunity for more people throughout the state. That is the world that re-allocation of water will create in the new American West, a world different than the one we currently inhabit, complete with a new set of winners and losers. To determine who will be whom, just watch who grabs and holds onto the water jug.

About the Authors

Jesús F. de la Teja is an associate professor of history at Southwest Texas State University and former director of Archives and Records of the Texas General Land Office. He is the author of the award-winning *San Antonio de Béxar: A Community on New Spain's Northern Frontier* (1995).

Shelly C. Dudley has been a senior historical analyst for the Salt River Project in Phoenix, Arizona, since 1983. She researches and analyzes water- and power-related topics with an emphasis on water rights and has published articles in *Western Legal History* and *Casa Grande Valley Histories*. She is currently cowriting a history of the Salt River Project for its centennial celebration.

Mark Harvey is an associate professor of history at North Dakota State University. He is the author of numerous articles as well as *A Symbol of Wilderness: Echo Park and the American Conservation Movement* (1994).

Donald C. Jackson teaches at Lafayette College and is the author of *Building the Ultimate Dam: John S. Eastwood and the Control of Water in the West* (1995), which was selected by *CHOICE Magazine* as an outstanding academic book for 1996.

Bonnie Lynn-Sherow is an assistant professor of history at the Kansas State University. She earned her doctorate at Northwestern University in 1998. Her dissertation, "Ordering the Elements," was awarded the Best Dissertation in Oklahoma History Prize by the Oklahoma Historical Society. Her most recent publication is "Beyond Winter Wheat," *Kansas History* (Spring/Summer 2000).

Sandra K. Mathews-Lamb is an assistant professor of history at Nebraska Wesleyan College. In addition to teaching the history of the American West, she has done historical contracting on water and land rights as well as other topics for both private and public organizations.

Daniel McCool is currently professor of political science and director of the American West Center at the University of Utah. He is the author of *Command of the Waters* (1994), coauthor of *Staking out the Terrain* (1996), and author and editor of *Public Policy Theories, Concepts, and Models* (1995) and *Contested Landscape* (1999).

Char Miller is professor and chair of the Department of History at Trinity University. He is coauthor of *The Greatest Good: 100 Years of Forestry in America* (1999), editor of

American Forests: Nature, Culture, and Politics (1997), and coeditor of *Out of the Woods: Essays in Environmental History* (1997).

Alan S. Newell is cofounder and a principal in the firm Historical Research Associates, Inc., in Missoula, Montana. Over the past twenty years, he has engaged in historical studies of water development on Indian reservations throughout the West and has served as an expert witness on Native American issues, including federally reserved water rights, in federal and state courts.

John Opie was one of the founders of the American Society for Environmental History and its journal, *Environmental History.* He is the author of *The Law of the Land* (1987), *Ogallala: Water for a Dry Land* (1994), and *Nature's Nation: An Environmental History of the United States* (1998), among many other works.

Donald J. Pisani is Merrick Professor of Western American History at the University of Oklahoma and past-president of the American Society for Environmental History. His research interests include natural resources, the law, and public policy, with particular emphasis on water. His most recent book, a collection of essays, is *Water, Land, and Law in the West, 1850–1920* (1996).

Brad F. Raley is a doctoral student at the University of Oklahoma, where he is writing a dissertation on irrigation in Colorado's Grand Valley.

Hal K. Rothman is a professor of history at the University of Nevada–Las Vegas, where he edits *Environmental History.* He is the author of *Saving the Planet: The American Response to the Environment* (2000), *Devil's Bargains: Tourism in the Twentieth-Century American West* (1998), *The Greening of Nation? Environmentalism in the U.S. since 1945* (1997), *"I'll Never Fight Fire with My Bare Hands Again"* (1995), and *On Rims and Ridges: The Los Alamos Area since 1880* (1992); the editor of *Reopening the American West* (1998); and coeditor *Out of the Woods: Essays in Environmental History* (1997).

Raúl M. Sánchez was formerly an associate professor at The St. Mary's University School of Law in San Antonio, Texas. He is now Special Assistant to the President for Diversity and Human Rights at the University of Idaho, Moscow.

Thomas C. Schafer is an assistant professor of geography at Fort Hays State University. His research interests include the changing geography and production patterns of farming in southwestern Kansas.

James E. Sherow is an associate professor of history at Kansas State University. He is the author of *Watering the Valley* (1990) and editor of *A Sense of the American West: An Anthology of Environmental History* (1998).

John P. Tiefenbacher is an associate professor of geography at Southwest Texas State University. He has an interest in spatial patterns of chemical contamination of the environment and has published papers on the mapping and identification of hazards posed by accidental spills of toxic chemicals, industrial emissions of toxins, and the use of agricultural chemicals in *Urban Geographer* and *Environmental Practice*.

Index